*building construction estimating*

McGRAW-HILL BOOK COMPANY, INC.
*new york   toronto   london   1959*

*George H. Cooper*
mechanics institute, new york

SECOND EDITION

# BUILDING CONSTRUCTION ESTIMATING

BUILDING CONSTRUCTION ESTIMATING

Copyright © 1959 by McGraw-Hill Book Company, Inc. Copyright, 1945, by McGraw-Hill Book Company, Inc. Printed in the United States of America. All rights reserved. This book, or parts thereof, may not be reproduced in any form without permission of the publishers.

Library of Congress Catalog Card Number: 58–11977

# preface

The material in this textbook has been used for many years with success by the author's classes at Mechanics Institute, New York City. In the present edition, other schools should find the book well arranged and useful for their purposes.

This is intended to be a standard textbook for a full course in technical and vocational schools and also for intensive study by properly qualified students in technical institutes and colleges. Such students should be high-school graduates and should have at least two years of architectural drafting to their credit.

Many books on estimating consider in great detail the design and construction of buildings, mathematics, specification writing, mechanical trades, and elementary plan reading. The present book is based on the principle that a textbook should be complete only in the matters before the class and should not treat in detail anything that is regularly handled in other classes. Because there is no other book on estimating that meets this test, the present book should be found acceptable by teachers and students desiring a thorough textbook treatment without cumbersome detail.

The practical requirements of the classroom have been given consideration throughout. The book forms a complete working plan for the instructor and contains all the required material, including two sets of plans, many specimen estimates, hundreds of construction terms, and all essential reference data. The numerous illustrations will enable the student to visualize better what is being discussed. The exercises are not too many for a regular class to handle completely.

Chapters 1 to 10 discuss contracting as a business, building codes, plans, specifications, contracts, and the general technique of estimating. This material comprises the first half of the course. Chapters 11 to 18 deal with the estimating of work that is customarily performed by the general contractor's own men. Chapters 19 to 28 treat of the work usually done by the subcontractors' men.

The aim is to present in orderly sequence a well-rounded course covering the everyday work of the building contractor's estimator. The final chapter deals with the terminal aspect of the course. The challenge "Are you an estimator?" gives the student a strong motive for learning. The comprehensive examination, with plans, provided in this chapter will enable him to place a definite and practical value on his attainments. This final examination might instead be based on a set of selected blueprinted plans. In that event, the plans in the chapter should form the basis of classwork or homework during the final months of the course.

A full set of specifications, as for a modern business building, would require many pages and might confuse the student. For these reasons it seemed best to include only the essential extracts, which may be elaborated upon as desired by the instructor. Several complete sets of plans and specifications should be available in the classroom for additional drill and test purposes. These plans will last longer if they are printed on cloth or mounted. Complete estimates for several types of buildings should be available to show their general form and arrangement. Copies of the local building codes also should be available. The final month or two of the course should be spent on these additional plans, and every student should be required to prepare at least two complete estimates, each for a different type of building.

It is also suggested that full-size samples of many building materials should be on hand in the classroom where the students may see and feel them, especially if they are not provided in concurrent classes. Models of different forms of construction and large-scale drawings will also be found helpful and should be used to illustrate the instructor's lectures.

Although the arrangement and contents of this book are original, and the line illustrations all especially made for it, the author wishes to acknowledge his indebtedness to all the books and manufacturer's catalogues which he examined for ideas. No names of manufacturers or of their products are mentioned either in the text or in the illustrations, nor has any effort been made to favor the use of any particular materials.

*George H. Cooper*

# bibliography

GENERAL

FREDERICK S. MERRITT, editor: *Handbook of Building Construction*, McGraw-Hill Book Company, Inc., New York, 1958.

FRANK E. KIDDER and HARRY PARKER: *Architects' and Builders' Handbook*, 18th ed., John Wiley & Sons, Inc., New York, 1931.

H. G. RICHEY: *Richey's Reference Handbook for Builders, Contractors, Architects, Building Materials Dealers, Carpenters, and Building Construction Foremen*, Simmons-Boardman Publishing Corporation, New York, 1951.

CHARLES G. RAMSEY and HAROLD R. SLEEPER: *Architectural Graphic Standards*, 5th ed., John Wiley & Sons, Inc., New York, 1955.

ESTIMATING

JOSEPH E. KENNEY: *Blueprint Reading for the Building Trades*, 2d ed., McGraw-Hill Book Company, Inc., New York, 1955.

CHARLES F. DINGMAN: *Estimating Building Costs*, 3d ed., McGraw-Hill Book Company, Inc., New York, 1944.

ROY W. WHITE: *Building Practice Manual*, D. C. Heath and Company, Boston, 1952.

ROBERT L. PEURIFOY: *Estimating Construction Costs*, 2d ed., McGraw-Hill Book Company, Inc., New York, 1958.

FRANK R. WALKER: *The Estimator's Reference Book*, 14th ed., Frank R. Walker Co., Chicago, 1957.

CONSTRUCTION

WALTER C. VOSS: *Construction Management and Superintendence*, D. Van Nostrand Company, Inc., Princeton, N.J., 1958.

J. RALPH DALZELL: *Simplified Masonry Planning and Building*, McGraw-Hill Book Company, Inc., New York, 1955.

EMANUELE STIERI: *Concrete and Masonry*, Barnes & Noble, Inc., New York, 1956.

JOHN A. MULLIGAN: *Handbook of Brick Masonry Construction*, McGraw-Hill Book Company, Inc., New York, 1942.

NELSON L. BURBANK: *House Carpentry Simplified*, 6th ed., Simmons-Boardman Publishing Corporation, New York, 1958.

E. A. LAIR: *Carpentry for the Building Trades*, 2d ed., McGraw-Hill Book Company, Inc., New York, 1953.

# contents

|   | PREFACE | v |
|---|---|---|
|   | BIBLIOGRAPHY | vii |
| 1 | Introduction | 1 |
| 2 | Construction Relations | 4 |
| 3 | The Architect | 11 |
| 4 | The Contractor | 20 |
| 5 | Plan Reading | 30 |
| 6 | House Plans | 41 |
| 7 | Contracts | 80 |
| 8 | The Technique of Estimating | 90 |
| 9 | Quantity Surveying | 106 |
| 10 | Expense and Summary Sheets | 115 |
| 11 | Excavating | 134 |
| 12 | Concrete Foundations | 143 |
| 13 | Concrete Floors and Roofs | 158 |
| 14 | Masonwork | 175 |

| | | |
|---|---|---|
| 15 | Rough Carpentry | 205 |
| 16 | Cement Work | 235 |
| 17 | Plastering | 245 |
| 18 | Finish Carpentry | 262 |
| 19 | Steel and Iron | 282 |
| 20 | Roofing and Sheet Metal | 296 |
| 21 | Stonework | 305 |
| 22 | Fireproof Doors and Windows | 310 |
| 23 | Tile, Terrazzo, and Marble | 318 |
| 24 | Painting and Glazing | 323 |
| 25 | Hardware | 332 |
| 26 | Plumbing | 337 |
| 27 | Heating and Air Conditioning | 346 |
| 28 | Electrical Work | 352 |
| 29 | Are You an Estimator? | 362 |
| | **INDEX** | 393 |

# ready reference

### PLANS

Plot plan, 43
Suburban house, 41–79
Two-car garage, 113
Bank and office building, 379–391
Architectural, 41–79, 379–387
Structural, 32, 388–391
Mechanical, 33, 340, 344, 360
Plumbing, 33, 340, 344
Electrical, 360
Masonry, 180–198
Rough carpentry, 210–230
Finish carpentry, 264–274
Plastering, 252–257
Architect's office, 12, 13
Contractor's office, 21, 26
Job office, 121
Job layout, 120

### SPECIFICATIONS

General conditions, 368
Excavating, 142
Steel and iron, 370
Concrete, 157, 368
Masonry, 203, 369
Rough carpentry, 233, 371
Cement work, 244
Plastering, 260, 273
Hollow metal and kalamein, 372
Finish carpentry, 279, 371
Cabinet work, 372
Architectural metal, 372
Roofing and sheet metal, 303, 375
Stonework, 309, 370
Marble, 374
Tilework, 321
Terrazzo, 374
Painting, 375
Glazing, 331, 375
Plumbing, 343
Heating, 377
Electric, 359
Elevator, 377

### CODES AND REGULATIONS

Definitions of terms, 38
Excavating, 141
Concrete, 156, 169–174
Masonry, 199–202
Carpentry, 231–233
Plastering, 259–260
Plumbing, 339–341
Heating, 350–351
Electrical, 356
Labor laws, 37, 129
Safety regulations, 119, 129
Public utilities, 354
Tile association, 319
Union regulations, 120, 353
Performance bond, 126

xi

## SPECIMEN ESTIMATES

General job expense, 133
Excavating, 140
Concrete, 146
Masonry, 178
Carpentry, 214
Plastering, 247
Overhead expense, 128
Estimate summary, 132

## LABOR COSTS

General job expense, 115, 133
Overhead expense, 127
Concrete, 24
Masonry, 25, 179
Rough carpentry, 25, 217
Sheathing, 219
Rough flooring, 219
Finish carpentry, 275
Wage rates, 9
Cost-per-hour analysis, 99
Cost records, 24

## TABLES

Reinforcing bars, 149
Nail sizes, 220
Wire gauges, 287
Wide-flange beams, 285
Standard beams, 286
Steel angles, 287
Steel channels, 286
Plastering terms, 258
Hardware finishes, 335
Electrical symbols, 355
Wattage requirements, 358
Waste allowances, 219
Job personnel, 9, 28, 116

*building construction estimating*

CHAPTER 1

# *introduction*

Building construction estimating requires a working knowledge of all phases of building work. The serious student of estimating may therefore be said to have an excellent opportunity to make a good general study of building work. As nearly every man will find this knowledge useful at some time or other, whether or not he is directly connected with the building industry, it will readily be seen that the practicality of a course in estimating can hardly be overemphasized.

**The Estimator.** In a well-organized contractor's office, the estimator is usually the center of activity. From the time the office makes the first contact regarding a proposed building until the job is well under way, the estimator is in control. He may even have to make the contacts. His first duties include quantity surveying, interviewing subcontractors, obtaining quotations on materials, and preparing the estimate. He then submits the bid and follows through to secure the contract. He has to make adjustments quickly and therefore must have his records in fine order and his wits about him all the time. His work is important at all stages.

Before the work at the job can begin, the estimator prepares the working estimate, the material lists, and the construction schedule. He is in constant touch with sources of supply, as it is his duty to award the subcontracts and to supervise the purchasing of materials. Unless these are properly and promptly taken care of, there is much waste energy all along the line afterward. The saying "a job well begun is half done" applies very aptly here, for many a job has been bungled merely because it was given into the superintendent's hands before it had been properly started in the office.

When the job gets under way, the general superintendent has control and the estimator gradually relinquishes his hold, in order that he may put his time to use in securing other contracts. A good estimator will see that complete plans and specifications, copies of all subcontracts, and all other required data are given to the superintendent. He will make sure that every adjustment has been made in the records up to the time that he thus turns them over to the superintendent. He will clarify all points about which he thinks there may be any question or which may tend to slow up the progress of the job. He will insist upon his records being so complete at the time of their final release by him that nobody can hold him responsible for what happens afterward.

**Purpose of the Course.** This course is for the practical purpose of training students in the everyday work of the building contractor's estimator. The plan is to cover the entire ground as thoroughly as possible in the time given to the subject. The textbook, which contains all the material required for a complete presentation of the principles underlying this subject, is planned for classroom use and for home study.

Two complete sets of plans are included and will be referred to in the text. These plans illustrate many kinds of materials and several types of construction. Detail drawings are included in the chapters in which the work shown by the details is discussed.

Specimen estimate sheets will be found in all the chapters dealing with the estimates. These illustrate the recommended method of making the entries required in each of the various lines of work. Each student individually should do all the actual measuring of the plans, and each should make all the entries in the estimates. In this way the plan indications, units of measure, and methods of scaling the plans will be brought out forcefully.

The hundreds of construction terms treated in the text are those commonly used by architects, contractors, and others concerned with building work everywhere. The student should realize that these form the heart of the construction man's language. As they are tied in with the text and the estimating work throughout the course, every serious student will have an opportunity to enrich his vocabulary by including in it these new words and expressions. They will probably be of more practical value than the same number of words learned in a class in English literature.

Exercises, given at the end of each chapter, are based upon the work of the chapter. They should, therefore, be worked out thoroughly by every student while the chapter is being studied. In this way the weak points of the instruction or of the study will appear and can be reviewed before the next chapter is started. The last chapter in the book contains final examination directions and problems and represents the goal of the course.

Many unit costs are given in the text and in the specimen estimates. These are for the purpose of making the work more realistic and may be employed for pricing the estimates. However, it should be borne in mind that costs vary, for many reasons that will be discussed in the text; and these unit costs should not be treated as permanent references for actual pricing of work.

This is a practical course and the students should react by asking questions and looking up information because of their interest in the practical nature of the work. Thus will be fulfilled the conditions that facilitate the acquisition of knowledge: a determination to know, an active mind, and an effective method of study. As the student increases in power, imitation must gradually decrease and reason play an ever-increasing role. The students are urged to discuss the work among themselves as well as with the teacher; in fact, they should work in pairs, if possible, always being sure, however, that each member of the pair goes through all the motions of the estimating process individually.

■ EXERCISES

1. Briefly, what are the duties of an estimator?
2. Up to what point is an estimator in control of a building operation?
3. What should the estimator do for the good of the job and for his own safety when relinquishing control?
4. Look over some of the books on estimating, such as those listed in the Bibliography. Write at least 100 words of notes on an introduction to estimating. Name the books that you have used for this purpose.

CHAPTER 2

# *construction relations*

The construction of a building involves many kinds of administrative and technical skills. This is a fact that is not often fully appreciated by the general public or even by businessmen in other lines of work. They see a building going up, but seldom have an opportunity to look behind the scenes and observe the working of the system that makes possible the construction of the building—the activity in the various departments of the architect's, the contractor's, and the subcontractors' offices.

In a contractor's office much careful planning and scheduling must be done, equipment must be arranged for, and materials must be purchased and coordinated as to sequence of delivery. The field working force must be organized anew for each project, and definite arrangements must be made for payrolls, accounting, insurance, and tax records, in connection with this floating population of workers on the job.

It is the contractor's skill and the use of his business and technical organization that the owner buys when he signs a construction contract. A contractor who attempts to operate without proper skill and without a proper business and technical organization is under a very great handicap.

Figure 2-1 shows the usual relationship that exists between the men who are concerned with the design and the construction of an average building job. This study is from the general contractor's viewpoint, and therefore the main line of action is shown running through the contractor's organization, as the chart indicates. The contractor looks

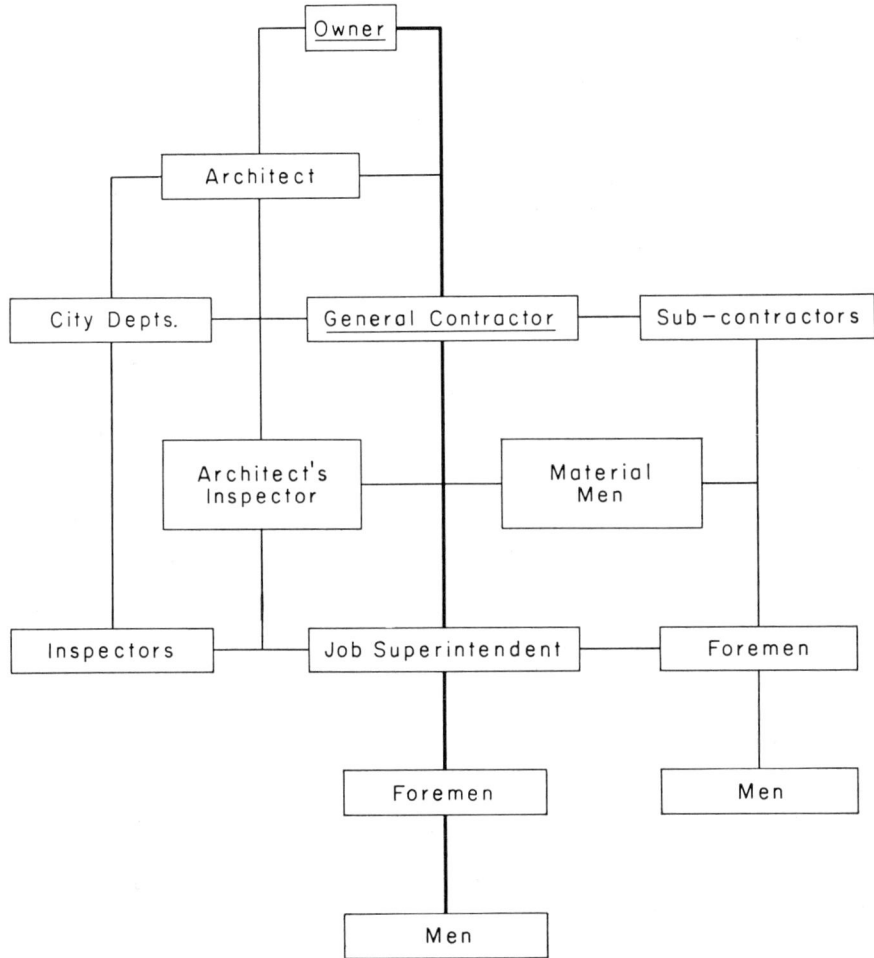

*Fig. 2-1  Construction relations.*

to the owner as the man who pays him and as the one he wants to satisfy regarding the particular job.

**The Architect.** The man who makes the plans is the architect. He also writes the specifications that accompany the plans and that further describe the work. He is engaged and paid by the owner to do this planning and to look after the owner's interests in connection with the building work.

At the beginning, the architect has dealings with the owner alone preparing preliminary sketches and then making the regular plans and writing the specifications. After this is done, the general contractor comes upon the stage. He is asked to submit a bid for the job of constructing the building. Usually, the architect invites the contractors to bid, although sometimes the owner does this.

Contractors are very likely to know about the job before it is time for bids, because they are always following up trade reports and other leads toward jobs and are constantly in touch with a number of architects and others who frequently have work to be estimated. As public contracts have to be legally advertised, one can always know about this class of work at least 10 days before the bids are due.

The architect lends each invited contractor a set of plans and specifications, in order that each one may prepare his estimates on which to base bids. After the contract between the owner and the successful contractor is signed, this contractor looks to the architect for further information regarding the work to be done and for detail plans, etc. He soon finds that he has both the owner and the architect to satisfy.

The architect sends a man from his office to act as his inspector on the job. This man is sometimes referred to as the "architect's superintendent." The contractor has his own superintendent on the job to handle the various trades and to lay out the work, to order material as required, and to manage the subcontract work, etc. The contractor's superintendent has continual contact with the architect's man and with the inspectors from the municipal departments that have jurisdiction over the work of constructing buildings.

**Municipal Departments.** The municipal departments are visited by the architect, as he has to file applications and plans in order to enable the contractor to secure a permit for the erection of the building. The contractor also visits them to get this and various other permits that contractors are required to have. After the job starts, it will be visited by the inspectors from these departments—and the contractor will have them to satisfy too!

Local laws always include regulations governing building work, which are generally referred to as "building codes." The building code states the minimum requirements regarding the design, construction, and use of buildings and other structures. Safe and sanitary buildings are required. The minimum requirements for the size and quality of materials for walls, floors, roofs, plumbing, etc., are given in the code.

**Subcontractors.** Men or firms who do work for the general contractors are called subcontractors. This may be work that the general contractor does not wish to do; or work that he has not the facilities to do with his own men; or, perhaps, work for which he has not the required license. Steel, iron, roofing, sheet-metal, plumbing, heating, and electrical work are some of the lines that are commonly sublet in this way. The subcontractors send their men and materials for the work they do on the job. The general contractor's superintendent sees that they do their work on time and in harmony with the other lines of work.

In small towns, where there are very few inspectors from the municipal departments, one man often acts as the whole personnel of the building department. Similarly, in small places there are fewer subcontractors than in cities. In such cases the general contractors employ men directly for most of the trades, instead of giving the work out to subcontractor specialists. In very large cities there seem to be subcontractors without end—for wood flooring, for just scraping and finishing wood floors, for doing only the cement finish work, for caulking, and for many other items of work that would ordinarily be done by the general contractor's own men in a smaller place.

Separate contracts are sometimes awarded by the architect or the owner, especially for the plumbing, heating, and electrical work, and on large jobs also for the structural steel, air conditioning, and elevators. The general contractor usually has no jurisdiction over such work but is expected to cooperate with the men who do the work. This cooperation may involve considerable expense to the general contractor and must be considered by his estimator in preparing the estimate. The owner saves the general contractor's supervision costs, overhead expense, and profit on contracts separately awarded but at the same time is himself burdened with the supervision and overhead expense connected with handling matters he may not be organized to handle.

Separating the work in this manner results in confusion on the job because the general contractor, who is experienced in coordinating the work of all trades, no longer has the power over these separate branches that would enable him to control the work properly. Realizing this, architects usually try to persuade the owner to make one complete contract for the entire job. Architects sometimes charge the owner an additional fee for having to deal with more than one contractor because they are, in effect, acting as general contractors and are put to additional expense in their office and on the job.

**Material Men.** Material man is the term used to denote any of the men or concerns that merely sell the materials used in building work. The lumber dealer is a material man. He sends the lumber to the job but does not send the carpenters who install it. The mason-supply yard, the sand and gravel company, the hardware dealer, and all others from whom materials are bought are called *material men* and are thus distinguished from subcontractors, who do actual work on the jobs.

The subcontractors themselves use the material men—for supplying their own men with materials. Thus, it may happen that occasionally the same material men will send material to the same job, both for the general contractor's use and for use by one or more of his subcontractors. The same mason-supply man may send bricks and cement for the general contractor's masonwork and plastering materials for the subcontractor who is doing the plastering.

**Mechanics.** The mechanics are the men who work with tools on the job. Helpers or apprentices also are employed in some trades. "Laborer" is a broad term, used to apply not only to the men who do the ordinary heavy work about the job but also to helpers and apprentices. A bricklayer's helper, for example, is often referred to as a laborer.

A list of the employee classifications found on large jobs, together with their rates of wages, is shown in Table 2-1. The rates are given here merely so that the student may become somewhat familiar with them; naturally, they do not apply in all parts of the country. The whole field of industrial relations and business economics is undergoing a change. Workmen on construction jobs are being unionized in practically every part of the country. The rates shown are union rates per hour that apply in some localities.

Plasterers and electricians usually work only 6 hours per day, excavating workers, ironworkers, and plumbers usually 8 hours per day, and other trades usually 7 hours per day.

Most unions maintain funds which provide the men with various welfare benefits such as sickness and vacation payments and pensions

TABLE 2-1  EMPLOYEE CLASSIFICATIONS

| Classification | Rate per hour | Classification | Rate per hour |
|---|---|---|---|
| Air-tool operator | $3.35 | Painter | $3.45 |
| Asbestos worker | 4.20 | Plasterer | 4.35 |
| Barman (wrecking) | 3.50 | Plasterer's helper | 3.85 |
| Barman's helper | 3.20 | Plumber | 4.35 |
| Boilermaker | 4.25 | Plumber's helper | 3.00 |
| Boilermaker's helper | 3.85 | Sheet-metal worker | 4.10 |
| Bricklayer | 4.45 | Slate and tile roofer | 4.10 |
| Bricklayer's helper | 3.35 | Slate and tile roofer's helper | 3.27 |
| Carpenter | 4.10 | Steamfitter | 4.30 |
| Cement finisher | 4.20 | Steamfitter's helper | 3.00 |
| Cement and concrete worker | 3.25 | Stonecutter | 4.20 |
| Composition roofers | 3.85 | Stonemason | 4.25 |
| Composition roofer's helper | 2.75 | Structural ironworker | 4.40 |
| Dock builder | 4.00 | Tile layer | 3.87½ |
| Electrician | 4.18 | Tile layer's helper | 2.95 |
| Elevator constructor | 4.24 | *Equipment operators* | |
| Elevator constructor's helper | 3.18 | | |
| Glazier | 4.10 | Air compressor | 3.65 |
| Laborer, common | 2.90 | Hoist | 3.65 |
| Laborer, concrete | 3.25 | Concrete mixer, small | 3.27½ |
| Marble carver | 4.10 | Concrete mixer, large | 3.40 |
| Marble cutter and setter | 3.95 | Tractor | 3.40 |
| Metal lather | 4.20 | Truck | 2.75 |
| Mosaic and terrazzo worker | 3.30 | Bulldozer | 3.27½ |
| Ornamental ironworker | 3.90 | Shovel | 4.15 |

when they retire. These are in addition to the workmen's compensation and unemployment insurance coverages required by state law and the social security payments required by the Federal government.

Good contractors also carry public liability and property damage insurance and sometimes are required to pay other insurances and taxes which are based on the men's pay. Because of all these payroll costs, contractors frequently speak of all-inclusive costs per hour instead of merely the regular wage rate per hour. A man whose wage rate is $4 per hour might cost the contractor more than $5 per hour when all these added costs are included.

■ EXERCISES

1. Make a chart showing the relationship between men on an average job or, if possible, on a job with which you are familiar.

2. State the relationship of subcontractors to a job and name at least six lines of work that are commonly sublet.
3. What is a material man? Name several kinds.
4. State what the term *mechanic* means as used in building work, and name six trade classifications of mechanics.
5. Name 10 positions held by men on a building job who are not employed by the general contractor.
6. Write at least 100 words of notes on the men concerned with building work, taken from one or two of the books listed in the Bibliography. Name the books used.

CHAPTER 3

# *the architect*

The architect's duties have been partly set forth in Chap. 2, which shows his relations with the group of men concerned in a building operation. He furnishes the drawings and specifications necessary for the proper prosecution of the work and also any additional drawings, details, and directions required as the work proceeds, except the shop drawings.

*Shop drawing* is the general term applied to all plans and details prepared by the contractor and his subcontractors and material men. The millwork details submitted by the material man for the millwork are shop drawings. The fabrication details and setting plans prepared by the structural-steel and miscellaneous-ironwork subcontractors are shop drawings.

The architect inspects and passes upon the work and determines whether it is in accordance with the contract and with his plans and specifications. He usually has the power to make decisions upon all the contracting parties and upon all controversies arising under the contract.

The architect has the power to determine allowances for changes in the work ordered by the owner. He has the power to act as the owner's agent in emergencies affecting the safety of limb or property and to order any work necessary to meet such emergencies. If he thinks such action is necessary for the proper execution of the contract or for safeguarding the work from injurious weather conditions, he may suspend the work until the causes of suspension have been removed. He certifies as to the amount due to the contractor and has the power to withhold payments to protect the owner from loss that is due to failure on the part of the contractor to meet obligations under the contract.

**The Office.** The architect's office personnel usually consists of a head draftsman, several other draftsmen, and a secretary. The head

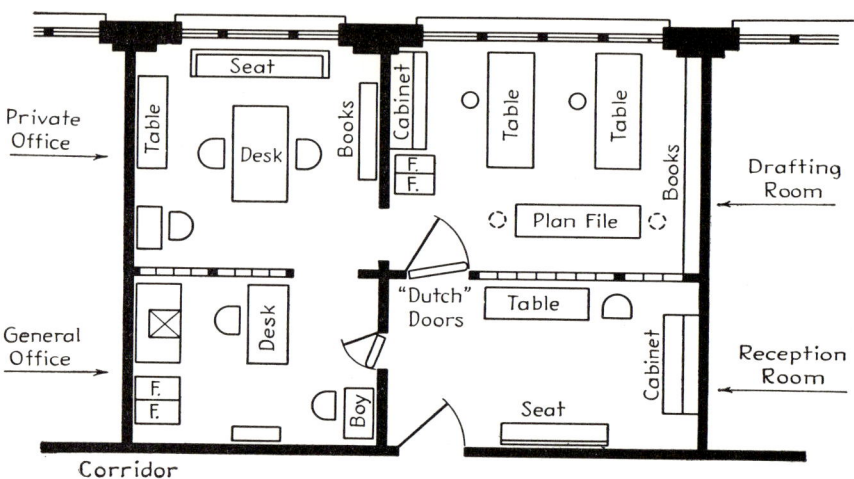

*Fig. 3-1  An architect's office.*

draftsman works in close conjunction with the architect in preparing the preliminary sketches and in studying the client's requirements, as well as in the general designing and planning. The other draftsmen complete the drawings and make the tracings.

In architectural offices there is always ample provision for drafting work. Even a small office will have several drafting tables. In addition, there are plan files, sample cabinets, bookcases, and other special equipment, as well as the usual desks, files, typewriters, etc., that are common to all offices.

Generally, either the architect or his head draftsman does all the specification writing. This is a sort of one-man specialty, as it involves not only continual contact with manufacturers' representatives but also the coordinating of all the building work described under the different specification headings (Fig. 3-1).

In large architectural offices there are designers, squad leaders,

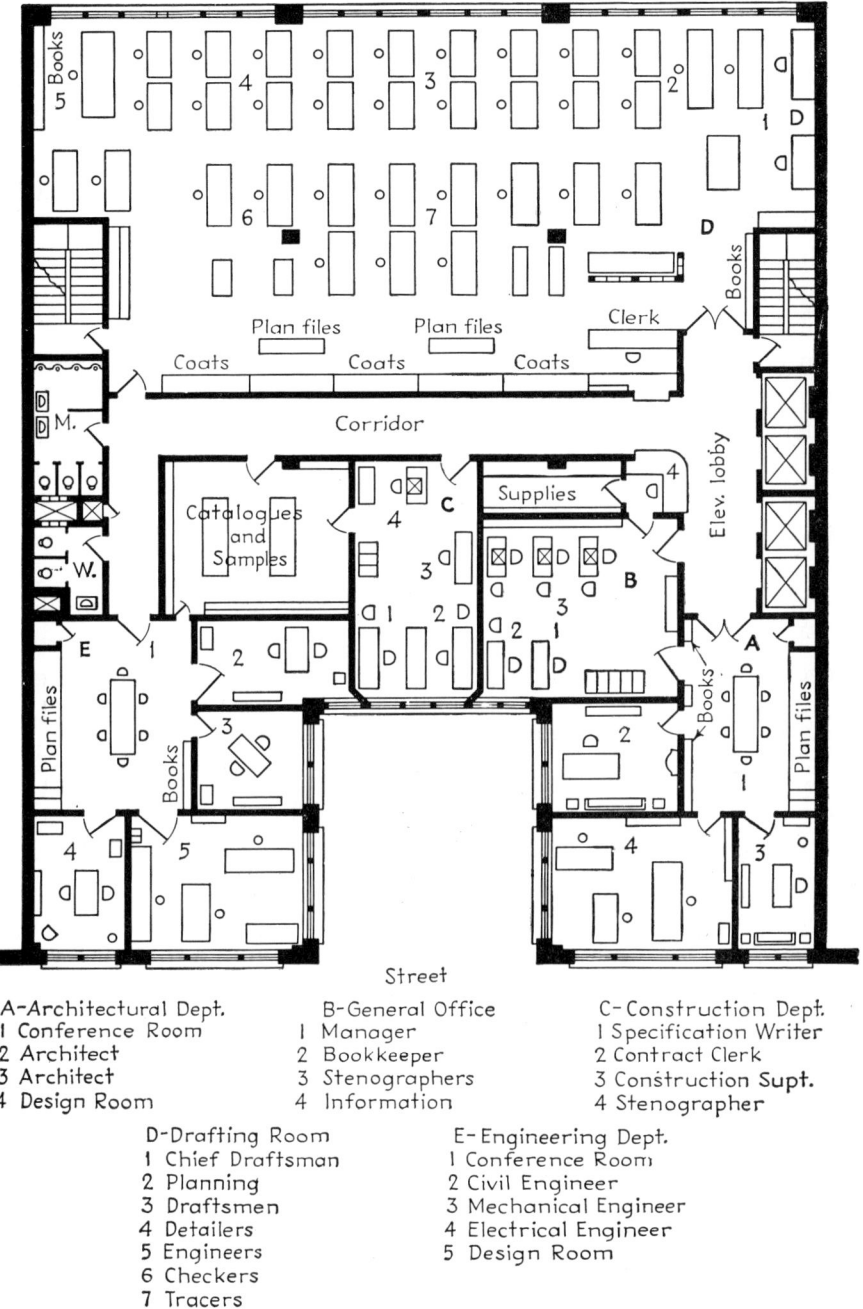

Fig. 3-2  *A firm of architects and engineers.*

A—Architectural Dept.
1 Conference Room
2 Architect
3 Architect
4 Design Room

B—General Office
1 Manager
2 Bookkeeper
3 Stenographers
4 Information

C—Construction Dept.
1 Specification Writer
2 Contract Clerk
3 Construction Supt.
4 Stenographer

D—Drafting Room
1 Chief Draftsman
2 Planning
3 Draftsmen
4 Detailers
5 Engineers
6 Checkers
7 Tracers

E—Engineering Dept.
1 Conference Room
2 Civil Engineer
3 Mechanical Engineer
4 Electrical Engineer
5 Design Room

draftsmen, tracers, detailers, engineers, structural draftsmen, mechanical draftsmen, checkers, etc. Such offices are often highly organized. Figure 3-2 shows the layout of offices for a firm of architects and engineers that handles the design of large industrial plants as well as the general run of architectural work.

**Licenses.** A license is required. In practically all states it is unlawful for any person to practice or to offer to practice architecture unless he has been duly licensed as an architect. Engineers, as such, may also obtain licenses, and they then have almost the same powers and duties as architects in connection with building work.

In New York State, licenses are issued by the State Education Department. To secure a license as architect the applicant must submit evidence that he is at least twenty-five years of age and a citizen of the United States. Besides, he must have completed an approved 4-year high school course or its equivalent, as determined by the State Education Department, plus the satisfactory completion of 2 years in an institution registered as maintaining satisfactory standards, conferring the degree of bachelor of arts or science, or the equivalent.

The applicant must submit satisfactory evidence to the Board of Regents of at least 5 years' practical experience in the office of a reputable architect, commencing after the completion of the high school course of study. The law in New York also provides that, notwithstanding all else, every applicant shall establish by written examination his competency to plan, structurally design, and supervise the construction of buildings and similar structures. Each complete year of study in a registered school or college may be accepted in lieu of one year of experience; in this case, the applicant must submit evidence of sufficient additional experience to give him a total of 8 years.

The examination in New York State is based on the four following subjects or groups:

## History of architecture

The candidate gives evidence in the examination, by means of clear descriptions, analyses of plan, construction, general expression, and ornament, that he understands the essentials that give character to the various historic styles of architecture.

## Architectural composition

The candidate must show that he understands the broad principles underlying the subject of architectural planning by the application of those principles to specific problems stated in the examination. The social, economic, and physical requirements of several architectural problems are outlined, and the candidate is asked to state the principal considerations that would guide him in the choice of an arrangement of plan that would most adequately express and fulfill the conditions suggested.

## Architectural engineering

In this subject the candidate's handling of the examination must give evidence that he has a thorough understanding of the appropriate use of the various materials used in buildings. He is required also to solve certain technical problems, such as the calculation of the proper economic dimensions of various structural members common to buildings in the several materials noted. The use of handbooks is permitted. Questions are asked relating to structural design, use of materials, heating and ventilating, electric equipment, plumbing and fire-protection equipment, and elevators.

## Architectural practice

In the examination the candidate must give evidence that he understands the moral and legal responsibilities of the architect in the proper performance of his duties. He is required to outline or draft clauses of contracts and to show that he understands the major provisions of state, county, and municipal laws and ordinances and the way they affect the different classes of buildings. Questions under this heading will be asked relative to the following topics:

**Business and professional functions of architects.** Professional relation of clients and contractors. Responsibilities of architects and methods of conducting their business.

**Building Laws.** State, county, and municipal. Filing plans and specifications. Obtaining permits.

**Contracts.** Drawings, specifications, and agreement, as essential parts of the customary contract between owner and builder. Provisions as to bids, letting contracts, requisitions, certificates, and payments.

**Specifications.** General conditions, purposes, and scope. Principles that should be observed in writing specifications.

**Drawings.** Purposes, use, and limitations of preliminary drawings. Essentials that should be embodied in contract drawings.

**Inspector.** The architect's inspector watches the men while they work and reports back to his office as to what is going on at the job day by day. If he is a practical and experienced construction man, his worth is appreciated on the job. His presence there is equal to that of an additional superintendent or engineer. He will be able to offer practical suggestions that will help toward making the job a happy place in which to work. No honest contractor objects to this sort of inspection or supervision on behalf of the architect or the owner. Some architects, however, send young draftsmen to the job to act as their inspectors, perhaps having them stop there only occasionally, if it is not a large job. As these young men have not had much to do with construction work, and as they do not keep in close contact with the work, they can merely act as messengers between the architect and the job. If they do just that much, the contractor's men are usually satisfied. The trouble is, a young man of this type may be overambitious and become a thorough nuisance in many ways; yet he must be treated courteously, even when he deserves rather to be thrown off the job. Specifications are usually based on average conditions, which are not always present. Unexpected circumstances may arise as the work progresses. Sometimes it is impossible or impractical to meet the specification requirements in every detail. In such cases an experienced inspector can work out a compromise.

Figure 3-3 shows a typical daily report as sent by an architect's inspector, employed full-time at the job, to his home office. The contractor's superintendent daily sends to his office a somewhat similar report.

**Specifications.** Specification writing involves a knowledge of the materials and methods of construction in detail and of the various customs that apply in all the trades. This implies that the specification writer must have experience in all these matters. Renderings for the client and well-prepared working drawings for the contractor should be followed by specifications that are as complete as the work warrants. The specifications are what guide the whole work in regard to the quality of materials and workmanship and the relations between the many parties concerned with the job. Specifications are legal documents as well as technical treatises and should always be written with these facts in mind.

A *good specification* is written in the same sequence, generally speaking, as the trades commence work on the job, following the section devoted to the general conditions. Thus, the first work heading would be Demolition, Clearing Site, or Excavating, depending upon which of these would be appropriate in the particular case. Suitable

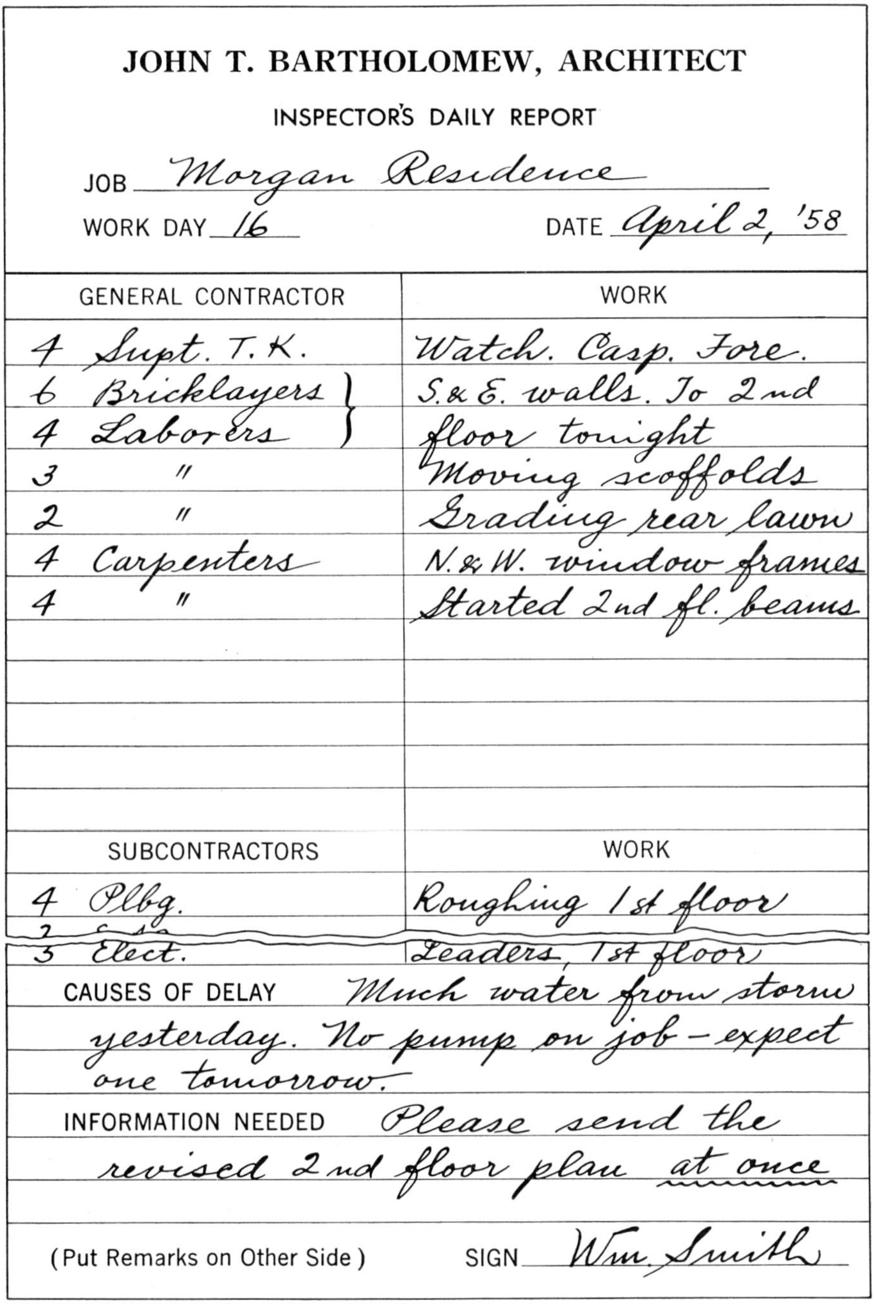

## JOHN T. BARTHOLOMEW, ARCHITECT
### INSPECTOR'S DAILY REPORT

JOB _Morgan Residence_
WORK DAY _16_    DATE _April 2, '58_

| GENERAL CONTRACTOR | WORK |
|---|---|
| 4 Supt. T. K. | Watch. Casp. Fore. |
| 6 Bricklayers } | S. & E. walls. To 2nd |
| 4 Laborers } | floor tonight |
| 3 " | Moving scaffolds |
| 2 " | Grading rear lawn |
| 4 Carpenters | N. & W. window frames |
| 4 " | Started 2nd fl. beams |

| SUBCONTRACTORS | WORK |
|---|---|
| 4 Plbg. | Roughing 1st floor |
| 3 Elect. | Leaders, 1st floor |

CAUSES OF DELAY  _Much water from storm yesterday. No pump on job — expect one tomorrow._

INFORMATION NEEDED  _Please send the revised 2nd floor plan at once_

(Put Remarks on Other Side)    SIGN _Wm. Smith_

Fig. 3-3  An architect's inspector's report.

headings for the rough and finish trades would follow. The so-called "mechanical trades," however, are generally placed at the end of the specification. These mechanical lines are plumbing, heating, electrical work, elevator work, and other piping, wiring, and machinery work.

The specifications are theoretically supposed to take precedence over the plans, what is described in the specifications being used in place of a differing plan indication for the same item. Similarly, large-scale drawings take precedence over small-scale drawings; therefore, estimators should always study all the detail drawings that are supplied with the general plans. Specifications, contracts, and plans must all be studied with great care; often a single sentence in a specification or a contract or a simple note or symbol on a plan will mean the loss of the profit on a job.

*Estimators* like a specification so arranged that they can follow its order in their own work and use the specification as a complete check on the items required. It must be said, however, that nearly all specifications show an overlapping of items. The estimator is, perhaps unfairly, expected to see that every item is provided for in the estimate, to ascertain whether all items are specified, and to straighten out any duplications in the specifications or in the estimate.

Contractors and estimators learn to beware of loose phrases in specifications, especially if they do not know the architect from previous experience with him. Someone has described specifications as an architect's dream, a contractor's nightmare, and a material man's dilemma. Such phrases as "to the satisfaction of the architect," "as directed," "in the opinion of the architect," "as approved by the architect," "if required," etc., depend for their meaning upon the character and the whim of the architect and his inspector on the job and are, therefore, obviously unfair to the contractor. The estimator is at a loss as to what prices to put on items so worded, because the architect does not specify just what is wanted.

Estimators should make a list of all vague wording found in specifications, especially if the indications are that their firm is to be awarded the contract, and then, if possible, should have a clarification put into the contract before it is signed. At least, they should have all such items discussed before the contract is signed, and notes kept of the clarifications.

**The Architect's Attitude.** Some architects set themselves up as a kind of court of last resort in regard to the judgment of quality of material, even going so far as to overrule the official grade markings of lumber inspectors, who are supposed to be unbiased. An estimator is forewarned by a specification in which an architect deliberately states that the architect's decision will be final. This usually means that he will be hard to please and will probably act as if he were paying for the building himself rather than serving as architect, or third party, to the contract. Despite the so-called ethics of the architectural profession, which claim that architects are fair-minded men, contractors know that many among them are very hard to please and technical and, in general, far from what one regards as fair-minded. Fortunately very few have the reputation of "breaking" the contractor or of "rubbing it in" on every job.

**Grounds for Dispute.** Most of the disputes in connection with building work arise from the fact that too little care has been given to the writing of the specifications. These disputes often lead to expensive lawsuits and arbitrations, which, although final, are unsatisfactory to all parties. It ought to be the duty of the architect to take care that when a contract is entered into, no disagreement may ever occur as to how much work the builder has to perform, what is to be the quality of that work, and how much money the owner must pay for it.

■ EXERCISES

1. Briefly, what are the functions of an architect?
2. Describe the work of an architect's office personnel.
3. Make a sketch plan of an imaginary architect's office.
4. What is a shop drawing? Name some types.
5. What are the four subjects or groups upon which the New York State architects' license law examination is based?
6. What is the general form and arrangement of a specification?
7. What lines of work are commonly referred to as the "mechanical trades"?
8. What is the order of precedence of specifications, general scale plans, and detail drawings?
9. What should an estimator do regarding vague phrases that he knows may bring disputes if they are not clarified?
10. What are four vague phrases often used in specifications?
11. Write at least 100 words of notes on the architect and his work, from one or two of the books listed in the Bibliography. Name the books used.

CHAPTER 4

# *the contractor*

**The Contractor's Duties.** The contractor (general contractor) organizes and is responsible for the entire job. He supplies the tools, equipment, and material for doing the work that is to be done by his own men and he makes subcontracts for the work that he does not intend to have done by his own men. He establishes the job office and provides the superintendent, job clerks, layout engineers, and watchmen. He usually has to obtain the building permits and to provide the temporary safeguards, temporary toilet facilities, and water supply. He builds the architect's office on the job (when one is called for) and on large operations he provides telephone service, heat, and water supply in it. The contractor pays all the bills and payrolls for his own work on the job, and he also pays the subcontractors. He has to satisfy a great number of Federal, state, and local laws, which call for taxes and insurance of many kinds, and he must make the whole job comply with the requirements of labor department, building department, and other public inspectors.

The term *contractor*, as used in building work, generally refers to the man (or firm) who undertakes to construct a complete building or to make a complete alteration of an existing building for an owner. The chart in Fig. 2-1 shows this relationship. He is sometimes called the *general contractor* or the *builder*, to distinguish him more clearly from the subcontractors. Subcontractors may be called contractors also, in their own special lines of work. Thus there are plumbing contractors, heating contractors, painting contractors, etc. Material men are occasionally referred to as contractors or subcontractors, too—millwork contractors, hardware contractors, etc., for example.

The building business is romantic and challenging, but it becomes increasingly complex with the passing years. No longer can the bright young foreman blossom out into a contractor by merely having a sign painted with his name on it. At least, this is not one of the important steps. Now, he must first organize to suit many laws and regulations, union requirements, financial arrangements, etc. Modern contracting

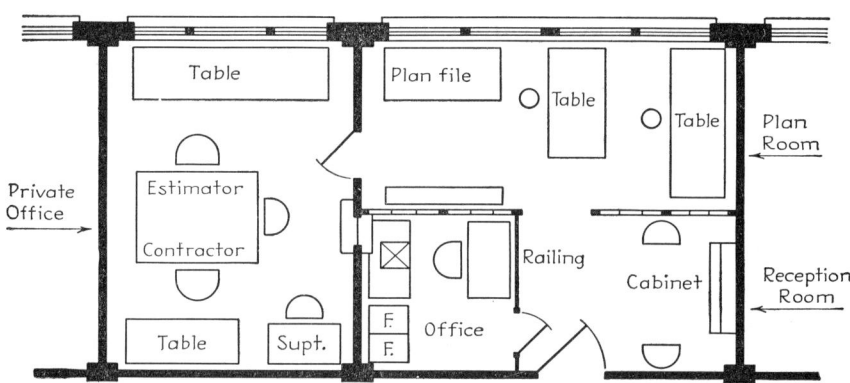

*Fig. 4-1*  A *contractor's office.*

is a business to be learned well before one undertakes it. It is not a poker game where the inexperienced may think they can jump in occasionally and win. Sometimes even an experienced contractor errs in thinking that he will be a wizard at anything he attempts. Only a small proportion of the men who start in business as contractors ever become really successful. Many of those who fail are expert mechanics, some are good foremen, and others are estimators who are lacking in experience. Perhaps the main reason for all failures is a lack of the understanding that the expert estimator possesses. Such an estimator is the logical man to start a building business; yet—by virtue of his deep insight into the many problems that are involved—he knows the importance of being organized for efficient management of every detail of the business, in addition to being strongly organized financially.

Figure 4-1 shows a plan of a typical contractor's office. The staff organization in such an office consists of the contractor, the estimator (or "office man"), the superintendent (or "outside man"), and the bookkeeper-stenographer. Where there is enough work, a regular bookkeeper and a timekeeper are employed. If the jobs handled are large enough in size, a superintendent is placed on each one and the so-called "outside man" becomes the general superintendent.

**Job Management.** Some men are peculiarly fitted for the planning and planting of construction work. Other men, who are deficient in this faculty, are capable of managing and executing plans. It almost goes without saying that unless a job is well planned, even the most conscientious management will all too frequently not be able to save it from loss. One is beaten from the start. Time could not be spent more wisely than in making a close study at the beginning to discover the best possible way of doing everything on the job. Obviously, hit-or-miss, slovenly ways of working should be avoided. Economy and savings usually follow in the wake of method and orderliness. In addition, the morale and interest of employees are raised by an atmosphere of order and method. Construction men like clean-cut, decisive instructions, but only with a well-thought-out program can such instruction be given with confidence.

A job is on comparatively safe ground when it is under the direction of a real superintendent, when it has a sensible but simple cost system, and when it has been properly planned and coordinated before being started. Good men cannot be held for such purposes in an organization, however, unless loyalty, ability, and earnest interest are rewarded. "Cheap" men are often very expensive in the end. A real superintendent is a man with experience, judgment, alertness, and a disdain for carping criticism. He is considerate and tactful in handling men, although he demands results from them. He avoids all favoritism. He is thick-skinned but always fair. He keeps informed as to new methods and equipment.

Jobs are affected by dissatisfaction among subcontractors. Too close bargaining or poor supervision on the job on the part of the general contractor's men may cause this. When a contractor holds the subs down to the last cent, things seldom go along as smoothly as they otherwise would. Bargains can be too close, and there is no tonic so stimulating to subs as making money. If they are squeezed too tight, by and by the general contractor himself gets squeezed, often in most unexpected ways. The law of compensation seems to bring this about.

A *progress schedule* should be made out for every job. This shows

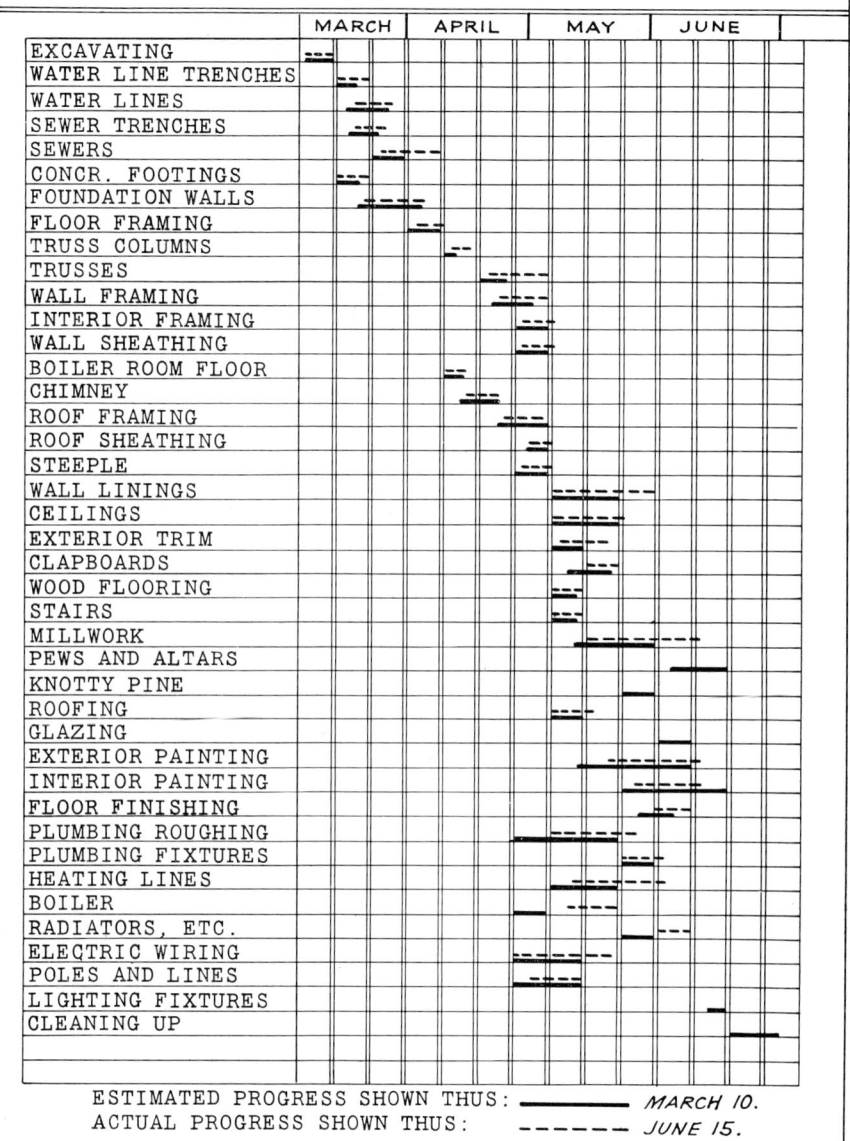

Fig. 4-2 *A progress schedule.*

the dates for the starting and finishing of each branch of work. In this way, control is better and the work of the various trades is less likely to conflict. The foundation, the rough superstructure, and the work of the finishing trades are the three general divisions of the schedule. Each of these divisions is subdivided to include the work of every different group of men that will be required on the job. The progress schedule is made out by the contractor's estimator, after consultation

| COST RECORD | | READY-MIX CONCRETE, PER CU. YD. | |
|---|---|---|---|
| Cost of concrete foreman, concrete gang, runways, stacking materials, and insurances. Add cost of delivered concrete, admixtures and hoisting. | | | |
| 1957 | Job | Cost | Remarks |
| June | Brenner res. | 4 59 | 12" fndn. walls |
| Sept. | Stewart Bldg. | 4 81 | 14" fndn. walls |
| Oct. | Do. | 4 46 | 6" floor on grade |
| 1958 | | | |
| Jan. | Dover Q.M.C. | 5 14 | 10" walls. Winter |
| June | Do | 5 01 | 7" reinfd floor, screeded |
| Nov. | Johnson res. | 4 54 | 12" fndn. walls |
| 1959 | | | |
| Feb. | Dover R.R. Sta. | 5 08 | 16" fndns. Winter |
| April | Do | 5 22 | 6" reinfd floor Trans |

Fig. 4-3   *A cost sheet for concrete work.*

with those subcontractors upon whose work the progress of the job will mainly depend. Copies of the schedule are given to the architect and to all the main subcontractors. The other subcontractors and material men are notified as to when they will be required. Figure 4-2 shows a typical progress schedule. This one was used to report and compare the actual progress.

*Cost records* should show all the items on the job—at least, all those in any given main division—and not just a sample day's work or a single item of work. The cost books should preferably be made to balance, just as an accountant's ledgers are balanced and closed. In this way, the incidental costs of foremen, hoisting, handling materials,

| COST RECORD | | | ROUGH FRAMING, PER MBF | |
|---|---|---|---|---|

Cost of carpenter foreman, carpentry gang, handling and stacking, nails, and insurances. Add cost of delivered material and any special hoisting.

| 1957 | Job | Cost | | Remarks |
|---|---|---|---|---|
| June | Brenner res. | 124 | 17 | Average, - all framing |
| Oct. | Stewart Bldg. | 120 | 91 | Floors, 3×12 |
| Nov. | Do. | 160 | 14 | Roof, 3×10, 3×8 |
| 1958 | | | | |
| Feb. | Dover Q.M.C. | 146 | 90 | Roof, 3×8, 10, 12 |
| July | Johnson res. | 120 | 80 | Average, - all framing |
| 1959 | | | | |
| March | Dover R.R. Sta. | 151 | 23 | Roof, 3×14 |

**Fig. 4-4** *A cost sheet for carpentry.*

| COST RECORD | | | COMMON BRICK, PER M | |
|---|---|---|---|---|

Cost of mason foreman, masons and laborers, handling and stacking, mixing mortar, and insurances. Add cost of all materials, scaffolds, cleaning, pointing, and hoisting.

| 1957 | Job | Cost | | Remarks |
|---|---|---|---|---|
| Oct. | Stewart Bldg. | 62 | 15 | 8" & 12" back-up |
| 1958 | | | | |
| Jan. | Dover Q.M.C. | 78 | 37 | 12" walls, Winter |
| July | Johnson res. | 76 | 20 | Chimney |
| Dec. | John's house | 75 | 90 | Chimney |
| 1959 | | | | |
| Feb. | Dover R.R. Sta. | 61 | 88 | 8" back-up |
| Mar. | Do. | 75 | 12 | Chimney |

**Fig. 4-5** *A cost sheet for mason work.*

insurance, etc., will not be ignored. The final analysis of the job-cost records should be very carefully compared with the estimate, and every possible lesson should be learned from this comparison for future estimating. Figures 4-3 to 4-5 present typical cost-record sheets. Nothing quite takes the place of one's own records when the matter of costs

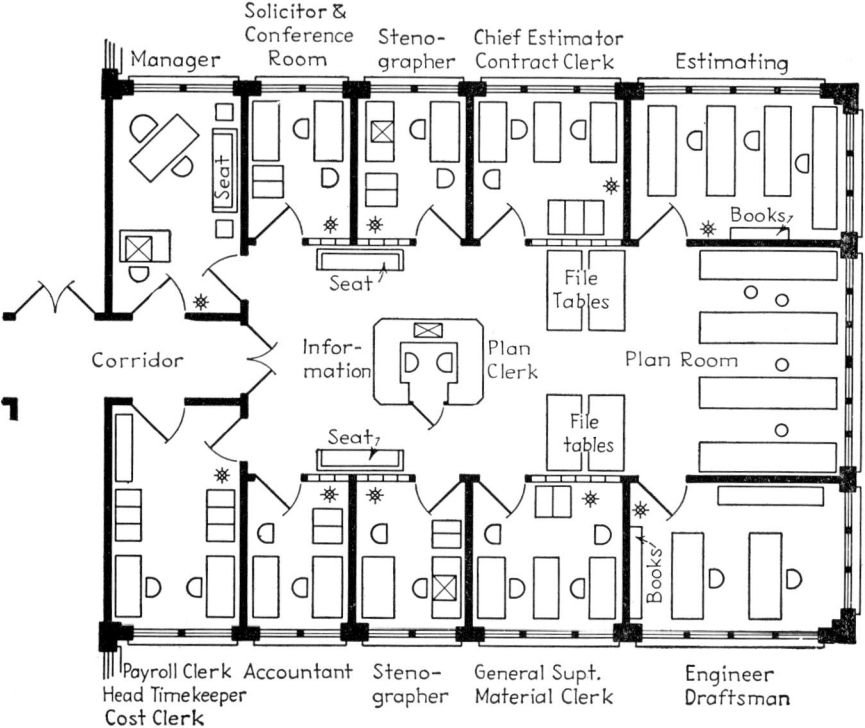

*Fig. 4-6* *A construction company.*

is concerned. Actual detailed accounts and analysis of work done under a contractor's own supervision give him a keen insight into the cost of the work performed that will develop in him (and in his estimator) the confidence required to estimate future items of the same kind.

**Construction Companies.** The construction company is simply an enlargement of the building contractor. There are many varieties and sizes of these companies. Some are national or even international as to the territory in which they operate. Some offer to build any kind of structure, while others specialize in certain types of structures or in certain kinds of construction. Often they maintain complete design-

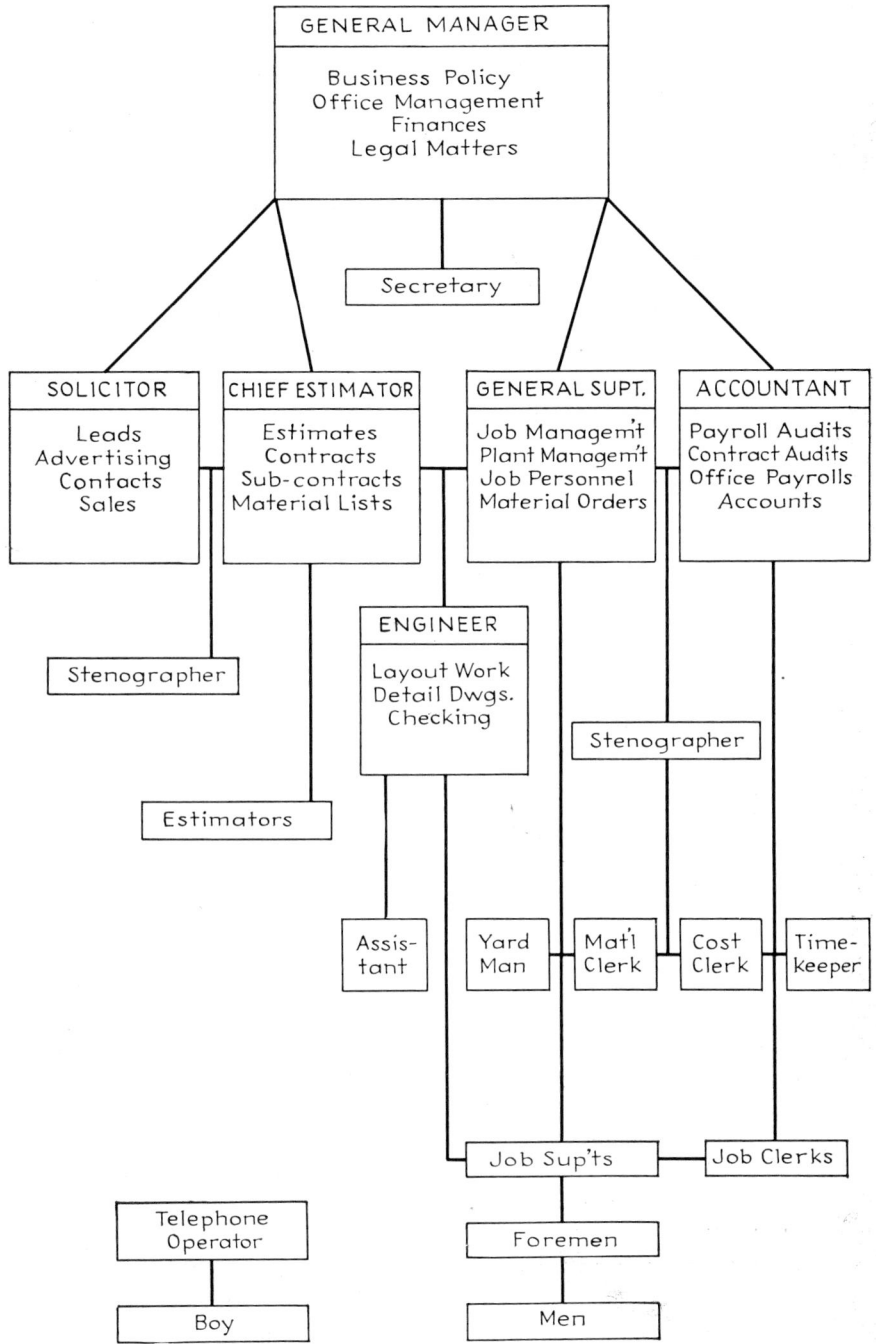

*Fig. 4-7  Organization of a construction company.*

ing and drafting departments and may go so far as to have financing departments, as well. A few companies have no workmen on their own payrolls and sublet all the various lines to other contractors. A few, at the other extreme, have all the workmen on their own payrolls and sublet nothing. The majority, however, do all the laying out and superintending and some of the work with their own men. Figure 4-6 shows the plan of one construction company's home office. Figure 4-7 is the organization chart of the same company.

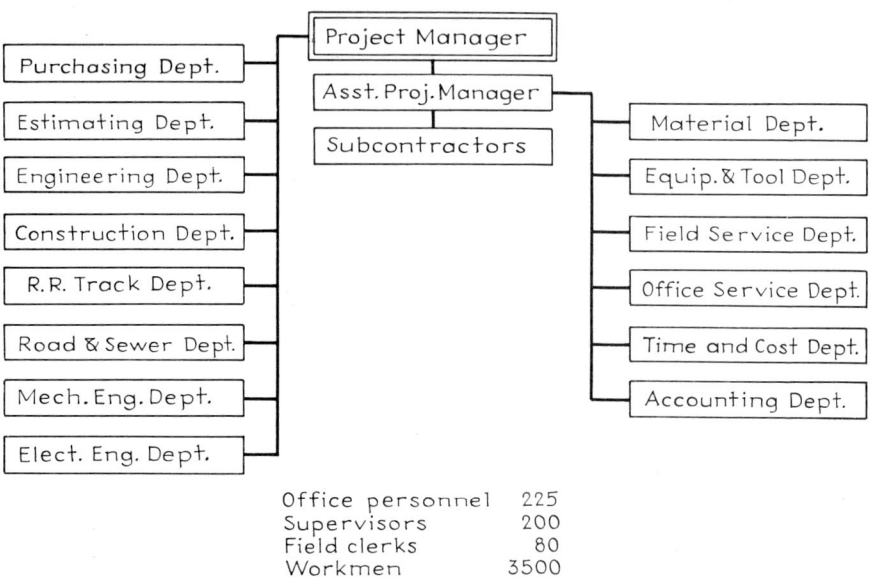

Fig. 4-8 *A large job organization.*

Very large operations are usually organized with full business facilities in the job office itself. The job office then becomes really a branch office of the company. Figure 4-8 shows the organization of a large and important job, employing about 4,000 men. Chapter 10, dealing with the cost of job expense and office expense, contains detailed lists of the personnel in each department. A study of these charts will enable the student quickly to get a grasp on the business of contracting and an idea of the items that go into the general divisions carrying such names as Job Expense, Job Overhead, General Expense, and Office Expense.

This chapter has discussed points that are of interest to everybody concerned with building work. They are of interest to students of estimating because of their effect on estimates. The general contractor's own personality, training, and character are very important factors that must be kept in mind by the estimator. The contractor's personality has a psychological bearing on the office organization, on the job management, and on the subcontractors. His training, if it is thorough and practical with respect to building work, can be turned to advantage in a very beneficial way; but if it is not based upon practical experience, it must be seriously considered in connection with the pricing of estimates. His character, as is the case in that of any businessman, is of prime importance. If he is honest and has a good reputation among his own employees and among subcontractors and architects, then the whole work in the office and on the job runs along in a pleasant way, without undue friction. If he is of poor character, you would better get away from him as soon as possible, for fear of becoming contaminated. Character and reputation—both more valuable than many young men realize—can make for happiness. Keep your character and your reputation *clean.*

## ■ EXERCISES

1. Briefly, what are the functions of a general contractor in connection with building work?
2. Make a sketch plan of an imaginary general building contractor's office.
3. What are the qualities of a good superintendent of construction?
4. Make a progress schedule for an imaginary job.
5. Name the departments that might be found in a highly organized job.
6. Write at least 300 words of notes on the contractor and the business of contracting, taken from two or three of the books listed in the Bibliography. Name the books used.

CHAPTER 5

# *plan reading*

*Blueprints* are copies of drawings that are made in a blueprinting establishment by a photographic process in which chemically treated white paper is exposed to a strong light. The tracing is placed over the sensitized paper, which turns blue where the light strikes it through the tracing, but remains white beneath the black lines and other markings on the tracing, which do not allow the light to pass through. The prints are the same size as the original drawing from which the tracing was made; thus, when anyone reads a blueprint, he is really reading the architect's drawing. Other kinds of prints are also used, some with blue or black lines on white paper. As many prints as are desired can be made from one tracing. Usually, about eight sets of the regular plans are required to provide enough for the contractor's office, for the subcontractors, and for other purposes.

## Plans

*Building plans* are generally understood to mean those that are prepared by the architect. These, together with the accompanying specifications, are intended to show or to describe all the work thoroughly. What is not shown on the plans should be fully described in the specifications, and what is not described fully in the specifications should be plainly shown on the plans. In general, dimensions, shapes, and other items of information that can best be shown graphically are put on the plans. Long descriptions of the way the work is to be handled and other matters that are best treated in written form are covered in the specifications. In some ways, the specifications are more important than the plans. Realizing this, a good architect takes as

*Fig. 5-1* Title boxes, etc., on drawings.

much care in the specification writing as he does in the preparation of the plans. Specification writing involves a detailed knowledge of the materials and methods of construction and of the various customs that apply in building work.

*Plan reading* involves specification reading as well, for although plans may be complete in every respect, the materials to be used and the workmanship desired are generally found only in the specifications.

In order properly to understand what is shown on the plans, therefore, it is usually necessary to read the specifications, too. Even the plans and the specifications together do not give every little detail of building work. It is assumed that the builder knows how buildings are constructed and, for this reason, he and the estimator must have training in the practical use of plans and specifications. They should also have had considerable experience with actual building work of the general nature of that shown on the plans being used. Figures 6-1 to 6-10 comprise a complete set of architectural plans, except that the borders, titles, etc., have been omitted in order to make use of the entire area of the textbook page to show the drawings to better advantage. Figure 5-1 shows two typical title boxes, etc.

Besides architectural plans, sometimes separate structural plans and sometimes separate mechanical plans are furnished. Structural plans, or framing plans, as they are often called, show the columns, girders, beams, and other framing members more clearly than they could be shown on the general architectural plans. These plans are used when the framework of the building is to be of the skeleton steel type or when the portions of the structure of the building are so complicated in design that a large amount of information is required on the plans. In like manner, mechanical plans are used when it is desired to show the plumbing, heating, and other branches of the mechanical equipment better than they could be represented on the regular plans. Special plans, such as these, may be prepared in the architect's own office or they may be made by outside engineers. The latter is usually the case when the special problems involved are very intricate and require the services of a specialist. Figure 5-2 shows a portion of a structural plan and Fig. 5-3 a portion of a mechanical plan.

## Types of drawings

The term *plans* is loosely used to refer to all the drawings, including not only floor plans but elevations, sections, and other drawings as well. Most building plans are drawn to the scale of a quarter inch to the foot. This means that every quarter inch measured on the plans represents one foot, or twelve inches, at the building. The scale rules are intended to save time in measuring the plans, although, of course, an ordinary rule can be used and the number of quarter inches measured can be counted. Other scales in common use for regular architectural drawings are ⅛" to the foot and, especially for details, ½", ¾", and 3" to the foot. Full-size and half full-size details are occasionally made for special features of buildings.

Some draftsmen have a bad habit of using several different scales in a single set of plans, where one scale could be used, or of using different scales for several drawings that appear on one sheet. They may even indicate one scale in the corner of a sheet and then make all the work or part of the work on the sheet at a different scale. Estimators learn to be very careful and to apply the proper scale to each drawing separately, regardless of the scale that may have been indicated.

**Floor Plans.** Floor plans, generally speaking, are pictures of the floors of buildings, such as one would get were he to imagine a building cut through on a horizontal plane, just above the window-sill level, and the upper part of the building removed. This view of the building shows the arrangement of the rooms and the location of the doorways, halls, stairs, etc., and gives the thickness of the walls and partitions and the size of the various parts of the building as measured horizontally. In making a floor plan, the draftsman uses conventional symbols or ways of indicating objects. These conventional indications are fairly well established, but there is no law making it necessary for any particular form of indication to be used. It should be noted, also, that draftsmen take liberties, so to speak, and show on floor plans items that would not be seen in a strict interpretation of the floor plan as a picture. Electric ceiling outlets, wall brackets, and kitchen wall cabinets, for example, are shown on the floor plans of a building. Other ceiling and wall features, also, are frequently included on floor plans. Figures 6-6 to 6-8 are floor plans.

**Elevations.** Elevations are views of the exterior of a building. The front elevation shows the outside of the front wall as one would see it when standing across the street. Note, however, that it is not a perspective or photographic view of the building. This is because the draftsman assumes that every point is directly opposite him or that he is always directly in front of every feature or point shown on the drawing. This is necessary in order that the drawing may be measured correctly with the scale rule. Figures 6-2 to 6-5 in Chap. 6 are elevations.

**Sections.** Sections show the interior of a building. A floor plan is really a section—a horizontal section, showing the floor layout, etc.—but the term *plan* is always applied in this case. Sectional views are

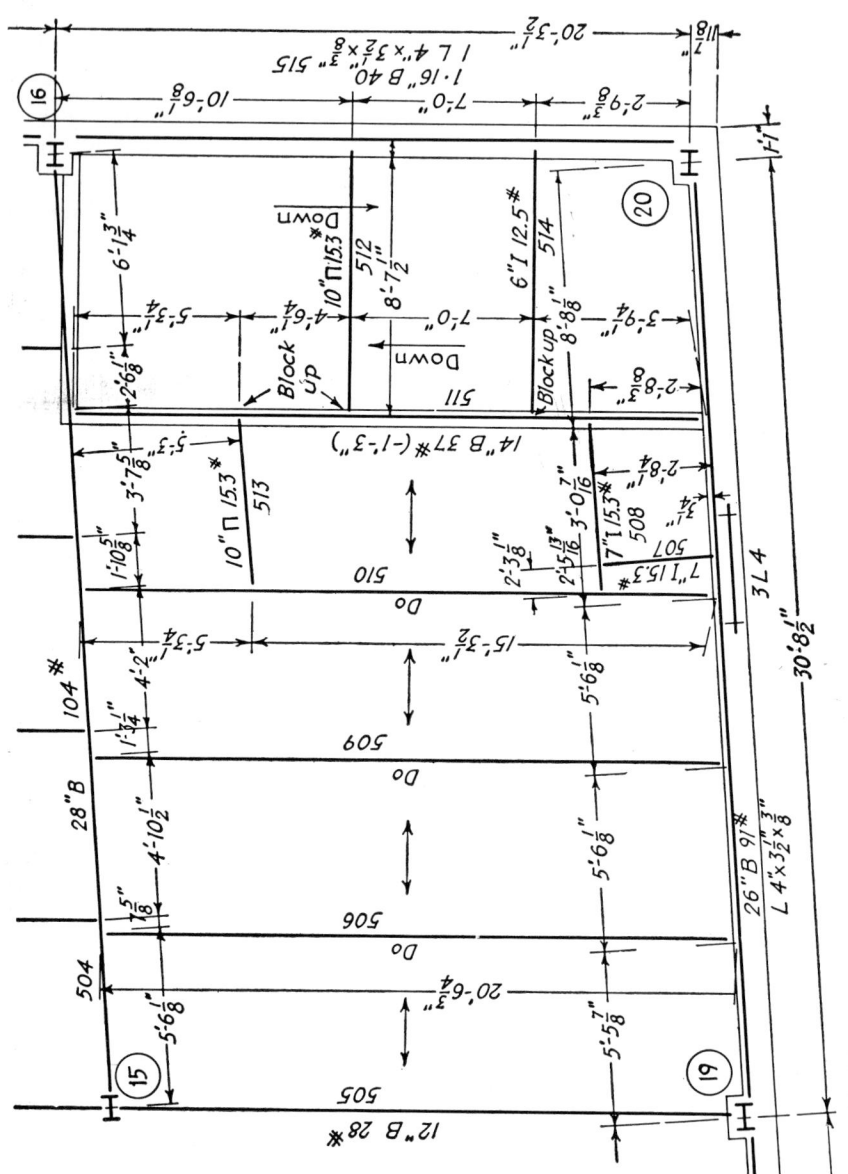

Fig. 5-2 Portion of a structural plan.

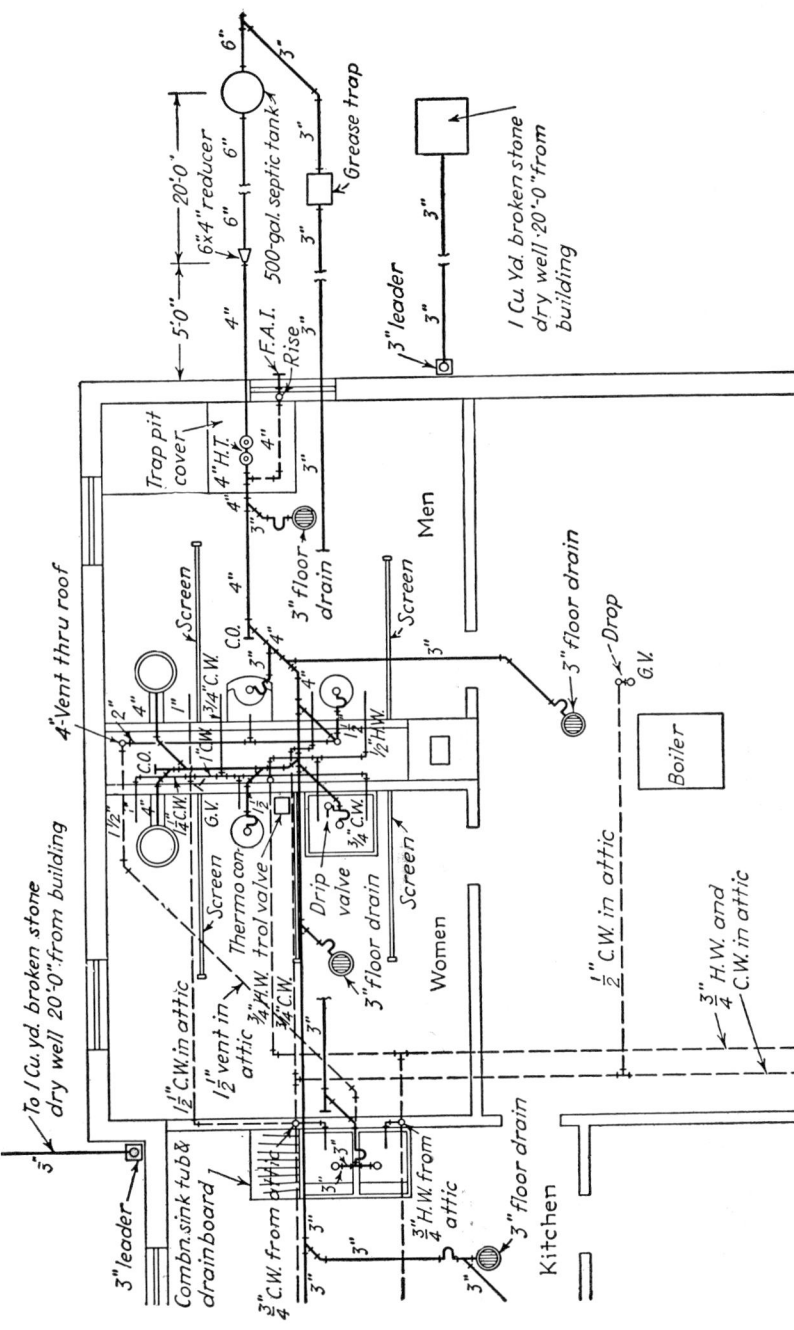

*Fig. 5-3* Portion of a mechanical plan.

35

those obtained by imagining a building cut through the other way—in a vertical plane instead of a horizontal plane—with one section of it removed. The picture thus presented shows the various floors, one above another, and the roof at the top. It shows the thickness of the floors and the heights of the basement and other stories, as well as their relation to the ground level at the building. Figure 6-9 is a section.

**Detail Drawings.** Detail drawings are made of the parts of the building that require to be developed in this way in order that the exact construction or shape of the parts may be seen more readily. This is done when these details cannot be shown well enough on the regular plans or when they cannot be readily described in the specifications. Details are thus made of doors, windows, cabinetwork, cornices, etc., when these are to be made up specially instead of in stock patterns. Special details of floor construction and of decorative work are often furnished along with the regular plans or are made up afterward. Figure 6-10 could be called a *sheet of details*, but the illustrations in Chaps. 12 through 23 are more truly detail drawings.

**Symbols.** Symbols are used on plans to save time in making the plans and in reading them. Unfortunately, like the conventional ways of indicating materials, etc., they are not absolutely established as to use. Good architects do indicate brickwork, terra-cotta blocks, and other regularly used forms of construction by crosshatching lines that are all but accepted as the standard method of indicating them. Doors, casements, and double-hung windows are generally shown on plans in such ways that persons familiar with plan reading seldom have trouble in discovering what is intended. National trade associations have endeavored to standardize the electrical and plumbing symbols. Despite all this, however, the estimator and all other men using plans have to be careful not to jump to conclusions regarding the symbols and other indications that they see on plans. What seems to be meant for gypsum blocks may mean terra cotta, instead; and what looks like brickwork may be concrete—just to mention two examples. Figure 5-1 shows a typical key to the symbols used on a plan.

**Abbreviations.** Abbreviations are used to a considerable extent on plans and less frequently in specifications. There is usually no difficulty about them, although at times it is hard to interpret some of them, and there are often two possible meanings that could be applied to one abbreviation. It is recommended that as few as possible be used and that those that are used be made easy to understand. Specifications should describe all work fully, with only the most common abbreviations, in order that disputes may be minimized.

## Laws and regulations

There are many laws and regulations pertaining to building work. The national labor laws and the social security and old-age pension laws have to be considered, both in the preparation of estimates and in the actual work of erection of buildings. State laws very directly affect the design and construction of buildings and the use of buildings after they have been completed. A thorough study of the state labor laws and the local building code would constitute an excellent review of the whole field of building design, materials, methods of construction, and the proper care and use of buildings of all kinds. However, an understanding of the more involved parts of these laws can be gained only by means of a background of technical training and experience in building work. Union regulations are becoming more definite and understandable, and these also will have to be considered by the estimator.

## Building codes

Building codes originated in the desire of people for fire protection. Later, safety of life and health was looked for and the present-day codes cover the design, construction, equipment, and occupancy of structures of all kinds. Codes are intended to be schedules of minimum requirements and should never be used as specifications of good design and construction. In many instances, they certainly fall far short of describing the best.

A building code is a technical treatise on a highly involved subject and is not readily understandable by the layman. The purpose of the New York City code, as stated in it, is to provide standards, provisions, and requirements for safe design, methods of construction, and sufficiency of materials in structures, and to regulate the equipment, use, and occupancy of all structures and premises. It deals with the heating and ventilating of buildings. Stairways and other means of egress are given consideration. The important subject of sanitation is treated in a separate section, and fire-resistive construction, fire sprinklers, and standpipes for fire fighting are provided for. Elevators, in which many thousands of people travel every day, are governed by what is probably the longest and most elaborate set of elevator regulations ever produced anywhere.

Codes usually give definitions of the terms used in them. One should keep in mind that the use of terms in a code is not always the same as that in trade language or the same as among real estate dealers and others having to do with property. For example, the terms *basement, cellar, fire wall, penthouse,* and others given in the code may have different meanings if used in other connections.

Some terms defined in the New York code are the following:

**enclosure wall.** An exterior nonbearing wall in skeleton construction, anchored to columns, piers, or floors, but not necessarily built between columns or piers, nor wholly supported at each story.

**curtain wall.** A nonbearing wall built between piers or columns for the enclosure of the structure, but not supported at each story.

**panel wall.** A nonbearing wall in skeleton construction built between columns or piers and wholly supported at each story.

**apron wall.** That portion of a panel wall between the window sill and the support of the panel wall.

**spandrel wall.** That part of a panel wall above the window sill and below the apron wall.

**parapet wall.** That portion of a wall extending above the roof.

**fire wall.** A wall provided primarily for the purpose of resisting the passage of fire from one area of a structure to another and having a fire-resistive rating of at least four hours.

**partition.** A nonbearing interior wall one story or less in height.

**fire partition.** A partition provided for the purpose of protecting life by furnishing an area of exit or refuge and having a fire-resistive rating of at least three hours.

**fireproof partition.** A partition other than a fire partition, provided for the purpose of restricting the spread of fire and having a resistive rating of at least one hour.

**fire door.** A door and its assembly capable of resisting fire, as specified in the code in detail.

**fire window.** A window with its sash and glazing having a fire-resistive rating of three-quarters of an hour.

**protective assembly.** An opening protective including its surrounding frame, casings, and hardware.

**floor area.** Any floor space within a story of a structure enclosed on all sides by either exterior walls, fire walls, or fire partitions.

**horizontal exit.** The connection of any two floor areas, whether in the same structure or not, by means of a vestibule or an open-air balcony or bridge, or through a fire partition or fire wall.

**place of assembly.** A room or space 2,500 square feet or more in area, or a space capable of or designed for containing 250 or more occupants and used for educational, recreational, or amusement purposes.

**private dwelling.** A structure occupied exclusively for residential purposes by not more than two families.

**human occupancy.** The use of any space or spaces in which any human does or is required to live, work, or remain for continuous periods of two hours or more.

**steel joist.** Any approved form of open-webbed beam or truss less than 20″ in depth, made from rolled- or pressed-steel shapes by welding, pressing, riveting, or expanding.

**controlled concrete.** Concrete for which preliminary tests of the materials are made and the water-cement ratio is made to correspond to these and other tests. All controlled concrete work must be inspected by the architect or engineer responsible for its design and a complete progress record kept.

**average concrete.** Concrete for which the preliminary tests are omitted and the work is without the inspection required for controlled concrete. When concrete structures are designed to receive their full loads at periods of less than 28 days after the concrete is installed, controlled concrete must be used.

**shaft.** A series of floor openings, consisting of two or more openings in successive floors, or a floor and a roof, shall be deemed a shaft and shall be enclosed. Such shafts shall be constructed of materials of three-hour rating, except dumbwaiter shafts and except vent shafts in nonfireproof residence buildings.

■ EXERCISES

1. Briefly, what is each of the following?
    a. A floor plan
    b. An elevation
    c. A section
    d. A detail drawing
    e. A plot plan
2. What do the following abbreviations usually stand for?
    a. Bldg. Line
    b. TC
    c. Fin.Flr.
    d. Kal.

e. Brk.
   f. DH
3. Describe briefly architectural, structural, and mechanical plans, and state what each shows.
4. What are blueprints?
5. Give three types of crosshatching used on plans, and state after each what it may be used to represent.
6. Give three symbols used on plans, and state after each what it may be used to represent.
7. What is meant by ¼" scale? Give an example of its use.
8. Make a sketch plan of a grocery store, using a full page, and show the things listed below as they are generally shown on floor plans:
   a. Brick exterior walls
   b. Toilet room, enclosed with TC partitions
   c. Show windows, counters, shelves
   d. Doors, DH windows
   e. Plumbing fixtures, electric lights
9. Write at least 300 words of notes on plans and plan reading, taken from two or three of the books listed in the Bibliography. Name the books used.

CHAPTER

# *house plans*

Figures 6-1 to 6-10 comprise a complete set of architectural plans for a house. These will serve for further drill in plan reading and will also be used considerably in the discussion of construction work and the study of estimating.

## The plot plan

A plot plan is shown in Fig. 6-1. This one indicates the location of the house on the property and also gives the ground grades and the relation of the floor levels to these grades.

The term *existing grade*, or *present grade*, means the ground surface before any change is made in it. The term *finish grade*, or *final grade*, means the surface that is to be produced by the changes. In this case the street surface at the driveway has been established as *datum*, which is the reference grade, or *bench mark*, from which the other grades are taken. The architect has marked this as zero on his plot plan.

There are five trees shown, and two of these are to be removed because they are in the way of the house. The existing ground levels, or grades, are to remain for the most part. However, the specifications for the job require that 9 in. of topsoil from the entire width of the property, and for a distance of 60 ft back from the sidewalk, is to be removed, piled at the rear of the property, and finally spread over the front lawn and terrace and elsewhere on the property as directed by

41

the architect. None of this material is to be removed from the premises. Topsoil is the rich soil usually found at the surface of the ground, sometimes as much as 18 in. deep.

**Excavation.** The excavation for the house will be about 4 ft deep below the removed topsoil, and most of this excavated material will have to be hauled away. The removed topsoil will be ample for any terrace grading required. Grading means lowering or raising the ground surface to bring it to the desired level or slope.

**Lot Lines.** Lot lines, or property lines, define the extent of the property but the driveway extends out to the street. The front-lot line is often referred to as the *building line*. This is a legal term meaning the line between public and private property. A corner plot would have a building line on each street. Local laws and regulations generally state how close to this line one may build. In sections zoned by the authorities for business use, one may build out to this line and, within specified limits, construct steps, areaways, and cornices which extend into the public property. In residential sections, however, it is likely that not only *zoning restrictions* but also *setback restrictions* are established by law, and possibly *deed restrictions*, established by the present or former owners of the property, may apply.

The house shown in Fig. 6-1 is required by law to be set back at least 20 ft from the building line and at least 5 ft from the side-lot lines and to be on a plot at least 7,000 sq ft in area. The deed restrictions require that the house be a one-family house, two stories high.

**Surveys.** Surveys are often used as plot plans, or information taken from a survey may be incorporated in the plot plan. Surveys are generally made by licensed surveyors. On them are accurately shown the property lines and the angles and directions of these lines, also streets, curbs, lamp posts, sewers, and other pertinent information. Frequently the contractor is required to furnish the owner, for his records, a survey showing the exact position of the building after it has been built.

### ■ EXERCISES

1. What is a plot plan?
2. What do the terms *present grade* and *finish grade* mean?
3. What is topsoil?
4. What do the terms *building line* and *lot line* mean?
5. What are *zoning restrictions* and *deed restrictions*?

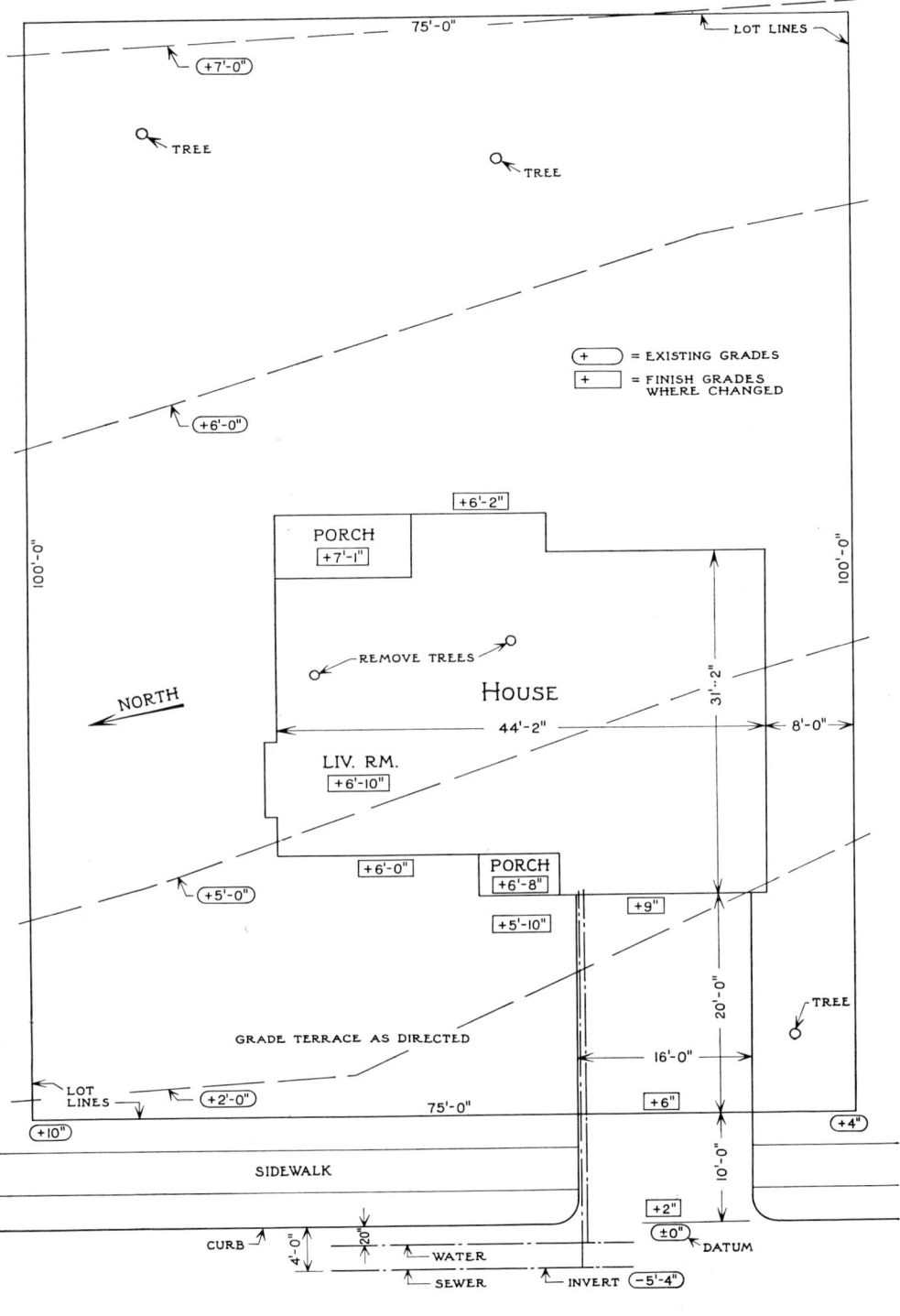

Fig. 6-1  A plot plan.

## Front elevation

Figures 6-2 to 6-5 are exterior views of the house. Figure 6-2 shows the front of the house and is referred to as the *front elevation.*

Elevations are sometimes named according to the way they face. This one could be called the west elevation because it faces west, as can be verified on the plot plan.

This is a frame house having a portion of the front and left side walls veneered with brick. The piers, or pedestals, under the flower boxes at the left side, and the chimney are also of brick on concrete-block foundations.

**Other Drawings.** The other drawings show that there is other brickwork involved. The left side, which shows the brick veneer and the brick chimney, also shows brick walls under the front porch. The basement plan shows four brick piers in the garage. The first-floor plan shows a brick floor, or hearth, in front of the fireplace. The detail drawing shows the brick veneer and also the fireplace wall and brick at the top of the wall between the crawl space and the playroom. This will point out the need to refer to all the drawings when looking for one material or one kind of work as, in this case, the brickwork. All the drawings must be examined and the specifications carefully read to obtain complete information.

**Garage Door.** The garage door is the overhead type, swinging up on tracks suspended below the ceiling of the garage. Sufficient space must be provided above the door opening for the door clearance.

**Main Entrance.** The main-entrance door is 3'0" wide, as shown on the first-floor plan. The specifications describe it and state further that it is 1¾" thick and 7'0" high and is of white pine. The frame for this door is also described in the specifications and, of course, the hardware and glass would be found noted under proper headings in the specifications.

**Living-room Window.** The living-room window is a triple window consisting of a plate-glass stationary picture window flanked on either side by a regular double-hung window. All three are contained in a special frame which projects out about 6" and has a copper-covered roof. The term *double-hung* means the commonly used type of window in which a lower sash and an upper sash slide past each other. The other front windows are regular double-hung windows, the one over the garage door being a triple window.

**Exterior Walls.** The exterior walls are faced for the most part with wood shingles but reference to the two side elevations will show that the gable areas have redwood siding set vertically. Also note that the portions of the foundation walls which extend above the ground are indicated on the side and rear elevations to be stuccoed.

**Rafters.** The rafters are 2 × 8's 24" O.C. This means that they are 2" × 8" in cross section and they are to be set 24" on centers, 24" apart from the center of each rafter to the center of the next one. The roofing, while not noted on this drawing, is indicated on the cross-section drawing and described in the specifications as composition asphalt shingles made in strips of four shingles each, laid on 1" × 6" T&G (tongued-and-grooved) roofing boards.

**Specifications.** The specifications must always be used in conjunction with the drawings. It will usually be found that some items are specified but are not shown on the drawings, and in many instances the specifications give important additional information regarding what the drawings do show.

The specifications for this house will be found in the exercises at the end of the chapters dealing with the various trades involved in the construction.

■ EXERCISES

1. With what materials is the front wall faced?
2. What is a DH (double-hung) window?
3. How many DH windows are there in the front wall?

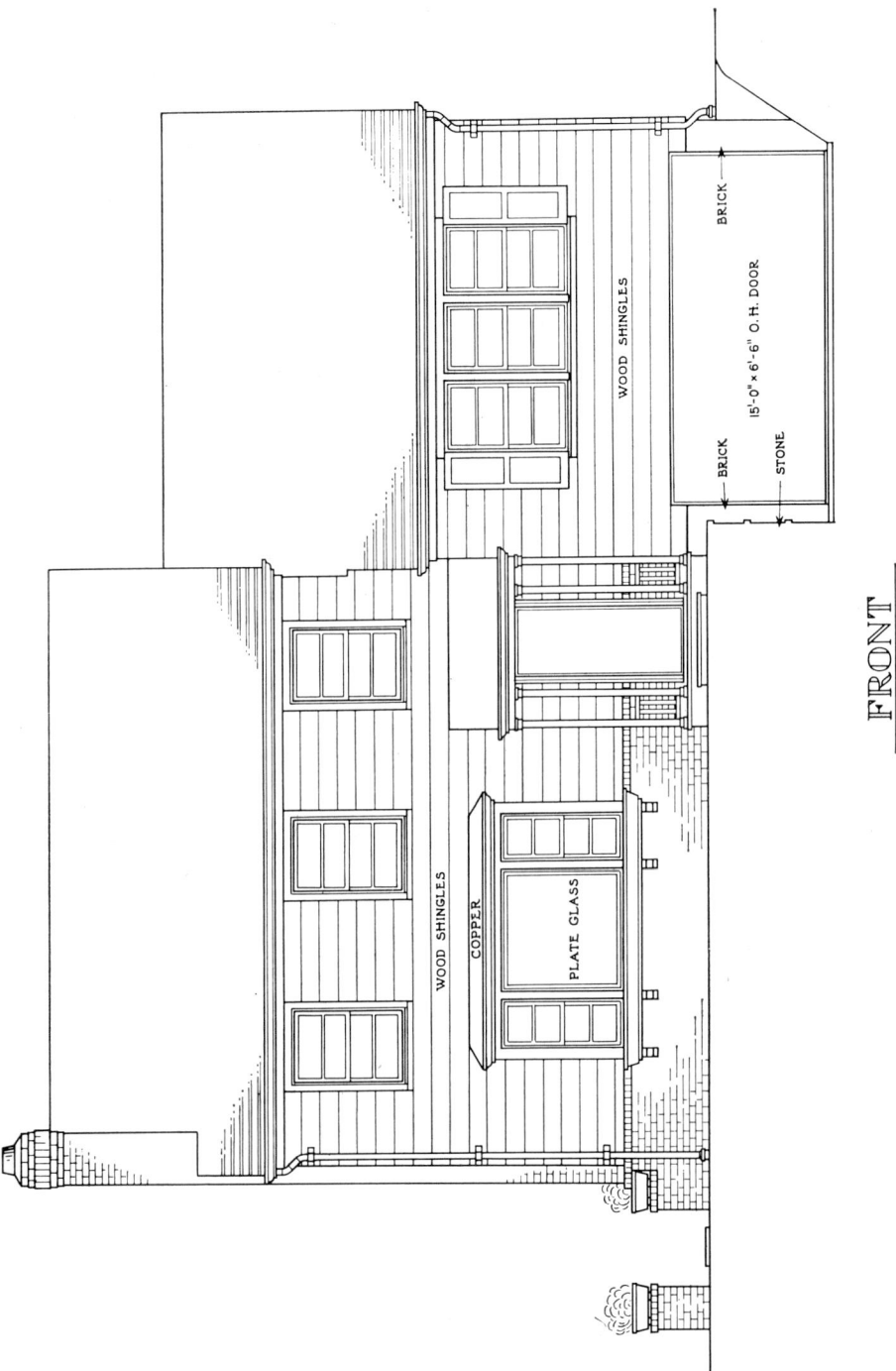

*Fig. 6-2* Front elevation.

## Left elevation

The left elevation is shown in Fig. 6-3. This can also be referred to as the north elevation because it faces north.

**Brick.** The brick chimney and the brick veneer on either side of it may be seen here. Two TC (terra-cotta) flues project from the top of the chimney, one leading from the boiler and the other from the fireplace. These flue linings are clay pipes and come in 2-ft lengths which are built in as the chimney construction progresses. The one serving the boiler in this house is round and has an inside diameter of 8″. The fireplace flue is larger, 13″ × 13″ on the outside, to provide a better draft, which is required for fireplaces. Flues are obtainable in round and rectangular shapes of various sizes. The outside dimensions are used for rectangular flues and the inside diameter for round flues. The chimney is broader at the bottom to accommodate the fireplace and is shaped above this to relieve the appearance.

**Gable.** One gable end of the house appears on this side. The gable is faced with redwood vertical siding and contains a DH window, a triangular gable louver, and false (decorative) rafters. The slope of the roof may be seen on this drawing. This is called a gable roof to distinguish it from a hip roof and a flat roof.

**Porches.** The *front-porch* floor is of flagstones (see first-floor plan), supported on a concrete slab and brick walls. The other portions of it are wood with asphalt shingles on the roof and an aluminum gutter along the front with a short spout at one end of it. The *rear porch* is all of wood, including the floor and the roof deck and railing. The roof deck is covered with heavy canvas for walking upon.

**Dining-room Window.** The dining-room window is a box window composed of a stationary picture window with an outswinging casement window at each side. It projects out about 18″ and has a window seat on the inside. The projecting portion is supported on two wooden brackets and the roof over it is covered with copper. See the rear elevation, the first- and second-floor plans, and the cross section for other views of this box window, or bay window, as it might be called, and the roof over it.

**Subsurface.** Subsurface conditions are sometimes shown on elevations. This one shows, in dotted lines, the wall footings and foundation walls at this side of the house and the foundation under the front porch. All foundations in northern climates should extend down far enough to escape the heaving action of frost. The law in these areas, through the state or municipal building code, generally requires a minimum depth of 4 ft for all foundations except those under steps and other minor parts.

The wall footing shown in Fig. 6-3 is "stepped" where the levels change between the playroom and the crawl space. This stepping is carefully constructed so as to provide level surfaces on which to construct the wall. Furthermore, if the excavation here or under any other footings is cut deeper than the plans show, it is required that the excess depth be filled with concrete, not with dirt.

Note that no footings are called for under the front-porch foundation walls nor under the foundation for the stone wall at the driveway.

All the foundation walls for this house are of hollow concrete blocks on 20″ wide concrete footings. The blocks under the chimney and the top course of all the other foundation walls are specified to be filled in solid with concrete. Brick piers are used to support the concentrated loads of the steel girders in the garage. See the basement plan for the thicknesses of the walls and the size of the piers.

■ EXERCISES

1. What size TC flues are specified?
2. What material is used for the front-porch flooring? For the rear-porch flooring?
3. Using a scale rule, what is the total height of the chimney from the top of the footing to the top of the brickwork?

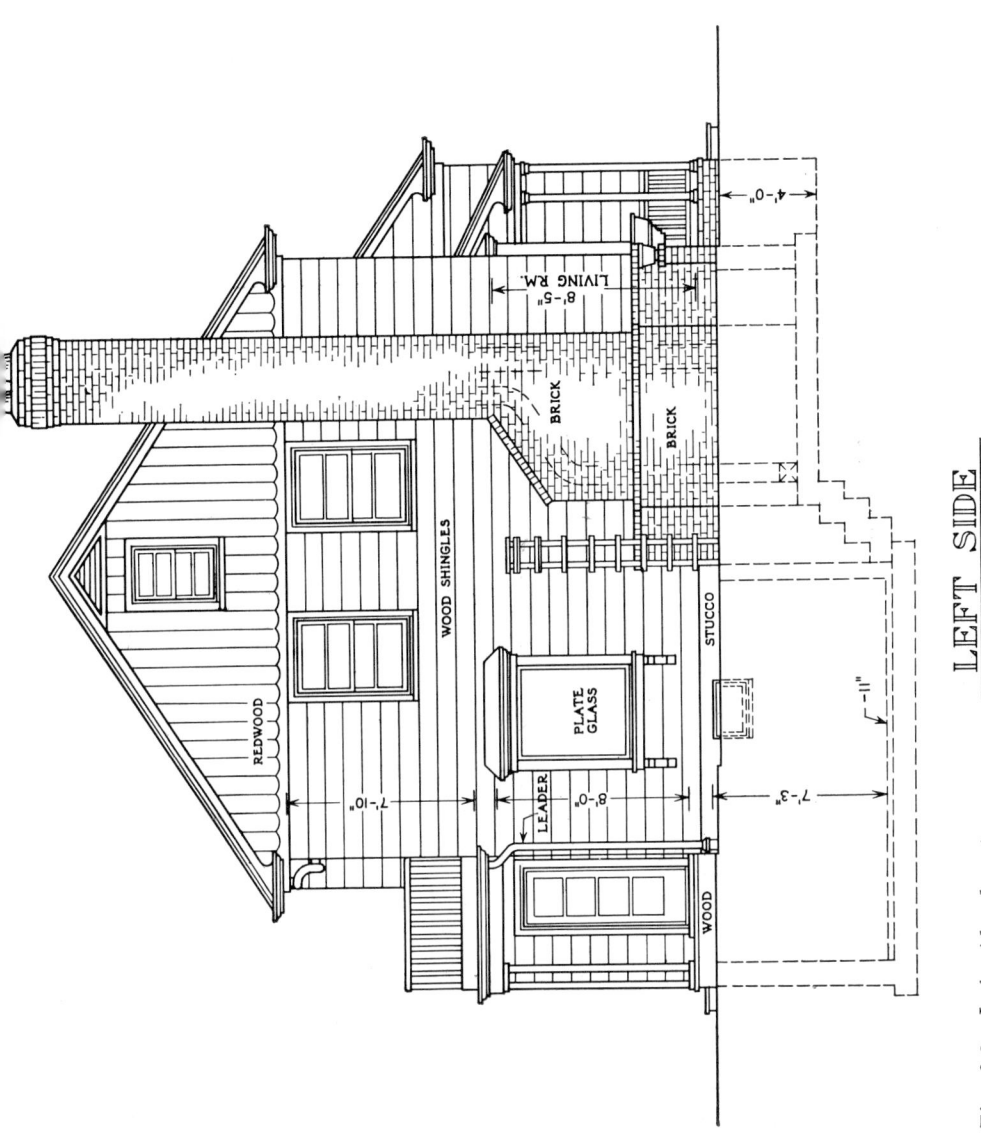

*Fig. 6-3* Left side elevation.

## Right elevation

The right elevation, or south elevation, is in Fig. 6-4. Stone steps and a stone wall are shown at the front of the house. The steps are curved and the specifications call for them to be carefully constructed over rough concrete steps, the concrete to be reinforced with ⅜" steel rods placed 6" O.C. both ways. This means the rods run crosswise every 6" and longitudinally every 6" and thus form a steel grid embedded in the concrete. The steps are specified to have rough local-stone risers and 1½" thick one-piece bluestone treads, all laid in cement mortar. The wall, which is at the north side of the driveway, is also of local stone, in random ashlar formation, with one-piece bluestone copings 1½" thick. The wall rests on a foundation of hollow concrete blocks filled with concrete and extending down 24".

At the right side of the drawing may be seen the portion of the kitchen which extends at the back of the house and, above this, the sun deck over the rear porch, which is beyond the kitchen extension. One of the kitchen windows, a casement, is shown. The window below the kitchen window is also a casement, or, more precisely, a pair of casements. This opening is in the areaway at the back of the house and provides ventilation and some light for the playroom. The two garage windows have single top-hinged outswinging sash.

The rain gutters are "returned" on the side walls about 24" to improve the appearance. These are moulded aluminum gutters which, with the fascia boards, soffit, and wood mouldings, form the cornice of the house. See the cornice detail in Fig. 6-10. The gable trim, or rake cornice, is composed of a false beam with shaped pieces at the lower ends and a moulding under the edge of the roof shingles.

The upper gable is on the two-story portion of the house. The lower gable is on the south-end wall at the storage room. The headroom is only about 6'8" under the rafter collar beams in the storage room.

**Collar Beams.** Collar beams are ties between rafters on opposite sides of a roof. Sometimes, as a supposed aid in tying the building together, collar beams are introduced part way up on the slope of the rafters. They should never be depended upon as ties unless the rafters are quite steep, because the nearer the tie is to the top, the greater the leverage action upon it, with consequent tendency to pull out the collar-beam nails, and also to bend the rafters. The proper function of collar beams is to stiffen the roof.

If the action of a collar beam is analyzed, it will be seen that it probably acts as a strut rather than a tie. When secured against spreading at the base, rafters tend to sag. The collar beam will, therefore, act as a strut or prop which, to be most effective, should be at the halfway point. Sagging of a rafter on one side tends to push out the one opposite. This reinforcement will prevent both from sagging.

The collar beams over the storage room in this house are placed high to provide headroom under them. Additional support for the rafters is provided by $2'' \times 4''$ stud bracing as shown in Figs. 6-8 and 6-9. Collar beams at the midpoint of the rafters, and also similar stud bracing, are specified for the attic over the main portion of the building.

■ EXERCISES

1. What size reinforcing rods are used in the slab supporting the front steps? How are they spaced?
2. What type of windows are used in the garage? What size are they?

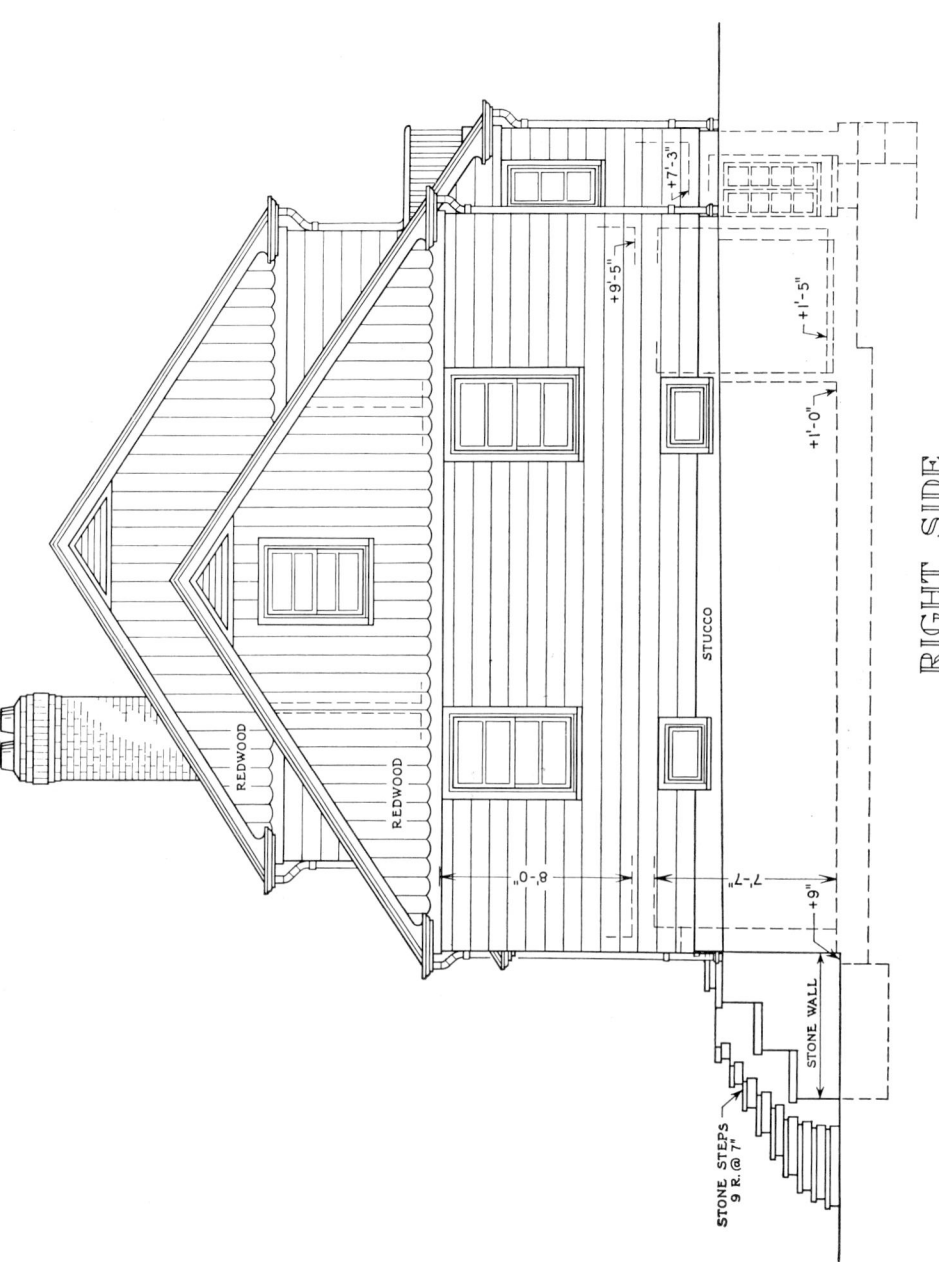

Fig. 6-4  Right side elevation.

## Rear elevation

The rear elevation is in Fig. 6-5. The pair of windows near the left-hand side are spoken of together and called a mullion window. The dividing member between them is called the mullion. Three windows, placed with mullions between them, are also taken together and termed a triple window. The narrow members dividing the glass panes are called muntin bars. Note, however, the term *pane* is never employed in building work. We speak of pieces, or *lights*, of glass, not panes of glass.

Areaway steps are indicated by dotted lines on the drawing. The areaway door leads to the laundry and garden room in the basement. It has glass in the upper part and is flanked on either side by a small DH window. The kitchen window is a unit containing a stationary window with an outswinging casement at each side.

At the right side of the drawing is a rose arbor with a pair of gates in it and, beyond this, the brick piers and flower boxes at the front of the house may be seen.

**Framing.** This house uses the western, or platform, type of framing. Light frame construction may be classified into three more or less distinct types: balloon frame, western frame, and braced frame. The principal characteristic of balloon framing is the use of studs extending in one piece from the foundation to the roof, the beam ends being nailed to the studs. The upper floors' beams are also supported by a ribbon or ledger board let into the studs. Platform framing is distinguished by floor platforms independently framed, the upper floors each being supported by studs one story in height. The braced frame is the oldest type and is still used, in modified form, in some sections of the country. In it heavy corner posts are used, often with heavy intermediate posts between, all extending from a heavy foundation sill to an equally heavy plate at the roof line and with heavy girts to carry the upper-floor beams. The studs in this type serve merely as fillers to support the finish materials. In modified forms lighter members are used and the studs support the floors and roof as in the other types.

In partitions parallel to the floor beams, the entire weight of the partitions may be carried on one or two beams. Ordinarily these beams are already carrying their share of the floor load and, for this reason, additional strength must be provided. The prevailing practice has been to double the beams under such partitions, that is, to put an extra beam beside one of the regular ones. Computation shows, however, that the average partition weighs nearly three times as much as a single beam should be expected to carry. At least two beams in addition to the regular ones should be placed under a non-bearing partition which is parallel to the beams, and at least three additional under a bearing partition. This is one that carries a portion of the load of the floor above.

For non-bearing partitions at right angles to the floor beams, it is ordinarily not necessary to increase the size or number of beams, because of the large margin of safety in the live loads for which beams are used. If, however, two such partitions parallel to each other cross the beams near the center of span, the beams should be spaced at intervals of 12" instead of the usual 16", or the depth of all beams increased 2", or the width of all beams increased 1".

■ EXERCISES

1. What is a mullion window? A triple window?
2. How many pieces, or lights, of glass are there in each of the two bathroom windows?
3. What is meant by the platform-frame type of house framing?

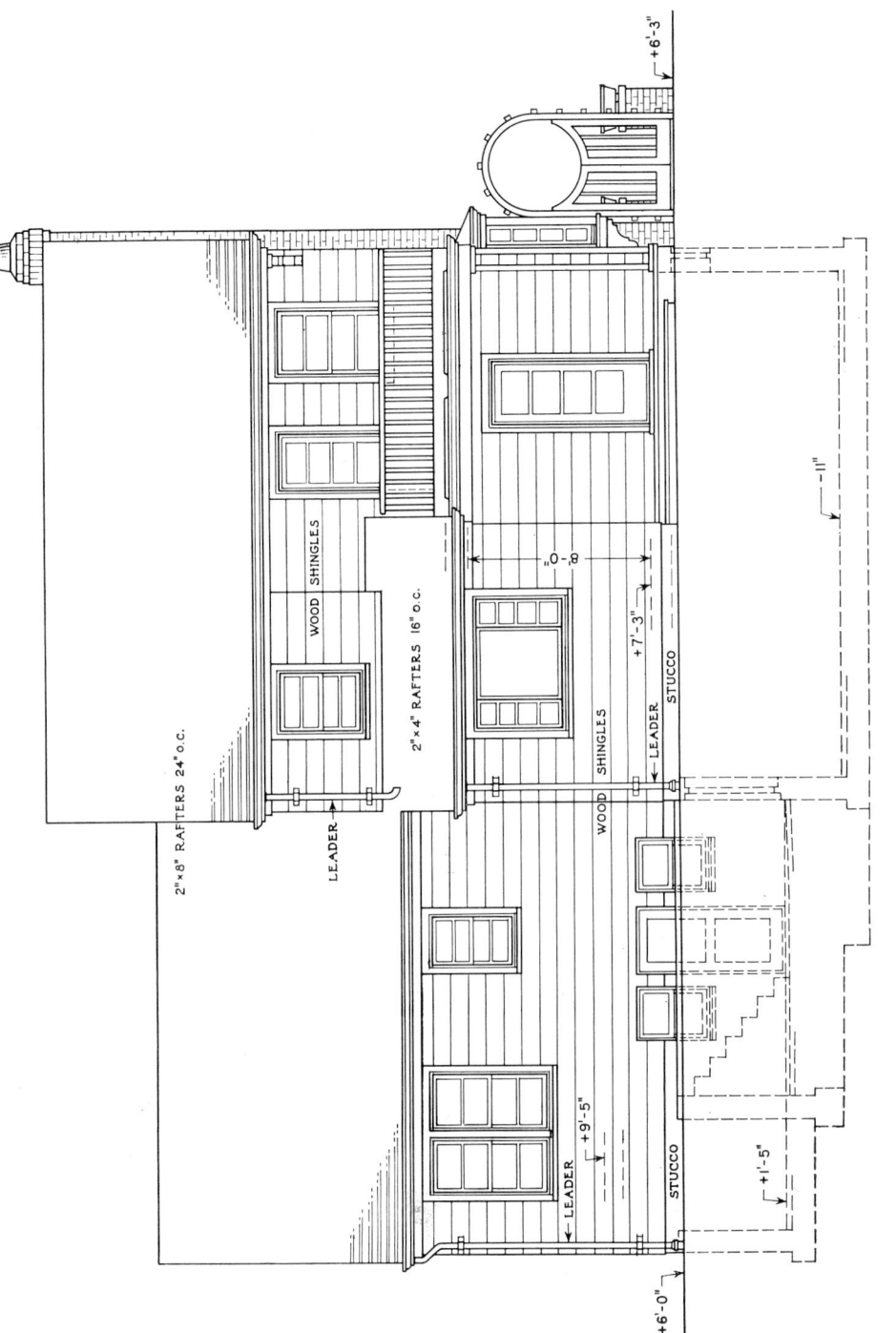

*Fig. 6-5* Rear elevation.

59

## Basement plan

The floor plans are shown in Figs. 6-6 to 6-8, Fig. 6-6 being the basement plan.

Basement is used broadly to denote any space below the main floor. Building laws, however, generally define the term to mean a floor space which is more than one-half above the ground. If more than half the height between the floor and the ceiling is below the ground level, it is then, in legal terms, a cellar.

**Walls.** The foundation walls (or basement walls) in this house are of hollow concrete blocks, some of which are filled with concrete. They are 12" thick in the main walls and 8" in the other walls, and all rest on concrete footings. The 4" cinder-concrete-block partitions start on the concrete basement floor.

The *heights* of the walls vary. It is necessary carefully to examine this drawing and the elevations and cross-section and detail drawings to determine these heights. Obviously, all those enclosing the playroom will be the highest walls because they start at the lowest level in the basement, and the beams over this room are 7'3" above the finish-floor level. However, the 8" block wall between the playroom and the crawl space extends up only to the underside of the living-room floor beams and is therefore about 10" less in height than the exterior walls of the playroom.

*Steel girders* are shown in the playroom and in the garage. The ends of those in the playroom rest on the block walls, and they are also supported by two Lally columns. *Lally column* is a term commonly used to designate a column composed of a steel pipe filled with concrete. The name is that of the originator, but many other manufacturers now make similar columns. The girders in the garage are carried on brick piers. The one at the front may be called a lintel. A lintel is any member placed over an opening in this fashion to carry the load above it.

The floor beams indicated on this drawing are those over the basement, and they support the first floor. These are all $2 \times 10$'s except under the back porch, where $2 \times 8$'s are shown. The dash lines indicate the direction in which the beams run.

This house has a back areaway in which there are steps leading up to the rear walk. Full advantage of the areaway is taken for windows in the laundry, toilet, and playroom and for a glazed door.

The laundry is planned to serve also as a garden workroom and as a passage between the stairway and the various rooms in the basement. All the rooms, except the garage, have asphalt tile on the cement floors and the concrete steps to the playroom have rubber treads.

**Construction.** However carefully a building may be framed, or however rugged it may be built, inadequate foundations will result in uneven settlement, cracking of plaster or tile, ill-fitting doors and windows, and other difficulties which will offset, to some extent, the advantages of good work in the superstructure.

Pier and column footings under girders of the average building carry loads much greater per square foot than do those under foundation walls. Such a footing may often carry 10 times the weight per square foot that is carried by the wall footing. The load on these footings should be figured and they should be made large enough to carry the load that will be placed upon them.

### ■ EXERCISES

1. What material is used for the basement and other foundation walls? What thicknesses?
2. How many steel columns are there? How many steel girders?
3. How many square feet of floor are there in the playroom?

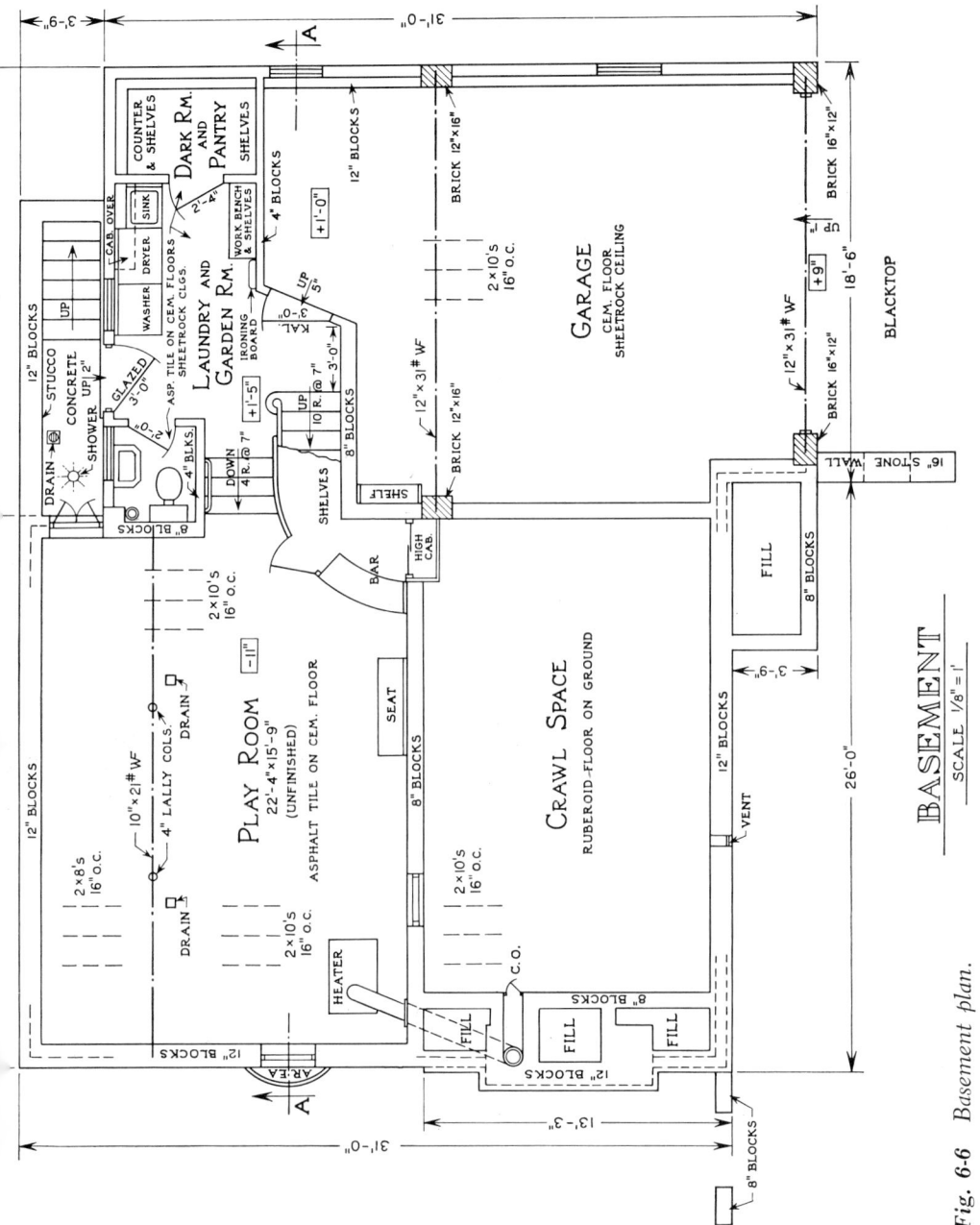

Fig. 6-6 Basement plan.

## First-floor plan

The first-floor plan is given in Fig. 6-7. This floor has three levels. The vestibule, foyer, and living room are on one level. The dining room, kitchen, and nook are up one step and the other rooms are four steps still higher. These are all "easy" steps, each one being more than an inch lower than usual in height.

**Vestibule.** The vestibule and foyer have quarry-tile floors laid on "deafening." This is cinder concrete 4" thick placed on wood subflooring. The floor beams here are set 4" lower for this purpose. The floor in the coat closet is stepped up to provide clearance, or headroom, in the garage below it. All the wide closets in this house have pairs of hinged doors and those in the bedrooms are fitted with full-length mirrors. Some people prefer wood or metal sliding doors or accordion-type doors.

The front-porch floor is of flagstones laid on a concrete slab. The posts, railing, and seat are wood.

**Living Room.** The living room has a fireplace, and the brick hearth extends also in front of the bookshelves. The balance of the living-room floor is ¾" thick plywood which is planned to be entirely covered with carpeting. Interior elevations of the living room are shown in Fig. 6-10, where the cabinets here and in the foyer may also be seen.

**Dining Room.** The dining room is one step up from the living room and the opening between the two rooms requires a strong lintel over it. This is composed of three 2 × 10's spiked together. A detail of this is shown in Fig. 6-10. This room will have oak flooring and there is a glazed door to the back porch and a double-acting door to the kitchen.

**Kitchen.** The kitchen and dining nook will have vinyl-tile flooring, plastic table and counter tops and splash backs, and enameled steel cabinets. The seat in the nook will have a leather-covered foam-rubber cushion and back and, above it, an open-front cabinet with plastic shelves. A door shuts off the kitchen from the hall and the basement stairway.

**Other Rooms.** The other rooms on this floor are set apart and at a higher level for privacy. The bathroom will have tile flooring and wainscoting and the other spaces oak flooring. The bathroom has a recess tub with a shower curtain, a lavatory contained in a plastic-topped cabinet, and a large medicine cabinet with a pair of doors. The small room is intended to be used as a study, a den, a guest room, or a child's bedroom.

The walks outside the house are shown on this plan. They are of flagstones laid directly on the ground.

**Construction.** Oak is considered the most desirable hardwood flooring for residences. It can be obtained either quartersawed or plain-sawed, both in red oak and in white oak. It is graded as quartersawed or plain-sawed, and these terms are further qualified as clear, select, No. 1 common, and No. 2 common.

Oak flooring comes in random lengths varying from 18 in. or less up to 12 ft or more, the average piece being about 4 ft long. Hardwood trees do not lend themselves to the manufacture of long lengths of flooring as do the softwood trees.

In general, narrow flooring is to be preferred to wider flooring because any tendency to open up at the joints is reduced in proportion to the width of each piece. In addition, there is less likelihood of cupping or squeaking, since the nailing comes at closer intervals.

### ■ EXERCISES

1. What is "deafening"?
2. What kind of flooring is used in the living room? Dining room? Kitchen?

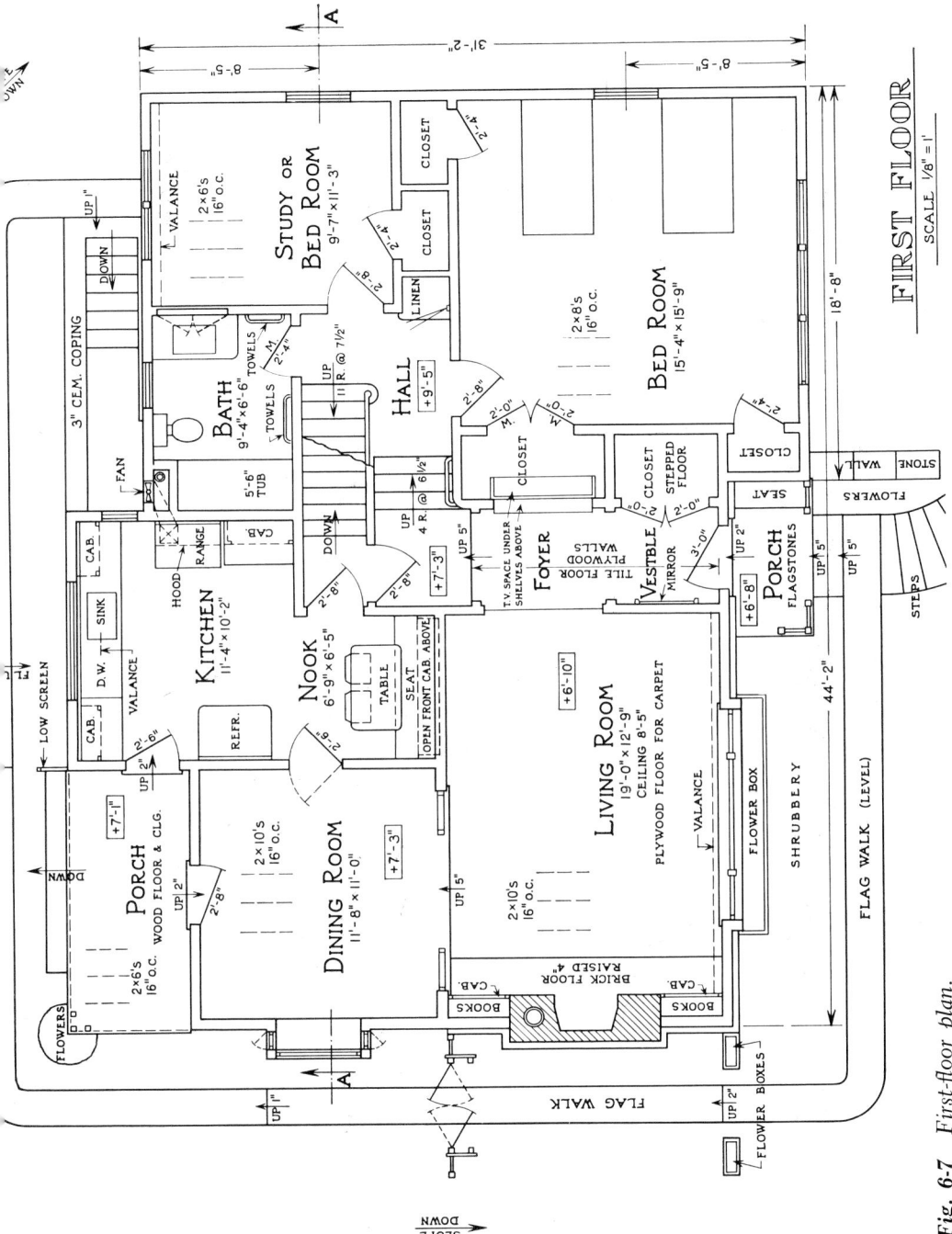

*Fig. 6-7* First-floor plan.

## Second-floor plan

The second-floor plan is in Fig. 6-8. The storage room on this floor is at a higher level than the other rooms. It is entered by stepping up on a ledge in the hall and then up one step at the door to the room. The room has a single plywood floor and is otherwise unfinished. This means that the wood studs and rafters will be exposed to view. No flooring is provided in the low-roof spaces adjoining the storage room, and therefore the ceiling beams over the rooms below will be exposed. The attic space above the main portion of the house is also unfinished, and it is reached by way of a ladder which leads to an access door above the linen closet. The specifications call for a catwalk, 24" wide, in this upper attic, from the access door to the north wall.

The chimney flues are shown on this drawing. The flue linings have shells about ¾" thick. Larger flue linings have thicker shells. All come in 2-ft lengths.

The roof is "broken" over the storage-room portion in order to show the interior construction. The roof over the main portion is not shown on this drawing. There is a canvas-covered flat roof, or sun deck, over the rear porch. The roof over the living-room windows is of sheet copper, as noted on the front elevation. The other three small roofs are covered with composition asphalt-strip shingles the same as on the main roofs.

Doors of various widths are noted on the plans. Those leading from the bedrooms are 2'8" wide to provide access for large pieces of furniture. The wide closets have pairs of doors, each leaf opening independently, to give full and easy access to the entire closet, and some are provided with mirrors. Note the way doors swing and refer to Chap. 18, Finish Carpentry, for a diagram that gives the "hand" of doors.

**Construction.** Note the way the various parts of the building are drawn. The walls and partitions are represented by two parallel lines scaling about 6" apart. The walls are constructed of 2" × 4" studs covered on the outside with insulating sheathing board and wood shingles and on the inside with plasterboard and plaster, the whole thickness being about 6". The 4" studs actually measure only about 3¾". The partitions are constructed of 2" × 4" studs with plasterboard and plaster on both sides.

For partitions, the common practice is to provide a double top plate. It might be mentioned here, however, that inasmuch as a plate is usually supported every 16″ and no weight is ordinarily imposed upon the plate between the studs, one $2 \times 4$ in most cases would be sufficient. Good practice, however, suggests that a double plate be used in the case of bearing partitions and that it is desirable to use double plates on walls to secure a good lap joint at corners and at partition intersections.

The best of materials and workmanship for plaster and interior trim will fail to give satisfactory results unless the underlying framework of a building is strong and rigid. Durability and wind resistance, freedom from cracks and settlement, all depend in part on good framework.

### ■ EXERCISES

1. In the two bedrooms, not including the closets and alcove, compute the following:
    a. Square feet of floor area
    b. Linear feet of baseboard
    c. Square yards of plaster ceilings

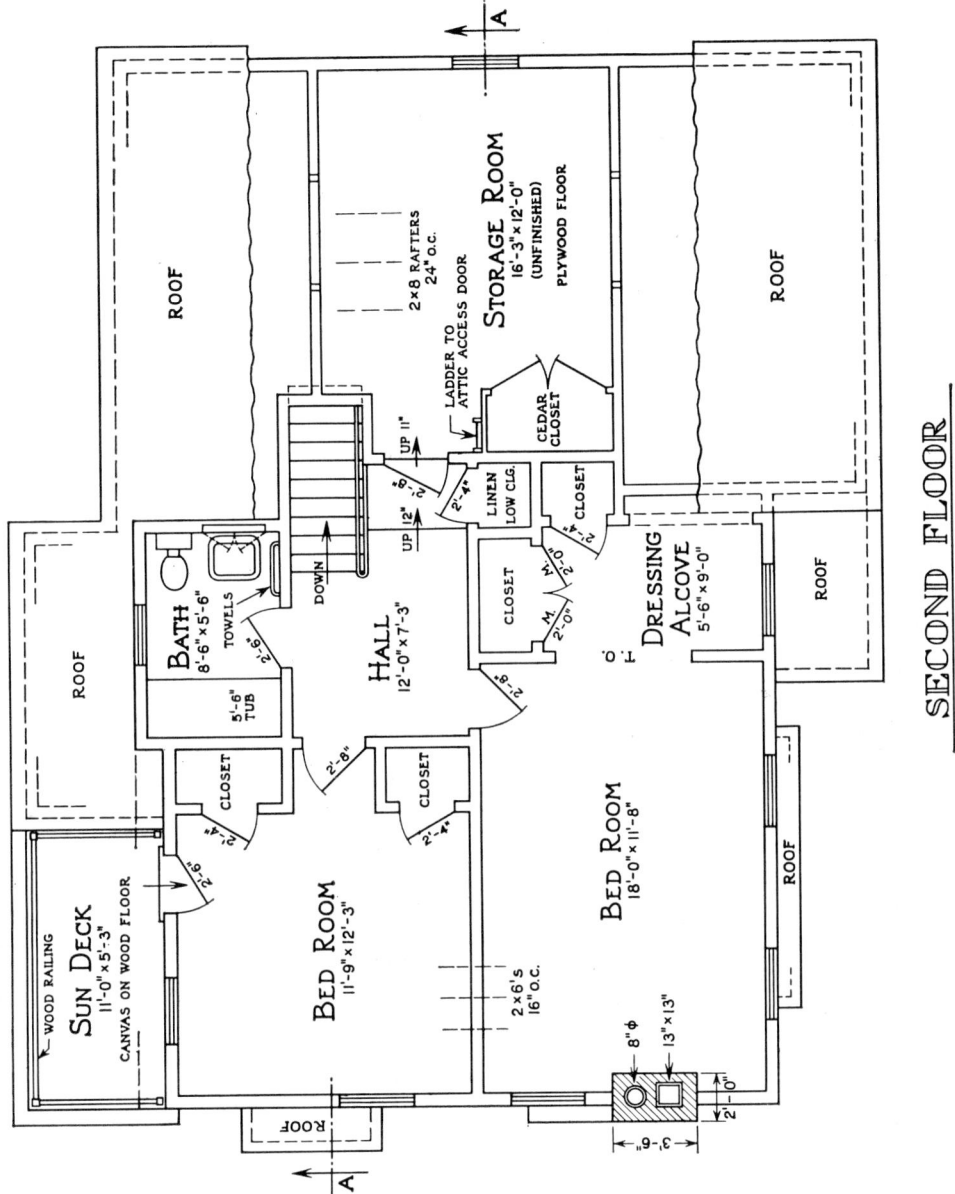

*Fig. 6-8   Second-floor plan.*

## Sections

Section A-A is shown in Fig. 6-9. Reference lines A-A, indicating where the section occurs, will be found on the basement and first-floor plans with arrows giving the direction of the sectional view.

This drawing is sometimes referred to as a cross section, but it is really a longitudinal section because it is taken through the length of the building. Other section drawings are sometimes provided by the architect, and the detail drawings in Fig. 6-10 may be referred to as sections.

The floor levels may be seen in this drawing. The playroom is at the lowest level, 11″ below datum, which, on these plans, is the street level at the driveway. This is shown on the plot plan in Fig. 6-1. The floor at the back of the garage is 12″ above the street level.

The stairs have $1\frac{1}{8}$″ thick oak treads and $\frac{7}{8}$″ thick white pine risers, all housed and wedged together.

**Dimensions.** Dimensions given on this drawing, as well as those on the other drawings, are nominal and will vary slightly depending upon actual sizes of the floor beams and actual thicknesses of flooring, etc. The $2 \times 8$'s are made from $2″ \times 8″$ rough lumber but actually measure only $1\frac{5}{8}″ \times 7\frac{1}{2}″$ after they are planed smooth. This is done in the mill before the lumber is shipped. The $1 \times 3$ oak flooring is made from $1″ \times 3″$ rough stock but actually measures only $^{25}\!/_{32}″$ or $^{13}\!/_{16}″$ thick. This flooring is T&G and, while being laid, the tongued edge of one piece locks into the grooved edge of the next piece. Thus the effective covering width is only $2\frac{1}{4}″$ instead of 3″.

**Construction.** The permissible minimum live load for purposes of design given in most building codes for residential occupancy is 40 lb per sq ft. There are occasional instances where codes permit 30 lb per sq ft on upper floors of single-family dwellings, and the same figure has been advocated by some authorities for general use throughout dwellings.

The permissible minimum live load given in building codes for business occupancy varies to some extent. If office space in buildings is taken as an example, the range in building codes recommended by various organizations for national or regional use in building-code standards is from 50 to 80 lb per sq ft.

The term *mercantile occupancy* is not used in all codes, but it covers such occupancies as stores, salesrooms, and markets. For these, the permissible minimum live load recommended ranges from 75 to 125 lb per sq ft.

Obviously, the *industrial* classification will contain widely varying examples of floor loading since it includes occupancies involved in manufacturing, fabrication, and assembly of all kinds of industrial products. For heavy manufacturing, some codes give values of from 125 to 150 lb per sq ft and others do not assign any particular value.

Safe building design requires the use of a combination of assumed loads and working stresses which will result in structures that will not be seriously overstressed in any part by the imposed loads. The first consideration, of course, is that there shall be ample strength. Many times, however, a beam which is strong enough may still be limber enough to permit a noticeable bending or vibration from walking. This not only is annoying but may be sufficient to crack the plaster of the ceiling below. Stiffness, therefore, must also be considered as a factor.

■ EXERCISES

1. How long are the Lally columns?
2. What size rafters are used in the main roofs? At what spacing are they?

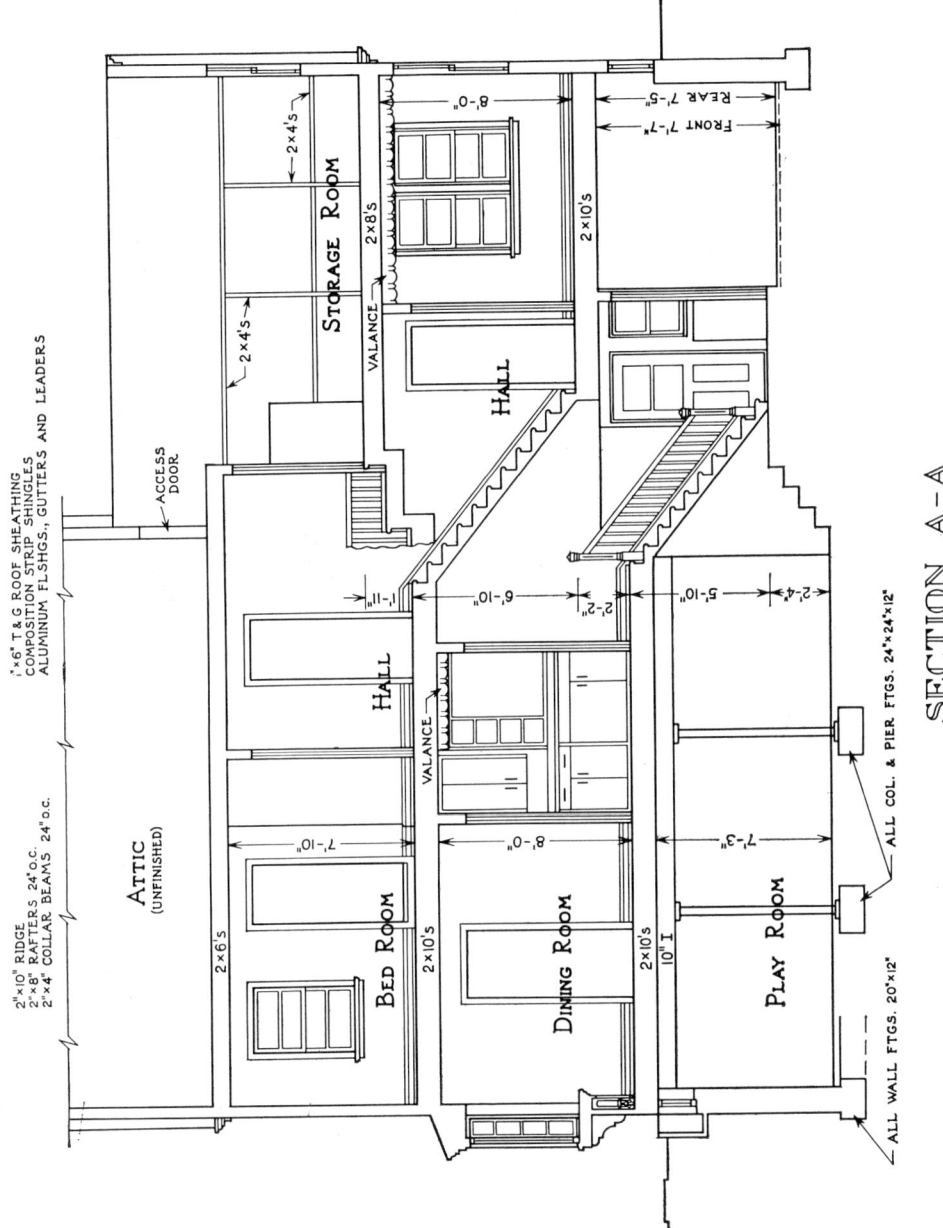

Fig. 6-9 Section A-A.

## Detail drawings

Detail drawings are given in Fig. 6-10.

**Fireplace.** The fireplace and the wall above it are faced with brick, the portion immediately around the fireplace projecting 1″. The mantel shelf is a plain 1¾″ thick oak board only 4″ wide and supported on four small oak brackets. The cabinet doors are of oak-faced plywood and the shelves and mouldings are also of oak.

**Railings.** Railings of ornamental wrought iron are provided in the opening between the living and dining rooms. This detail also shows the floor beams under the living room.

**Foyer.** The foyer-opening detail shows the space for a television set and shelves and cabinets for a record player and record albums. All the exposed wood in the foyer is oak plywood. The detail also shows sections through the dining nook and the front window and flower box. The nook contains an open-front cabinet above the upholstered seat. There is a convector enclosure under the front window and a drapery valance which extends along this side of the living room.

**Main Cornice.** A main-cornice section is shown in one of the details. The wall plate is a double 2 × 4. The ceiling beams or attic floor beams are 2″ × 6″ over the north portion of the building and 2″ × 8″ over the south portion, where they support the storage room. All the rafters in the main roofs are 2″ × 8″. The cornice outlookers are 2″ × 4″. All the exposed wood in the cornice is specified to be white pine and all the gutters and leaders are aluminum.

**Lintel.** The lintel over the dining-room opening is composed of three 2 × 10's spiked together and the second-floor beams rest on this and are spiked together.

**Sills.** Box sills are shown in the foundation details. The sills, or foundation plates, are 2 × 8's fastened on the concrete-block walls with ⅝″ × 18″ anchor bolts cemented into the blocks. The sill course on the brick veneer is a rowlock course of bricks cut to 5″ length and tilted slightly to shed water.

**Construction.** The length of the sill is determined by the size of the building, and therefore the foundation should be laid out accordingly. Dimension lines for the outside of the building are generally figured from the outside face of the sub-siding or sheathing. Where high winds are at all probable, it is important that the building be thoroughly anchored to the foundation. In fact, anchoring is good practice in all localities. It is accomplished by setting, at intervals of 6 to 8 ft, bolts that extend at least 18 in. into the foundation. They should project sufficiently through the sill to receive a good-sized washer and nut.

Estimators do not like plans drawn at a scale as small as $\frac{1}{8}''$ to the foot because this does not allow enough room for showing the work clearly, and they are thus forced to rely on the specifications which, in turn, may not be very clear. Some architects always use $\frac{1}{8}''$ for the regular plans and elevations but give also a number of large-scale drawings which show the important features properly.

Mechanical work is not shown on these drawings. Separate plans will be found in Chaps. 26 and 28 for the plumbing and electrical work.

## ■ EXERCISES

1. Compute the area of the brick veneer in square feet. Allowing seven bricks per square foot, how many bricks are required for the brick veneer?
2. How long are the second-floor beams over the living room? Over the dining room?
3. Describe the main-cornice construction.

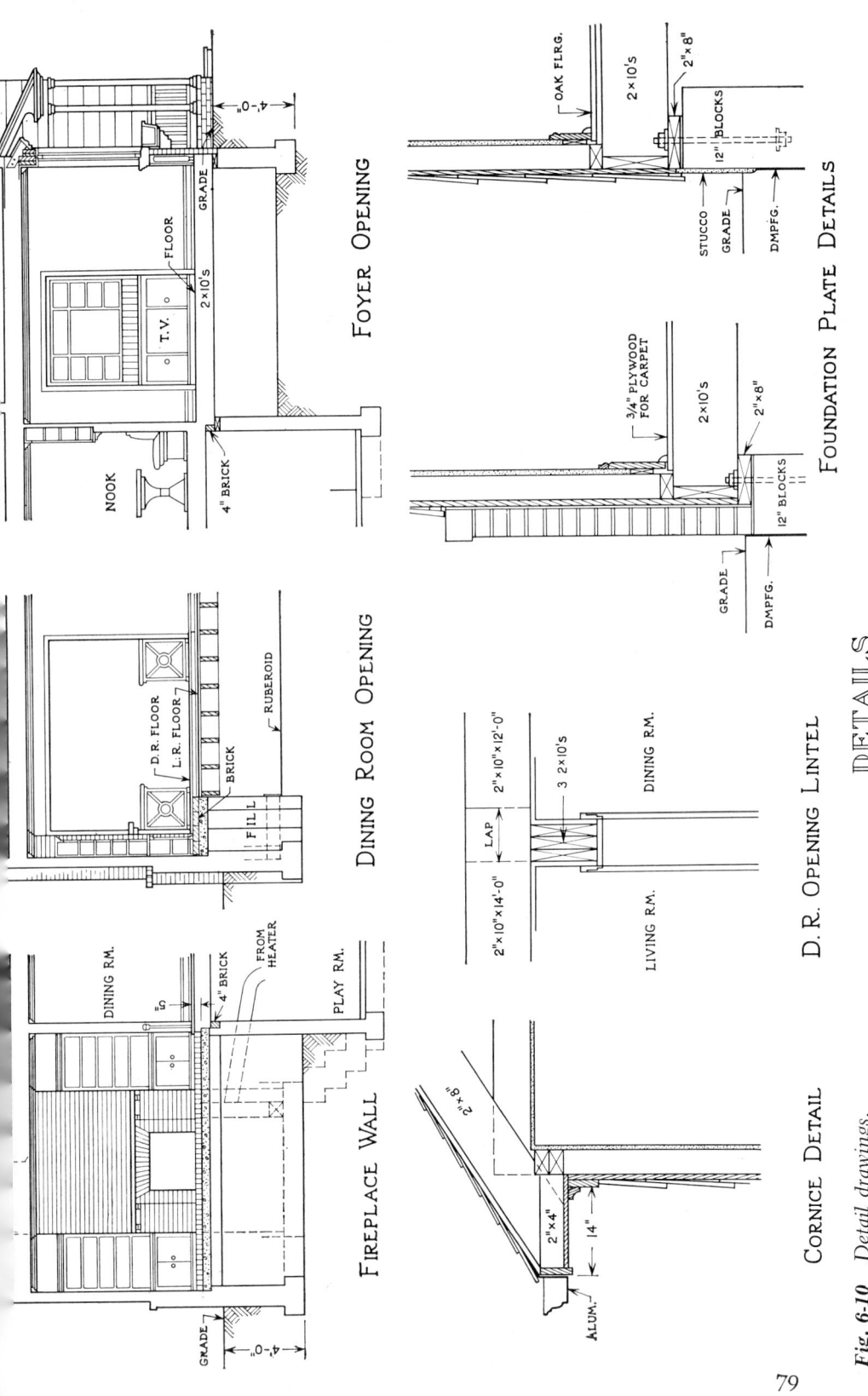

Fig. 6-10 Detail drawings.

CHAPTER 7

# *contracts*

A contract is an agreement between two or more parties. It must be properly made and must be for a legal purpose. It should be so clearly written that it merely forms the record of what is agreed upon between the parties and, therefore, can be filed away as soon as it is signed. This means that the contracting parties understand so well what is expected of them that no further reference to the contract should become necessary during the construction of the job. A contract should not review in detail items of material, workmanship, etc., that are already covered by the plans and specifications. It is merely the memorandum of the terms of the agreement between the parties that sign it. Because the contract may finally be interpreted in a court, it must be written in such form that it will be readily understood and have strength legally.

Architects generally use standard printed forms with which most contractors are familiar.

Contracts, or letters of acceptance, should clearly state:

> The full names and correct addresses of the parties and their relationship to each other.
>
> A brief description of the work to be done by the contractor and definite reference to the plans and specifications that are to be used.
>
> The amount to be paid to the contractor and a statement showing when the several progress payments and the final payment will be due.
>
> A clause providing for prompt arbitration of any dispute between the parties that cannot be settled directly by them and the architect.

Proper signatures, matching the names given in the contract, signatures of witnesses, and corporate seals where required.

The plans and specifications should be signed by both parties and this fact noted in the contract. Progress payments should be liberal, and prompt receipt of all payments should be insisted upon by the contractor. The arbitration clause should state that either party has the right to demand an arbitration at any time before the final payment has been accepted by the contractor.

**Management Type of Contract.** One form of contract provides for the contractor to furnish merely the services of his business organization to manage the construction work. This may be called the *management* type of contract. Under this form, the owner uses these services in much the same way as he would the services of a person whom he might employ and place on his own payroll. The contractor is the manager employed by the owner for the particular job mentioned in the contract. The contractor's responsibility is to run the job entirely for the owner's interests—if there is no guaranteed limit of cost involved.

This type of contract is used also in cases where the owner wants a guarantee as to the cost of the job. Here the contractor has a vital interest in the cost. He makes an estimate beforehand, upon the basis of information sufficient to enable him to do so; then, if changes are made in the information or in the amount of work to be done, a revision in the amount of the guaranteed cost becomes necessary. The contractor's compensation under the management type of contract may be a flat fee or it may be a percentage of the cost, as agreed upon when the contract is being written. This is then stated in the contract.

Another form of contract provides that the contractor shall furnish all materials, labor, and expense at unit prices that have been agreed upon. This is like buying an order of groceries; each item is priced per unit of measure. As the work progresses, it is measured and the contractor is paid at the prices stipulated in the contract. If any new features of construction are introduced, new prices are arranged for them as they appear.

**Lump-sum Contract.** The lump-sum form of contract is the one most commonly used. In it the contractor agrees to construct the job complete, in accordance with the plans and specifications, for a definite sum of money. The plans and specifications should be as clear-cut as possible, so as to minimize the chances of any changes becoming necessary. If changes are made, they should immediately be estimated upon by the contractor and a supplemental contract should be made to cover them, before any work on the changes is started.

In the lump-sum form, the architect and the owner are not directly concerned with the details of quantities of materials and labor, unit costs, etc. In the management form of contract, without guarantee of cost, the contractor has every incentive to work in close conjunction with the architect and the owner toward completing the job with the best materials and workmanship and in the shortest time, and it is possible for the work to proceed before the plans and specifications are completed.

In any form of contract in which the contractor is limited as to the amount that the job is to cost, it is natural for him to use every reasonable method to economize in both materials and workmanship. This economizing is very likely to extend through to the subcontractors and material men, as well.

As has been stated before, what the owner primarily buys when a construction contract is signed is the contractor's skill and the use of his business and technical organization. Regardless of his ability to secure contracts, a builder without proper skill and a proper business and technical organization is operating under a very great handicap.

**Financing Contracts.** Some contracts provide that the contractor shall pay all his bills every month before asking for a payment from the owner to cover them. Thus, if a contractor has 10 jobs under way at one time and they average an expense for one month of $25,000 each, he would have to pay out $250,000 before sending in his bills to the various architects for approval. Then, before he received his payments from the owners, another half month would go by and he would have about $375,000 outstanding. On this basis, the contractor would have to be a very wealthy man or else he would be in the banking business to a greater extent than in the contracting business. It is, therefore, part of the estimator's duties to warn the contractor against accepting contracts with such extreme wording regarding the method of making payments. If it is necessary to accept such a contract, the means of financing the undertaking should be worked out before the contract is signed, and proper interest and service charges should be included in the contract amount to take care of the unusual financing requirements. The more customary procedure is for the contractor to pay his bills immediately after receiving payments from the owner and, if necessary, to submit proof to the architect or the owner that he has actually paid these bills before his next payment is approved by the architect.

On an average job costing $500,000 and extending 6 months, the amount held back in the customary 15 per cent that is retained is a large sum, as will be seen from Table 7-1.

TABLE 7-1

| Month | Amount done | 15% withheld |
|---|---|---|
| 1 | $ 18,000 | $ 2,700 |
| 2 | 55,000 | 8,250 |
| 3 | 110,000 | 16,500 |
| 4 | 245,000 | 36,750 |
| 5 | 420,000 | 63,000 |
| 6 | 500,000 | 75,000 |

By the time the fifth payment is made, and only about $40,000 of work remains to be done, the amount being withheld is about $70,000. Thus, instead of 15 per cent, there is 175 per cent being withheld, based upon the amount of work yet to be done.

The 15 per cent is often referred to as the contractor's profit. It really represents the cost of his office rent, the salaries of his estimators and general superintendent, general office salaries, and expenses. The salaries have to be paid every week, and the many other items involved, at least every month. Then, whatever may be left is profit! It might be better if contractors would write out their bills to show general expense and office overhead and profit separately; the profit item would be only 2 or 3 per cent instead of 15 per cent.

The percentages and the time of payments often require thought on the part of the estimator. In some cases it may be necessary to refuse to figure a job or to accept a contract unless better terms of payment are offered. Architects seem to be forever guarding against overpayments to the contractors. Very few of them are good businessmen, and this fact may account for their fear of granting more liberal terms. Contracts should be worded definitely as to how long the contractor must wait for each monthly payment. If 20 or more days are to elapse, he will not receive one payment until it is nearly time to apply for another. Meanwhile, his payrolls have had to be met every week with cash. Inasmuch as the partial payments are based on estimates of the work done and not on detail measurements, it should be possible for the contractor to submit requisitions 3 or 4 days before the end of the month and to expect these to be checked through so that he will receive payment by the fifth day of the following month, instead of having to wait several weeks, as is often the case.

**Arbitration.** Arbitration should be an important factor in contractual relationships. This method of settling disputes chooses men who are familiar with the practical work and problems of building construction to take the place of courts and lawyers, who are likely to be all tangled up in legal technicalities and very often are blind in their judgment, because they have not such practical knowledge. Estimators would do well to insist that an arbitration clause be provided in every contract. By this method, usually three men are chosen to settle any dispute arising out of the contract or plans or specifications. Each party to the contract selects one man and these two select a third. Each side presents its case to this three-man board and, when two of the three agree on a basis for settlement of the dispute, the matter is ended (thus it should be stated in the contract) and their decision is binding on both parties to the contract. One of the main advantages of arbitration is that either party has the right to ask for an arbitration any time it wishes. Thus, disputes may be quickly settled while the matters are still fresh in the minds of those familiar with the trouble. Law cases have a habit of dragging out into months and even years, and it is often impossible to find the records and witnesses that are necessary for properly presenting the case. Besides, law cases are notoriously expensive, while arbitrations may involve only a few hours of work on the part of the arbitrators. The fee to be paid to the arbitrators and the decision as to who is to pay it should be left to the arbitrators, or else each side should pay for the arbitrator it retains, while the payment of the third member is decided by the three members acting together as a board.

## *Typical lump-sum form of contract*

MEMORANDUM OF AGREEMENT made this twentieth day of February 1958, by and between HENRY V. MALLIN, of 494 Fifth Avenue, New York City (hereinafter called the Owner), and ALSPACH CONSTRUCTION CO., INC., of 30 West 42nd Street, New York City (hereinafter called the Contractor),

WITNESSETH: That for and in consideration of the agreements herein contained, it is agreed by and between the parties hereto, as follows:

1. The Contractor agrees to furnish all the labor, materials, equipment, tools, and supplies for constructing and completing the hereinafter described house (hereinafter referred to as the work).

The Owner agrees to pay to the Contractor the sum of money hereinafter mentioned at the times and in the manner and upon the terms and conditions hereinafter set forth.

2. The house to be constructed under this contract is to be a one-family house constructed on the Owner's property at 110 Park Avenue in the Village of Florence, Nassau County, New York. The exact location, dimensions, and other characteristics of the house are given in the specifications and plans hereinafter mentioned.

3. The work is to be in strict accordance with the specifications dated February 5, 1958, prepared by John Edgar Smith, Architect, and the plans referred to therein, which specifications and plans have been signed by the parties hereto and are hereby made a part of this contract.

4. The Contractor will prosecute the work diligently at all times until completion and upon completion will leave the work complete and perfect. He will not have materials provided or obtained, nor will he have work performed or labor or means employed in the carrying out of this contract that would in any way cause or result in suspension or delay of, or strike upon the work.

5. The Owner will pay the Contractor for the work, subject to additions and deductions as hereinafter provided, the sum of FOURTEEN THOUSAND SIX HUNDRED DOLLARS ($14,600).

Payments to the Contractor will be made in current funds and in installments payable on the fifth day of each month in amounts not to exceed ninety percent (90%) of the value of the work done and materials substantially incorporated in said work on the last day of the month preceding. The other ten percent (10%) will be retained by the Owner as part security for the faithful performance of the work, this amount to be retained until thirty (30) days after the entire work has been completed and the Contractor has paid all claims for labor and material furnished on said work.

The right of the Contractor to receive any payment hereunder shall be evidenced by a certificate issued by the Architect. No payments made or certificates thereof will in any way be construed as an acceptance of any part of the work, nor will the same in any way lessen the total and final responsibility of the Contractor.

Prior to final payment and as a condition thereto, the Contractor will furnish the Owner with a verified statement that all bills and claims have been satisfied, and a release of all claims against the Owner, arising under and by virtue of this agreement.

6. This agreement, and the specifications and plans referred to herein, may be modified and changed from time to time, as may be previously agreed upon in writing between the parties hereto, in order to carry out and complete more fully and perfectly the work herein agreed to be done and performed.

The Contractor will submit a written proposal covering each modification or change, at the request of the Architect, within ten days after receipt of such request, and such proposal shall be accepted or rejected by the Owner within ten days. In the event of rejection because of the price quoted in the proposal such modification or change may be paid for, at the option of the Owner, on the basis of cost of labor and materials and insurance, plus twenty percent (20%).

7. The Contractor will procure all necessary permits and licenses, abide by all applicable laws, regulations, ordinances, and other rules of the States or political subdivisions thereof, or any other duly constituted public authority. He will assume, pay, and be responsible for any and all taxes and contributions under Federal, State, and Municipal tax laws arising as a result of this agreement.

8. In case the parties hereto cannot agree as to the interpretation of any part of this agreement, the same shall be determined in accordance with the arbitration laws of the State of New York. One arbitrator will be chosen by the Owner, one by the Contractor, and the third by the two so selected, and the decision of a majority of the said three persons will be binding upon the parties hereto.

IN WITNESS WHEREOF, the parties hereto, for themselves and their respective heirs, executors, administrators, and assigns, executed this agreement the day and year first above written.

Attest: _____          _____
      John W. Randolph                Henry V. Mallin
                               Alspach Construction Co., Inc.
Attest: _____          by _____
      George H. Cabot                William T. Alspach
         Secretary                    President

## Management, or "cost-plus," form of contract

MEMORANDUM OF AGREEMENT made this seventh day of March 1958, by and between Williamstown Manufacturing Co., Inc., of Williamstown, New Jersey (hereinafter called the Owner), and George H. Cooper, Inc., of 140 West 42nd Street, New York, N.Y. (hereinafter called the Construction Manager),

WITNESSETH: That for and in consideration of the agreements herein contained, it is agreed by and between the parties hereto, as follows:

1. The Construction Manager agrees to furnish the full use of its business and technical organization and skill for constructing the hereinafter described building (hereinafter referred to as the work).

The Owner agrees to pay to the Construction Manager the sum of money hereinafter mentioned, at the times and in the manner hereinafter set forth.

2. The building to be constructed under this agreement is to be a three-story factory building, approximately eighty feet wide and four hundred feet long, built on the Owner's property in Williamstown, New Jersey. The exact location, dimensions, and other characteristics of the building will be given in the specifications and plans which the Owner will furnish to the Construction Manager.

3. The Owner will pay the Construction Manager a fee equal to eight percent (8%) of the cost of the work as defined in Section 5. Payments will be made in current funds and in installments payable weekly and equaling eight percent (8%) of the amount of the expenditures made during the preceding week as hereinafter mentioned.

4. The Construction Manager will work in cooperation with the Owner's architects and engineers and will assume such responsibility as the Owner may delegate to it. The Owner will have full control of the work in all its phases.

The Construction Manager will prosecute the work diligently at all times, acting as the agent of the Owner.

The Construction Manager will make all payments for the work from funds advanced by the Owner for the purpose, and render weekly detailed statements of receipts and expenditures, supported by proper vouchers. All discounts and rebates are to be credited to the cost of the work.

5. The cost of the work above mentioned is the total of the "Job Cost" and includes the cost of all materials, labor, equipment, subcontracts, permits, taxes, insurance, contributions, and all other costs and expenses incurred directly in connection with the work.

6. The fee above mentioned includes the services of the entire head-

quarters office of the Construction Manager, as presently constituted, in the organization and administration of the work, as may be necessary.

7. This agreement may be terminated by either party hereto by serving a four days' written notice to that effect upon the other party.

IN WITNESS WHEREOF, the parties hereto, for themselves and their respective heirs, executors, administrators, and assigns, executed this agreement the day and year first written above.

WILLIAMSTOWN MANUFACTURING CO., INC.

Attest: By:

| Secretary | President |
|---|---|

GEORGE H. COOPER, INC.

Attest: By:

| Secretary | President |
|---|---|

Custom plays an important part in the interpretation of contracts and, to a lesser degree, of plans and specifications. If one can prove that a definitely established custom prevails in a matter under dispute, he has already won an advantage toward settlement in his favor.

All contracts, except those for very minor amounts, should be reviewed by the contractor's lawyer before being signed. It is also advisable to have an insurance agent look over contracts that contain *hold harmless* clauses. In such cases, the contractor agrees to indemnify and save harmless the owner for injuries to persons and damage to property. He thereby assumes liabilities under which his insurance carrier is prevented from holding parties at fault in the event of a claim. Policies generally have a clause stating that "the insured shall not voluntarily assume any liability, settle any claim, nor incur any expense except at his own cost, nor interfere in any negotiation, settlement or legal proceeding without the consent of the company previously given in writing." If such a contract must be signed, the insurance carrier may give this consent but will undoubtedly increase the premium to cover not only what may be reasonably expected to happen, but also what could possibly happen in the worst circumstances. It may be necessary for the contractor to charge the owner an additional amount to cover this extra expense.

Contracts must be honest. A party to a contract who discovers that major misrepresentations were made to him may either continue and sue for damages, or he may repudiate the contract and sue for damages.

Courts have held that when a defrauded party repudiates a contract he is entitled to recover damages incidental to the contract and caused directly by the fraud. One fraud that courts recognize is for an owner to conceal the known existence of adverse conditions or information by not disclosing them in the plans or specifications, and in such instances the contractor can sue the owner for damages. Many of these cases relate to subsoil conditions, but other types of concealment sometimes exist. An example may be found in the case of a contractor who once had a job on which other separate contractors were also engaged and who was led to believe at the time of bidding that all contracts would be awarded together. The other contracts were not awarded for a long time and this contractor was forced to work for a considerably greater time than would be expected. The court held that he was entitled to recover for the increased cost of the work resulting from the failure of the owner to award all the contracts at the same time. Failure to disclose that the other contracts had not been let constituted constructive fraud. The court stated:

Silence by the defendant as to a change in plans so important to the timely performance of the work served to mislead the plaintiff into a false assurance that the work contemplated by his bid would progress and be integrated with the work of the other contractors. Silence may constitute fraud where one of the parties to a contract has notice that the other is acting upon a mistaken belief as to a material fact.

■ EXERCISES

1. What are the essential requirements of a building contract? (60 to 100 words.)
2. What is a lump-sum building contract? (50 to 75 words.)
3. What is the management type of building contract? (50 to 75 words.)
4. What is meant by arbitration? (50 to 75 words.)
5. Write at least 200 words of notes on contracts, taken from one or two of the books listed in the Bibliography. Name the books used.

CHAPTER 8

# *the technique of estimating*

When one speaks of an estimate in connection with building work, he generally refers to the regular estimate that is used for computing the amount of the contract price.

Preliminary, or budget, estimates are sometimes made up to find the approximate cost of a proposed building. These are usually made before the plans and specifications are complete. Such an estimate enables the architect to gauge the extent to which he may go in developing the details of the work on the final plans and in the specifications. Financial arrangements also may require a preliminary approximation of the cost.

## Preliminary estimates

A preliminary estimate may be quite comprehensive in scope. If it is to be used as a guide for the owner in determining his total expense, a preliminary estimate may include such items as the cost of the property upon which the proposed building is to be built. The legal costs of transferring the ownership of the property from other parties may also be shown. The owner's cost of investigating into the proposition, his expense for architects' and engineers' fees, the interest on his money for temporary and permanent financing of the undertaking, and other items not commonly considered by builders' estimators may have to be included. Often a contractor is given a set of preliminary sketches that show the proposed building but do not include walks, roads, floor coverings, lighting fixtures, decorations, and other items that are obviously necessary to make the building complete and ready for use. If it is intended to serve for the owner's full guidance, an honest preliminary estimate should list all these items and any others that the person making the estimate can think of as possible items of expense. They should be listed by name without figures after them, if no figures are available. Among the items that might be included are taxes and insurance to be borne by the owner, besides an allowance for additional work that may be ordered and for furnishings, the operation of the building, maintenance costs, etc.

Preliminary estimates for buildings are sometimes made by merely computing the total bulk volume in cubic feet, including the portion below ground, and multiplying this by a price per cubic foot based on the cost of similar buildings. Obviously, this can give only a very rough approximation of the cost because no two buildings are built under exactly the same conditions. Other methods of approximating use an assumed cost per square foot of floor area or the cost per room or occupant. As preliminary estimates seem to have an annoying tendency not to agree with the final estimates, those to whom they are submitted should always be warned that the former are only rough approximations and that it is not safe to regard them as anything else.

## Regular estimates

The regular estimates are made up by the general contractor's estimator, while preliminary estimates are sometimes made up by the architect or by an agent of the owner, instead of in a contractor's office.

Some contractors make a habit of using what may be called *working estimates*. These are prepared by the estimator—or by the superintendent when another man's check is desired—after word has come that the contract is to be secured. They are made up in much the same style as the regular bidding estimates, but much more detail is placed on them, so that they form valuable aids in the management of the job. The plans and specifications are carefully checked over and all adjustments are made up to this point. These working estimates form the basis for awarding the subcontracts and for preparing the material lists, construction schedules, and other working data.

Regular printed forms for estimates and material lists are obtainable, as well as standard forms for contracts of all kinds. Although some contractors use these, most either do their estimating on stock-type journal-ruled pads, or else have estimating sheets made up to suit their own ideas.

In the main heading of the estimate is put the name or type of building, its location, the owner's name, and the architect's name. Some estimators give a brief description of the building, its size, the number of stories in height, the type of construction, the cubical contents, and other information.

The cost of doing different kinds of construction work under different conditions cannot be exactly determined, but if the plans and specifications are well prepared, there is no question that the costs can be approximated very closely. A great part of estimating, however, is done by very loose and unscientific methods; figures are based on limited rather than broad experience, and many factors of great importance are left out of consideration.

The first step in estimating any kind of building work is to determine in the simplest way possible, yet with a reasonable degree of accuracy, the number of units of work to be done. In some cases this may mean the amount of material required, but usually it is not so simple as that. In their effort to oversimplify the work, careless estimators are led into omissions.

Estimating, like accounting, is an established business routine and is an integral part of the business of constructing buildings. In a broad sense, it must be thought of as equal in importance to the preparation of the plans and the writing of the specifications. Estimating that is done by men not trained for it is usually based on hasty judgment or on guesses. Such estimators lack the thorough schooling in the analysis work that is necessary for the preparation of complete and reliable estimates. The methods used by some contractors in going over work preparatory to making bids are superficial almost beyond belief. In the main, these are responsible to a great degree for the losses later sustained by the contractors.

The estimate should be a complete story of the construction work—in builder's language. It should be so arranged and so worded that the man who made it or any other man familiar with building estimates may be able to obtain from it a clear mental picture of the proposed work in all its parts.

Estimates, schedules, and material lists are generally arranged in the same order as specifications, that is, in the sequence in which the trades commence work on the job. The mechanical lines, however, such as plumbing, heating, and electrical work, are more often put at the end.

The estimate is divided, in general, according to the branches of work that are required on the particular job under consideration. There is no definite arrangement to be followed. The divisions are made for convenience in pricing the work that is to be done. They do not conform to the divisions used in the architect's specifications; these may not suit the method of handling the various parts of the job from a construction standpoint.

The heading for the first part, or division, of the estimate is Excavating, as excavation is usually the first work to be done on the job. However, if there are on the property any old buildings or other structures that have to be removed, a preliminary heading Demolition, or Clearing Site, would be required.

A division of the estimate is not necessarily confined to the trade that the heading might indicate. The heading Concrete Work, for example, will very likely include some carpenter work (for installing the temporary forms) and some metal lathers' work (for installing the reinforcing steel rods or mesh) as well as that of the concrete workers themselves. The heading Plastering nearly always includes lathing too, as only established custom has made the name *plastering* the one to be commonly adopted. This heading is, in fact, sometimes extended to Lathing and Plastering. These are examples of divisions that embrace more than their names indicate.

Other headings may include less than the name indicates. Carpentry is one of these. Although under this heading are mentioned all the items that are customarily grouped there, not all work to be done by carpenters will usually be listed among these items. As has been mentioned before, the carpentry of installing the temporary forms used to support concrete is more often placed under the heading of Concrete Work. Another part of the job done by carpenters is the installation of kalamein work (metal-covered doors, etc.). This special branch of carpentry is usually set up under a separate heading in specifications and estimates. Sometimes, however, particularly in specifications, it may be included under Carpentry.

In spite of this apparent lack of definite rules, the experienced estimator finds little difficulty in arranging his estimate.

*Neatness and order* in the form of the estimate will help to eliminate errors. Besides, the student should try to be systematic and to develop a uniform method of working, arranging the entries so that the result will be both artistic and easy to read.

Some estimators use colored pencils to check off the items on the plans as they are listed on the estimating sheets. This is an aid in some lines of work and on some kinds of jobs. One danger of this practice lies in assuming that the check marks cover all the items required—in other words, in marking the items that were seen and listed and failing to look further for other items that may have been missed. Experienced estimators seldom mark plans.

Some estimators use separate sheets for pricing the quantities, which are first grouped on the take-off sheets and then carried forward to these pricing sheets. This is desirable on large jobs, where many sheets are necessary for the take-off for each line of work. In this case, an accurate check should be made to see that all the items are actually carried over to the pricing sheets. On smaller jobs it is perhaps better to price directly on the same sheets with the quantities, thus eliminating any possibility of failing to carry over all the items. This method also keeps all the work of an item in one place. Thus, if any corrections or adjustments are required, there is only one place, instead of two, in which to make the changes. For the same reasons, on the smaller jobs it is best to dispense with summaries and grand summaries.

Calculations should be made by the estimator and then checked by another person. In busy offices where adding machines and calculating machines are used, the process is reduced to a simple matter of mathematical checking, which can be done by any office assistant after a little practice. Every reasonable effort should be made to have the estimate entirely correct.

*Check lists* are helpful toward catching items that may be omitted, but, because of the vast number of possible items, it is impracticable to have complete lists. A standard job-expense sheet is generally used, however. On this all the items that are expected to be encountered in this category are already enumerated and, therefore, serve as a check. Here the items of scaffolding, hoisting, temporary protection, temporary heat, and others that are likely to be omitted entirely are picked up. Even if an item is included on another sheet there is no harm in having it appear here also, as it can easily be crossed out if not needed. Many contractors, in order to make sure that their insurance costs are not overlooked, carry these costs on the job-expense sheet, instead of attempting to include such large items on all the various sheets having to do with the divisions of the work.

It is impossible to use anything in the nature of standard costs for reference in the pricing of estimates. Even cost records kept by the contractor's own office are not reliable to the extent that they may be safely used for all estimates. So many factors enter into the actual cost of building work that each job is a problem in itself. However, accurate cost records should be kept of all work done, in order that the estimator may keep generally abreast of actual costs as far as possible, and also that they may serve the purpose of finding changes made in the work that have, perhaps, not been properly charged to the owner. To help toward better estimating thereafter, cost records should also be used to bring out errors and omissions made in the original estimate. Figures 4-3 to 4-5 show typical cost-record sheets.

**Plans and Specifications.** The plans and specifications are examined first, by the estimator, in order that he may become familiar with the job that is to be estimated. No definite procedure that will be applicable to all jobs can be given, but certain rules do apply. Generally speaking, in the arrangement of the estimate the same order is followed as in the construction of the building, beginning with the excavation and proceeding with the various divisions in the order that the work of the different lines starts at the job. A good specification follows this order, too, and when this is the case, it may be possible for the estimator to follow the exact order of the specification. However, as has been stated before, nearly all specifications show an overlapping of items and it is the responsibility of the estimator to see that every item is provided for and that none is duplicated. He must be experienced enough to feel that the prices he applies to his estimates will be correct for every important item and suitable for the conditions that will probably be required by the particular job under consideration.

An estimator likes a specification that states definitely just what is to take place on the job, exactly what materials are desired and what their quality should be, and exactly what will be done about inspecting the work in its various parts—in short, one that thoroughly covers the whole work. If the architect expects to see a thoroughly manned job, with ample job office and supervision facilities, instead of one that has only a superintendent (acting also as timekeeper, material clerk, layout engineer, etc.), he should so state in his specification, calling for what he wants. Otherwise, many contractors will be tempted to underman the job, in order to save expense, and then the quality of the work will of course suffer. It would not be a bad idea for architects to enumerate the nonworking positions and, perhaps, to state the minimum salary that each is to carry and the period during which each position must be filled, with credit to be made to the owner for any saving effected.

Plans and specifications must be checked carefully by the estimator. When a contract is signed, it is usually for all the work called for in the specifications or shown on the plans. This means that an architect does not have to include all the work in the specifications alone, but may show some of it on the plans without mentioning it in the specifications. For this reason, estimators double-check against the plans, and from their knowledge of construction work are always able to add a number of items in their estimates for work that is not in any way mentioned in the specifications. Architects are not closely enough in touch with builders' problems to enable them to arrange their specifications so that these may be relied on entirely, either in first figuring the job or in using them as a basis of direct reference in awarding subcontracts. There are many overlappings and omissions.

**The Site.** The examination of the site should be made before the estimate is very far advanced. An experienced estimator will investigate the conditions relative to the handling of materials, the storage of these materials, the distance to the railroad (especially if carload shipments are expected), the means of unloading and transporting from the cars, etc. Some materials require temporary protection and substantial covers, sheds, or platforms. These temporary conveniences should be listed and measured and priced on a unit basis, just the same as other carpenter work. The maintenance and dismantling of temporary work often is not given the right amount of consideration in estimating. Temporary water lines and temporary heating often entail extra expense to a large amount, because of the daily (and sometimes nightly) attention required for proper maintenance. Even the work of unloading and stacking materials may require the inclusion of a large sum of money in the estimate, especially if it involves the handling of heavy or fragile materials and especially also if the items must be carried or wheeled a considerable distance from the point of unloading to the places where they may be stored. Matters such as these reveal an estimator's practical knowledge of job problems; he has to visualize the working conditions in order to know what to expect and what to provide for.

Comparison has often to be made between the cost of materials delivered to the job by truck and the cost of the same kind of materials shipped by freight and then unloaded and hauled to the job. When they are shipped by freight, the uncertainty of delivery on time must be judged, as well as the availability of sufficient storage space for large lots of material. If carload shipments require that temporary storage sheds be built for protecting materials until they are needed, it may very well prove foolish to set down a low price and later have to face the cost of the sheds. If there is not an abundance of room around a job and if materials must be stacked close together and close to the work, it will probably be necessary to move some of them more than once—and every handling adds to the expense. Speaking of crowded conditions, it sometimes happens that a job is so cramped that men cannot work to advantage. In cases like this, the lower output of work must be carefully taken into account. In addition, overtime work may become necessary, entailing premium pay for those men who work overtime, either to move materials or to do construction work that must be accomplished in order that the job may run smoothly the following day. This overtime work, moreover, may all be made necessary by poor judgment in ordering the materials.

Labor conditions at the site must be looked into if they are not already known. The availability of sufficient men of the proper kinds is important to the smooth running and proper sequence of the work.

What is to be done with excavated material that is needed at the site for backfilling or for filling inside the building should be given thought in the pricing of an estimate. In some extreme cases, it is better to haul away part of this material and then haul it back or buy other material for filling, rather than to have piles of dirt in the way of the men for a considerable time, which may prove more expensive, if this dirt has to be moved out of the way several times. It may even be more economical to concentrate on the foundations of one part of the job, letting the other parts wait, and then to backfill the first part with the material taken from the other parts. All these things, of course, are considered under the heading of Excavating. They are discussed here because of their importance in relation to the inspection of the site.

The expected date of delivery of the structural steel is often the deciding factor in the consideration of the general handling of jobs involving steel. Even the handling of the excavating may be affected by it. If the steel is expected quickly, the plan would be to concentrate on those footings that are to receive the first steel deliveries and in this way enable the work to proceed continuously thereafter, with the structural frame of that part of the building, even before the other portions of the building have been started. Thus the steelwork affects excavating, concrete work, backfilling, and perhaps other work besides.

A general study of the method of laying out the job and running it is made at the time the estimate is being prepared. An approximate progress schedule will be found helpful in this connection. It may prove better economy to spend more money speeding up some parts of the work in order to be better prepared to do other parts. It may also be found necessary to speed up portions in this way, just to meet promised dates of completion that may be contained in the contract.

**Waste.** An element of cost is waste. Forms for concrete involve waste if they can be used only once or twice on a job, rather than a great number of times. Concrete may be wasted if the forms are made larger than called for, or if some of it has to be thrown away because too much has been ordered for a particular pouring, or because the forms are not ready to receive it. Mortar for brickwork drops to the ground and is wasted, or too much is made and the surplus is thrown away. Improper mixing of concrete or mortar at the job may involve the use of more cement, the expensive ingredient in these items, and thus both money and time are wasted. Estimators often wonder where all the cement goes on a job. Perhaps if they saw the work being done they would note this waste. One careless, wasteful man can throw away hundreds of dollars on a good-sized job if he has a position in which he controls the supply of an item that can readily be wasted. The fault, of course, lies with the foreman or superintendent in his not checking figures every few days and so discovering these leaks.

Time wasted is money wasted. Table 8-1, using assumed figures, shows the cost per hour of a typical workman.

If 20 per cent overhead expense and 10 per cent profit are added, the owner would be paying the contractor $8.05 for each hour of the man's time.

It will be seen that the estimator must take into account the type of men his company employs for supervisors. The saving effected by paying low salaries may be ten times offset by the inefficiency or lack of interest shown by men of poor quality or by men who are being underpaid.

TABLE 8-1

| | |
|---|---|
| Base pay, per hour | $4.00 |
| Workmen's compensation insurance | 0.38 |
| Public liability insurance | 0.05 |
| Property damage insurance | 0.01 |
| Social Security insurance | 0.10 |
| Unemployment insurance | 0.11 |
| Union welfare fund | 0.20 |
| Union pension fund | 0.12 |
| Holiday pay, prorated | 0.02 |
| Lost time, prorated | 0.02 |
| Insurance on holiday and lost time | 0.08 |
| Union funds on holiday and lost time | 0.01 |
| Direct cost | $5.10 per hr |
| Foreman's pay, insurances, funds, etc., prorated | 1.00 |
| Total cost of man | $6.10 per hr |

Face brick and glazed-tile blocks break in transit and while they are being moved about the job, lumber and scaffold planks break, doors and windows become marred so that they have to be rejected. All these and other items of waste must be considered by the estimator and provided for. On some jobs there will be very little such waste, but on others, especially where speed is essential or where carelessness and lack of system prevail, it will amount to a considerable loss.

A regular and recognized form of waste is always allowed for in figuring lumber, because of the manufacturing processes used and established customs in the lumber business. A board foot is said to be equivalent of a board one inch thick and one square foot in area, yet if a floor is to be covered with ordinary wood flooring it is found that instead of 100 board feet being required for a room 10′ × 10′, nearer to 135 board feet is required. In this case, 100 sq ft does not equal 100 board feet, even with boards 1″ thick. This is because the stock from which the flooring is made, and not the lumber actually received, is charged for. This is a trade custom and is inescapable.

Flooring that, when laid, covers a width of 2¼", is made from 3" width stock. Thus a loss of ¾" is entailed in every piece, owing to the forming of the tongue and groove alone. In addition to this, when the flooring is being laid, it is necessary to throw away bad pieces so that a further loss is sustained. Ends of pieces are cut off in order to fit the flooring boards into the size of a room and many small ends are thrown away. This is all called waste, although it is not all caused by the men on the job. Another source of loss in the matter of flooring is that one does not even get the thickness that is charged for. Flooring may come only ¾" thick, or even less, yet it is counted 1" thick. This is because another trade custom says that lumber less than 1" thick must be figured the same as if it were 1" thick.

Flooring and other lumber over 1" thick is figured at the stock thickness from which it is made. Even timbers, such as floor beams, are scant of the dimensions shown in estimates and in purchase orders. A so-called 2" × 10" timber means one that is made from 2" × 10" stock; actually it will measure only about 1¾" × 9½" when received at the job.

**Contingencies.** It is difficult for the estimator to provide for contingencies and yet have his estimate low enough to win the contract. Winter weather must be taken into account if the work is to extend into that season so that temporary enclosures and temporary heating will have to be provided, and so that the loss of efficiency caused throughout the job during bad weather will have to be met. Incidentally, if a job is expected under normal conditions to be closed in before bad weather comes, and if additional work is given by the owner or changes are made that hold up the work, the estimator, immediately upon finding this to be the case, should have compensation demanded from the owner for any protection or other extra winter costs thus made necessary.

If a job is to last several months, it may very well be that the wage rate of one or more of the trades will be increased before the work is completed. Rarely is it possible for the contractor to get any additional money from the owner for this increased cost. Some subcontractors always put into their estimates a clause stating that they will not be responsible for any increase that their men may demand during the life of the job. The general contractor is thus forced to carry not only his own risk in this regard but that of his subcontractors as well.

The time required for constructing a building is of great importance to a contractor, because his funds and also his working staff are tied up in the work. To become too involved in jobs may cause his financial ruin, especially if his funds are limited. For this reason, it is better for him to stop figuring new jobs for a while and to use the estimating talent in connection with jobs already on hand—toward securing better efficiency on them and toward completing them sooner than would otherwise be the case. Too little work on hand is bad, but too much work on hand may be worse.

It is always necessary to make an analysis of the specifications and plans before starting an estimate. Enough detail should be incorporated in this analysis so that it will form a complete check list. First the specification is analyzed and then items from the plans are added to the list, as well as other items, gathered from the estimator's own knowledge of construction work and of the special requirements that will be involved in the particular job being estimated. It is the estimator's duty to see that every item of expense is included in the estimate, either among his own items or among those of the subcontractors whose figures he incorporates in the estimate.

## *Subcontractors*

Subcontractors' estimates require careful study before the figures presented in them may be used. First of all, the work included by the subcontractor in each case must be ascertained. Even those estimates that read "according to the plans and specifications" may not, when studied, really cover all the work that at first glance may seem to be included. If a plumbing contractor, for example, says that he has included all the work in his line, this does not necessarily mean that he has accounted for all the items that the architect chose to specify under the heading of Plumbing. The architect may have mentioned in the plumbing section many items of equipment that are ordinarily hooked up by the plumbers, but the plumbing contractor can take the stand that these items, or some of them, are to be furnished by other parties and not by him, that he is merely to hook them up to the plumbing system. Thus the cost of furnishing all such items may be omitted from the estimate if care is not exercised. Similarly, if an electrical contractor says that he has included all the work called for in his portion of the specifications, he may not be counting in the lighting fixtures, if they have not been specified therein, even though they may be shown and perhaps listed in detail on the plans.

The lowest subcontract estimate in a given line is not always the best to choose. One that is far below the other estimates received should be checked very carefully before being incorporated in the contractor's own estimate. For one reason or another the subcontractor may not take the job if it is offered to him. He may have found out in the meantime that he made a mistake in figuring or that he omitted some of the work. Even if he signs a contract at the low figure he submitted, trouble may start while he is working on the job; he may find that he is losing money and ask for more than he contracted for, or he may be financially unable to proceed. At any rate, he probably will not be very enthusiastic about the work or give it proper attention, such as he would give if he had the prospect of a good profit. All these things must be in the estimator's mind as he weighs the pricing of his estimate for the whole job, and every reasonable precaution must be taken.

Subcontractors often wish to use a different material or one of a different grade from that specified. Sometimes they state this in their estimates, perhaps noting the substitute; and sometimes they take a chance and later state that they had anticipated using different material. If the substitution is perfectly fair, usually no trouble results; but in other cases a dispute may ensue. At the time of estimating there is no opportunity to take any action on what the subcontractor may have in mind to do afterward. If, however, a subcontractor in his estimate gives any intimation that he plans to make a substitution or if he verbally states any such intention, the wise estimator, before using that subcontractor's estimate in preparing his own bid for the job, will make a thorough check as to the possibility of a misunderstanding.

Occasionally, a subcontractor, at the time he is figuring the job, will tell the estimator of a substitution that he plans to make, but will neglect to state this in his estimate. Later he may claim that he told the estimator and then try to force an adjustment. While there may be no legal claim on his part, an embarrassing situation is created that should have been avoided by the estimator's taking proper steps when he first learned of the substitution plan. He should at least have given warning that it would be the architect's responsibility to accept or reject any substitutions.

## Unfair dealings

Architects and owners are often not fair in their dealings with contractors. For one thing, many estimates are made for building that is never done. Yet the cost to the contractors bidding on it is never considered by those who unhesitatingly ask for bids. If a code of estimating practice is ever written, one rule should be that no bids shall be submitted unless plans for the proposed building have at least been filed with the local building department. This will tend to eliminate some of the useless bidding, because it is not likely that an architect would file plans unless the job had a fair certainty of going ahead.

Architects sometimes specify that only approved subcontractors may be used. This gives rise to confusion in the selection of subcontractors by the contractor. It could be eliminated if architects would furnish full sets of plans and specifications to the subcontractors who meet with their approval at the time bids are first asked for—at least, in the main lines of work. These subcontractors could then send a copy of each bid to each of the contractors bidding on the job and there would then be no need for doubt as to their being approved later.

Another thing that annoys estimators is to find out, after a job is well under way, that an architect had in his possession surveys and other information regarding ground levels and conditions existing at the site and in its surroundings that he did not issue with the plans when bids were asked for. It should be a rule that every bit of pertinent information that an architect has shall be passed on to the contractors bidding and thus perhaps save them from gambling on uncertain conditions.

All work should be deemed to be union work, with no overtime work considered. The standard form of lump-sum contract of the American Institute of Architects, without alteration, should be understood to be used for all work. If any variation from these two requirements is desired by the architect or the owner, it should be clearly so stated in the original bidding specifications, in order that contractors and their subcontractors may take such variations into account at the time of making their original calculations.

Rock excavating, pumping, underpinning that is not shown on the architect's plans, foundation work that is not shown on the architect's plans, and other work or expense upon which it is impossible for the contractor to make a satisfactory estimate should be paid for separately, in addition to the work proper set forth in the contract.

A fair price should be chargeable for all drafting or engineering services that contractors or subcontractors are called upon to render, whether these are called for at the time of bidding or at any other time.

Neither the contractor nor any subcontractor should be held responsible for loss due to any delay in the execution of a contract, when such delay is in no way the fault of the contractor or of one of his subcontractors; nor should any payment be held up on this account.

The contractor and each subcontractor should be paid in current funds at least 90 per cent of the value of the work completed, every 2 weeks, or at least 95 per cent every month for the work done and the materials delivered.

Every contract should provide for prompt payments and also should require final inspection and payment in full within 30 days after the completion of the work of each division of the specifications, regardless of the final settlement of the job as a whole.

Differences arising between the contractor and other parties to the contract or concerned with it should be subject to and settled by arbitration.

No changes should be made and no work should be held up pending any changes, until an agreement is entered into for the adjustment of the contract price on account of the changes. This should apply to subcontracts also.

Unless specific provision is mentioned in the contract, an extra charge should be made for any variation from standard stock materials and methods of construction.

When liquidated damages are specified, a corresponding bonus also should be included.

All the foregoing are merely suggestions of practices that would make contracting more of a business or profession and less of a gamble. They will serve also to point out some of the business troubles encountered by the estimator in his daily work.

## ■ EXERCISES

1. What are preliminary estimates? (60 to 100 words.)
2. What should a complete builder's estimate contain, and what is the general arrangement of such an estimate? (100 to 150 words.)
3. Why cannot the unit costs given in this and other books on estimating be used for all jobs? (75 to 100 words.)
4. What is the general procedure in making an estimate? (100 to 200 words.)
5. State some contingencies that an estimator must consider providing for. (60 to 100 words.)
6. How are subcontractors' estimates handled? (75 to 100 words.)
7. Write at least 300 words of notes regarding estimating procedure, from several of the books listed in the Bibliography. Name the books used.
8. Prepare the first sheet of an estimate for the house shown in Chap. 6, with proper heading, etc.
9. Make an analysis, similar to a specification analysis, of the requirements for the house.

CHAPTER

# *quantity surveying*

The quantity-survey portion of an estimate lists every essential detail that will affect the cost of the proposed building, except that subcontract items are listed by name only. A quantity survey is not a material list, however, nor is it as complete as a working estimate. The essential details for computing the cost consist of the name and location of every item of work and the number of units of work involved in each.

Quantity surveying is the measuring of the work that is shown on the plans and specified and the listing of these items on the estimate make-up or quantity sheets. These give the amount of work under each classification, ready for pricing at so much money per unit of work. This unit price is the contractor's cost price. It includes the cost of materials, labor, equipment, scaffolding, labor insurance, and all the incidental expense entering into the items; it excludes only the portion of the general job expense that could be considered also as part of the items. The main-office expense and the profit are entered in the estimate separately from these work items.

Some builders, especially the larger firms, employ men as quantity surveyors alone, leaving the pricing and summarizing of the estimate to the builder himself or to the head estimator. One criticism of this method is that it is not conducive to the proper study of the unit prices and the application of them to the items. It is better for one man to carry the whole estimate through to completion and then have the whole thing looked over or checked in detail by another person.

Quantity surveying involves reading the blueprints and scaling the quantities of work from them. Some knowledge of architectural drafting is necessary for doing this well and readily. Architectural drafting gives a keener hold on the details of the making of plans and a feeling for the work of the architect and the draftsman, which will help toward easier reading of the plans.

A good quantity survey is one that can be used to advantage after the contract is signed—one from which the lists of materials can be readily made up. However, it is rarely practical to enter on the original quantity survey every detail, as is required for the material lists. Most contractors are satisfied to have in mind only the winning of the contract when their office is estimating a job. They use the quantity survey and the other sheets of the estimate to guide them and check them in making up subsequent lists and schedules. Many follow the plan of making a new, or working, estimate as soon as they receive word that the contract is to be awarded to them. The original estimate is used as a general guide in preparing this working estimate, but all the figures are taken from the plans anew. In this way, not only is a careful check made of the original estimate as such, but more details are entered and the entire job is thoroughly studied. This makes the office more familiar with the work to be done; in fact, some contractors, with just that in mind, have the general superintendent or the engineer make the new estimate. The estimator is already familiar with the plans. The working estimate for a large job should show the items broken down, floor by floor or portion by portion of the job, in order that it may serve as a guide for deliveries of material and for other uses.

Quantity surveying should show all the items used in the estimate. These items are shown in the customary way, by means of the customary units of measure. Thus, plastering, for example, is listed as so many square yards of the different types of plastering work and not (as in a material list) as so many bags of lime, cubic yards of sand, so much lath, so much hair or fibre, etc. A list for concrete work gives the number of square feet of forms required and the number of cubic yards of concrete, all item by item of work; not the number of board feet of different kinds of lumber for the forms nor the sand, gravel, cement, etc., of which the concrete is composed.

The quantity survey shows how many thousands of bricks of each different kind are required and indicates their location in the building. A material list, on the other hand, shows all this and also the amount of sand, cement, lime, mortar color, brick ties, and other incidental items of material that are needed for producing the amount of brickwork required. All this additional information is necessary for the writing of purchase orders and is therefore entered on the material lists, after the job is secured. It is not needed in the original estimate, because the unit prices there include the cost of all these incidental materials.

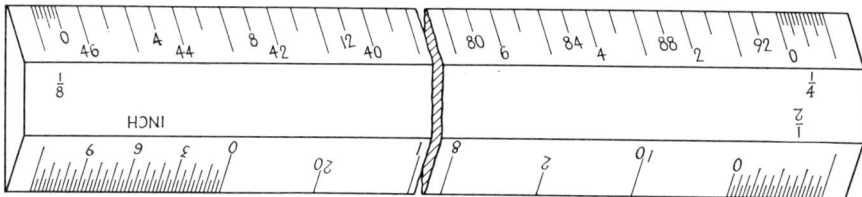

*Fig. 9-1* A 12-in. scale rule.

Chapter 5 treats of plans in general. Here the present chapter, making a deeper study of them, discusses the way to separate the work shown on them into the customary parts for estimating purposes. These divisions generally follow the same work as the trade name used in them may signify, but sometimes it is necessary to include work of more than one trade in a division of the estimate. Only by considerable practice is easy and familiar handling of such matters gained; but a knowledge of the general rules of scaling the plans is of first importance.

Symbols and conventional indications are quickly read by anyone experienced in building construction and the use of plans for this work. In fact, some of them, like the symbols for bathtubs and lavatories, are almost pictures of the things they represent and anybody will recognize them after a little practice.

**Scale Rules.** The scale rules in common use have two scales on each edge, and it is therefore necessary to be careful to read the scale that is intended to be used. The ¼" and ⅛" scales, the two most used, are usually both on one edge of the scale rule, one reading from each end of the rule. Figure 9-1 shows a typical scale rule.

As most of the dimensions of a building are given in figures on the plans, it is not necessary to use the scale rule to find them. In fact, it is better always to read the figured dimensions than to scale the plans; the figures are more accurate. In any event, care must be exercised in reading the dimensions. If the scale is used, care must be taken to read the right scale, as has been said. If the plan is made to ¼″ scale and the ⅛″ scale is used by mistake in taking a reading, the reading will obviously be twice as long as is actually required. If two dimensions, as for a floor area, are taken by the ⅛″ scale instead of the ¼″ scale, the error will be much greater; a 10′ × 10′ room would read 20′ × 20′, making a total of 400 sq ft instead of 100 sq ft!

The more familiar a student is with mathematics, the easier will he find calculation work. Higher mathematics, however, is not ordinarily used by contractors or estimators. Only in certain specialties is it necessary to have a good knowledge of solid geometry and trigonometry. Plain arithmetic, the use of fractions and decimals, and mensuration or plane geometry cover the mathematical operations used in building construction.

By cutting down the number of figures, whether in the quantities or in the pricing, the chance of making errors is automatically reduced and considerable time is saved. Therefore, an endeavor should be made to get the final results in the most direct way. Estimators, because of the limited time usually available for the preparation of estimates, find it absolutely necessary to reduce their work to the simplest number of operations consistent with accuracy. No one should attempt the study of estimating, however, until he can work readily the simple mathematical problems involved; this is a test of general ability, as well, and the general ability thus proved is also necessary. A good knowledge of the handling of fractions and decimals is an indispensable beginning.

Some estimators, immediately after reading the dimensions on the plans, turn the dimensions of feet and inches taken from the plans into feet and decimal parts of a foot, or feet and fractional parts of a foot. They then enter these figures on the quantity sheets, instead of entering feet and inches. The theory is that this makes it easier for an office assistant to extend the figures, as he would otherwise have to use his own judgment in handling the fraction of a foot. There are three objections to this procedure on the part of the estimator:

1. It forms another mental operation for his already busy mind, as the figures are either given in feet and inches on the plans for him to

read there, or else are scaled in feet and inches for him to read from the scale. Not only is he assuming the extra mental burden all through the process of reading and entering the figures, but he is adding one more opportunity for error in the handling of every entry.

2. Simplifying the work for the office assistant may seem all right, but if the assistant is of such poor quality that he (or she) cannot handle the simple calculations in feet and inches, it is to be seriously doubted that he has the ability to be trusted with the estimating sheets at all.

3. As the sheets will undoubtedly be used for later reference, the figures should be written on them in a way consistent with the method commonly used on plans and in verbal expressions among construction men—that is, in feet and inches.

The student who goes through this textbook and makes a complete estimate will probably be surprised to note the few formulas that are used in estimating. The mathematics of estimating is actually quite simple.

One rule that should be kept in mind, from a practical standpoint in estimating, is that when many items are to be extended from one set of figures, instead of using the figures for only one item, it is important that the figures thus used over and over should be right, as otherwise the error will be multiplied by the number of times that they are used. For example, if the size of a room is entered in length and width on the sheet for rough flooring and is later used, without being checked, for the finish flooring, and then perhaps for the floor scraping and finishing, and possibly for the ceiling plaster, it will readily be seen that, if the original dimensions of the room are wrong, many errors result in addition to the one regarding the rough flooring for which the dimensions were first intended. Some careful estimators make a habit of taking the measurements off separately every time they are required, and then later looking over the sheets and using the similar entries to check one another. At any rate, it is necessary always to watch out carefully for this type of error.

**Excavation.** Excavating work involves measurements of length, width, and thickness (or depth), in feet and inches; and then—computed from these three dimensions—the cubic feet of volume, and from that the cubic yards of volume. Feet multiplied by feet give square feet, and this result multiplied by feet again gives cubic feet. Thus, $10' \times 10' = 100$ sq ft, and this $100' \times 10'$ deep gives 1,000 cu ft. As there are 27 cu ft in 1 cu yd, dividing the number of cubic feet by 27 will give the cubic yards or, in this case, approximately 37 cu yd.

Excavating is measured in cubic yards. This is the customary unit of measure employed for that work. If the excavation for a cellar, for example, measures 25′ wide × 60′ long × 7′6″ deep, these three dimensions in feet (or feet and parts of a foot) are multiplied together, and the resulting number of cubic feet is divided by 27, in order to arrive at the number of cubic yards. The 7′6″ in this case is treated as 7½ or 7.5 ft. The answer is 417 cu yd. Figure out these cubic contents and find the same result. Then calculate the following cubic contents also:

$$20' \text{ wide} \times 50' \text{ long} \times 8' \text{ deep} = \text{―――} \text{ cu yd}$$
$$40' \times 32' \times 6'3'' = \text{―――} \text{ cu yd}$$
$$17'6'' \times 30' \times 7'4'' = \text{―――} \text{ cu yd}$$
$$39'8'' \times 83'9'' \times 11'5'' = \text{―――} \text{ cu yd}$$

**Concrete.** Concrete is, in most cases, measured in cubic yards also. Thus, to find the amount of concrete in a wall 23′ long × 10′ high × 1′ thick, follow the same procedure as in the above cases: $23 \times 10 \times 1 = 230$ cu ft; and this divided by $27 = 8^{14}/_{27}$ cu yd, or practically 8½ cu yd. It is not necessary to have the final answer in an odd fraction, the nearest ¼ or ½ is usually accurate enough for concrete work. The nearest whole yard is close enough for excavating work.

**Cement.** Cement floors are generally measured by the square foot, so the calculation is merely a matter of measuring the length of a floor and multiplying it by the width. Thus, a floor measuring 16′ wide and 34′ long would equal 544 sq ft. This is the required answer, as far as the quantity is concerned. This is then carried out to the pricing column and if it is priced at, say, 16 cents per sq ft, the cost extended would be $87. It is not customary to show cents in the final amount. This actually is figured out to $87.04, but is extended to the nearest whole dollar. In a long column of figures, there will be only an insignificant error if one gives and takes the cents in this way all down the line.

**Plaster.** Plastering is figured by the square yard. As there are 9 sq ft in 1 sq yd, find the two dimensions in feet of any area to be plastered, and multiply these two dimensions together to get the area in square feet. The result divided by 9 gives the number of square yards. Take a wall to be plastered, measuring 17′ long × 8′ high. This equals 136 sq ft; 136 divided by 9 equals 15⅑ sq yd. Enter this as 15 sq yd. If it were nearer to 15½, it would be called 15½ sq yd.

**Lumber.** Framing lumber used in building work is practically all figured in board feet, and waste has to be taken into account. It is easy to figure board feet. All that is necessary is to find the number of square feet, 1″ thick, or the equivalent. Thus, a board 10′ long × 1′ wide × 1″ thick has 10 board feet. If the same board were 2″ thick it would have twice as many board feet as the one that is 1″ thick, or 20 board feet. If the same board as first mentioned were only 6″ wide instead of 12″ wide, it would have half as many board feet, or 5 board feet. For any piece of lumber, multiply the three dimensions of it together and divide the result by 12; the answer will be the number of board feet in the piece of lumber. If more than one piece is concerned, multiply the amount in one piece by the number of pieces. The dimensions are always given in inches of cross section and length in feet; thus there may be a piece of 2″ × 8″ × 18′ (or, as spoken, 2 by 8 by 18). This piece measures 2″ × 8″ in cross section, and it is 18′ long. Multiplying the 2 × 8 × 18 gives 288, and dividing this by 12 gives 24 board feet. The rule is: multiply the standard dimensions together and divide by 12.

**Wood Flooring.** Wood flooring is sometimes listed simply by expressing the number of square feet to be covered by the flooring. In this case, it would be measured in the same way as the cement flooring mentioned above. For the ordinary types of wood flooring, the number of board feet is found by adding the waste allowance to the number of square feet of floor area. Ordinary flooring is 1″ or less in thickness and is therefore taken as 1″ for figuring purposes. All lumber less than 1″ thick is considered 1″ thick for figuring purposes.

Flooring boards have a tongue-and-groove formation cut on them, which takes away about ¾″ from the width of the boards when they are laid as flooring. Flooring boards with a face measurement of 2¼″ are made from 3″ wide stock, for example, and therefore 25 per cent of the width is lost or taken up by the tongue and groove. To provide for this loss, add the 25 per cent to the number of square feet of floor area, thus getting the answer in board feet. A floor 10′ × 20′, equaling 200 square feet of floor, would equal 250 board feet. This would be the amount of flooring to order, if every piece were good and of the right length so that no further waste would be encountered; but there

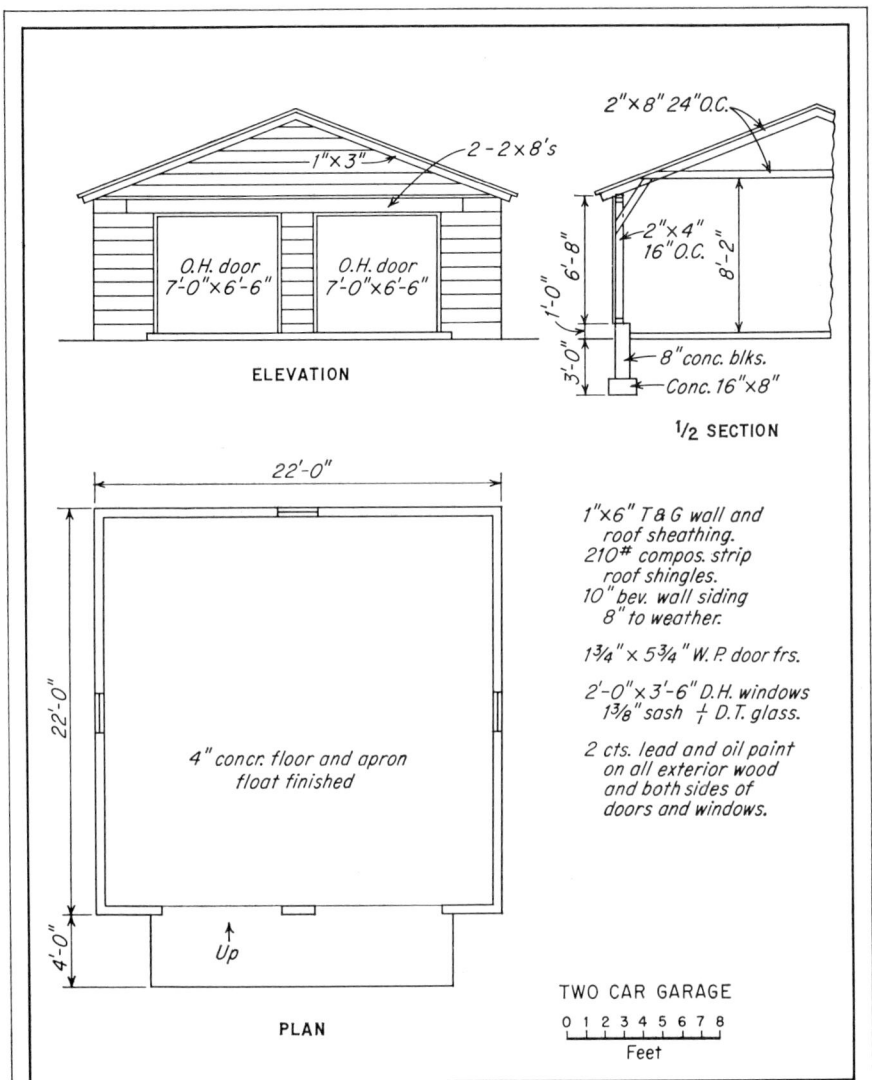

Fig. 9-2  Two-car garage.

is further waste, because some of the material is rejected, broken, etc., and because small ends have to be cut from many of the pieces. For this additional waste, add between 8 and 15 per cent more, depending upon the grade of flooring specified and the kind of lumber, etc. Therefore, 200 sq ft of floor area requires about 275 board feet of flooring.

All the plans must be used in listing the items and the specifications must also be kept in mind at all times. Inexperienced estimators have a bad habit of taking measurements from one drawing alone for an item when, by reference to the other drawings, the item may appear differently and may require different treatment from that which the one drawing indicates. Reference should be made to cross sections and detail drawings, especially, more often than is usually done. The preceding paragraph gives one reason for keeping the specification requirements in mind when measuring the plans—that the grade and kind of flooring affect the percentage of waste allowance. Other examples of the importance of reading the specifications carefully and keeping them in mind while measuring the plans could also be given.

Numerous specimen quantity-survey sheets are given in the succeeding chapters. The student should look these all over now and become generally familiar with the arrangement and grouping of the items, the wording used, the layout of the measurements, and the extensions of the figures. They will be discussed further as the various divisions are taken up.

■ EXERCISES

1. What is a quantity survey? (75 to 100 words.)
2. Name several items that are customarily measured by each of the following units: linear foot, square foot, square yard, cubic yard, pound, ton, thousand.
3. Make a quantity survey for the concrete, masonry, and cement work for the garage shown in Fig. 9-2.
4. Make a quantity survey for all the carpentry for the garage shown in Fig. 9-2.

CHAPTER 10

# expense and summary sheets

The items of cost that most directly and obviously enter into the total cost of any building are labor and materials. Besides these, there is a cost called *expense* that cannot be regarded as either labor or material. Cost items of this sort that can be applied against a particular job are entered under Job Expense. Items of expense that cannot be applied against a particular job are entered under Overhead Expense.

## Job expense

The heading Job Expense provides for the cost of supervision, temporary construction and protection, permits, and other items that are for the use of the entire job. In some instances it is difficult to determine whether an item should be put under this heading or under some work heading. For this reason, no arbitrary rules can be laid down. One rule for the estimator to follow, however, is to see that all the expense items are properly covered in some part of the estimate.

The superintendent's salary is one of the main items, usually, of the job expense. For example, if a job is to require 4 months to build and the superintendent is paid $200 per week, an expense of about $3,400, plus about $650 for insurances and taxes, is involved in this one item. This must be included as part of the job cost, and the proper place for it is under the heading for the general expense of running the job. In like manner, a watchman is generally needed on building jobs and, if the regular night watchman is paid $70 per week, there will be an item here involving about $1,400 more. A day watchman may be required for Saturdays, Sundays, and holidays, and this, in the example given, would add $400 more. Of course, some jobs will not require so much watchman service and some may not require any watchmen at all. On the other hand, there are jobs where a day watchman is needed every day, and some where two or more night men must be on duty every night.

Other members of the job office personnel may include a timekeeper and a material clerk, to keep records of the men's time and pay and of the supply of materials. On large jobs there are assistant superintendents, layout engineers, checkers, tool clerks, and others. These men are not actually engaged in the construction of the building but they are needed in connection with the work on the job, and the cost involved in having them there must be taken into account. On out-of-town jobs it is the custom to send the supervising personnel from the main office. In this case the expense will be increased, owing to the traveling expenses and, in some instances, to the living expenses of these men. Occasionally, workmen also are sent out in this way.

Table 10-1 gives lists of the personnel, by departments, on a large job such as that shown in Fig. 4-8. They show the activities attendant upon such work and serve to illustrate the elements of costs thus involved in construction work. The positions given in parentheses are held by men assigned from other departments.

TABLE 10-1

| *Project Manager's Office* | *Purchasing Department* |
|---|---|
| Project Manager | Purchasing Agent |
| Assistant Project Manager | Secretary |
| Control Clerk | Building-material Buyer |
| Attorney | Equipment and Tool Buyer |
| Secretary | Track, Road, and Sewer Buyer |
| Report Clerk | Mechanical and Electrical Buyer |
| Field Inspector | Field and Office Service Buyer |
| Plan Clerk | Buyers' Clerks (5) |
| Mail Clerk | (Voucher Clerk) |
| Stenographer | Local Expediter |
| Typist | Shipment Expediter |
| Subcontractors (20) | Traveling Expediter |

TABLE 10-1 *(Continued)*

*Purchasing Department*

Stenographer
Typist
Comptometer Operator

*Estimating Department*

Chief Estimator
Secretary
Estimator
Assistant Estimators (2)
Draftsman
Contract Clerk

*Engineering Department*

Chief Engineer
Secretary
Office Engineer
(Voucher Clerk ½)
Draftsman
Layout Engineer
Field Engineers (10)
Rodmen (10)

*Construction Department*

General Superintendent
Secretary
Assistant General Superintendent
(Voucher Clerk ½)
Night General Superintendent
Area Superintendents (5)
Superintendents (15)
Foremen (120)
Workmen (3,000)

*Track Department*

Chief Track Engineer
Secretary
Assistant Track Engineer
(Buyer ½)
(Storekeeper)
(Cost and Time Clerk)
(Tool Clerks 3)

*Track Department*

(Field Engineer ½)
Night Clerk
Track Superintendent
Foremen (10)
Mechanics (20)
Laborers (150)

*Road and Sewer Department*

Superintendent of Roads and Sewers
Secretary
Engineer
(Buyer ½)
(Timekeeper ⅓)
(Storekeeper)
(Tool Clerk)
(Field Engineer ½)
(Cost and Time Clerk ⅓)
Night Clerk
Foremen (2)
Maintenance Men (20)
Subcontractors (3)

*Mechanical Engineering Department*

Mechanical Engineer
Secretary
Plumbing Superintendent
Plumbing Foreman
Plumbers and Helpers (6)
Heating Superintendent
Heating Foreman
Steamfitters and Helpers (6)
(Buyer ½)
(Storekeeper)
(Tool Clerk ½)
(Cost and Time Clerk ⅓)
Night Clerk
Subcontractors (3)

*Electrical Engineering Department*

Electrical Engineer
Secretary

TABLE 10-1 (Continued)

*Electrical Engineering Department*

Assistant Electrical Engineer
(Buyer ⅓)
(Cost and Time Clerk ⅓)
(Storekeeper)
(Tool Clerk ½)
Night Clerk
Foremen (2)
Maintenance Men (20)
Subcontractors (2)

*Material Department*

Department Manager
Secretary
Night Manager
Head Checker
Checkers (20)
Truck Router
Chief Storekeeper
Storekeepers (8)
Lumberyard Manager
Lumberyard Foreman
Lumber Handlers (10)
Lumberyard Clerks (3)
Mill Manager
Mill Foreman
Mill Carpenters (40)
Mill Clerks (3)
(Buyer)
(Cost and Time Clerk)
Stenographer
Typists (3)

*Equipment and Tool Department*

Department Manager
Secretary
Assistant Manager
Stock Clerk
Salvage Clerk
Stenographer
Typists (2)
(Voucher Clerk)
(Cost and Time Clerks 2)

*Equipment and Tool Department*

(Buyer)
Equipment Superintendent
Equipment Clerks (2)
Equipment Operators (100)
Tool Superintendent
Tool Clerks (10)
Truck Superintendent
Truck Clerks (2)
Chauffeurs (100)
Field Checkers (6)
Repair Manager
Master Mechanic
Assistant Master Mechanic
Repair Men (4)
Oil and Grease Man
Gas-station Attendant

*Field Service Department*

Department Manager
Secretary
Safety Engineer
Canteen Supervisor
Rubbish Foreman
Rubbish Men (20)
Scrap Clerk
Medical Doctor
First-aid Men (6)
(Auxiliary First-aid Men 6)
Nurses (6)
Orderlies (2)
Ambulance Drivers (3)
Accident Clerk
Police Chief
Assistant Police Chief
(Auxiliary Chiefs 5)
Guards (75)
Fire Chief
Assistant Fire Chief
(Auxiliary Chiefs 5)
Firemen (10)
(Auxiliary Firemen 30)
Fire-truck Drivers (2)

TABLE 10-1 (*Continued*)

*Field Service Department*

Bus and Car Manager
Bus and Car Clerk
Bus and Car Drivers (80)
Field Messenger
(Storekeeper)
(Buyer ½)
(Cost and Time Clerks 2)
Chief Waterboy
Waterboys (75)
Stenographer
Typist

*Office Service Department*

Department Manager
Secretary
Night Service Manager
Assistant Department Manager
Stenographers (4)
Typists (4)
Comptometer Operators (4)
Mimeograph and B/P Operator
Personnel Manager
Personnel Clerks (4)
Labor Relations Manager
Receptionist
Messengers (8)
(Storekeeper)
(Cost and Time Clerk)
(Buyer ½)
Head Janitor
Janitors (4)
Matrons (2)
Head Switchboard Operator
Switchboard Operators (4)

*Time and Cost Department*

Department Manager
Secretary
Chief Timekeeper
Timekeepers (4)
Payroll Clerks (7)
Payroll Typists (4)
Payroll Comptometer Operators (5)
Chief Cost Clerk
Cost Clerks (3)
Time and Cost Clerks (6)
Field Checkers (10)
Cost Typists (3)
Cost Comptometer Operators (2)
Paymaster
Assistant Paymasters (4)
Termination Clerk
Night Clerks (2)
Bookkeepers (3)
Stenographer
Typist
Comptometer Operator

*Accounting Department*

Chief Auditor
Secretary
Assistant Chief Auditor
Auditor of Disbursements
Assistant Auditor of Disbursements
Voucher Clerks (6)
Cashier
Reimbursement Clerks (2)
Social Security Clerk
Bookkeepers (3)
Stenographers (2)
Typists (3)
Comptometer Operators (2)

**Offices.** Temporary job offices, tool houses, storage sheds, and other such conveniences are part of the expense of carrying on a job. Temporary fences, barricades, runways, bridges, stairways, and platforms are also to be considered as items of cost. The law demands that proper safeguards be constructed and maintained in good order to prevent accidents at open shafts, wellholes, and other places that are likely to be the locations of accidents if not so guarded. The law also demands temporary toilets, drinking water, and other conveniences

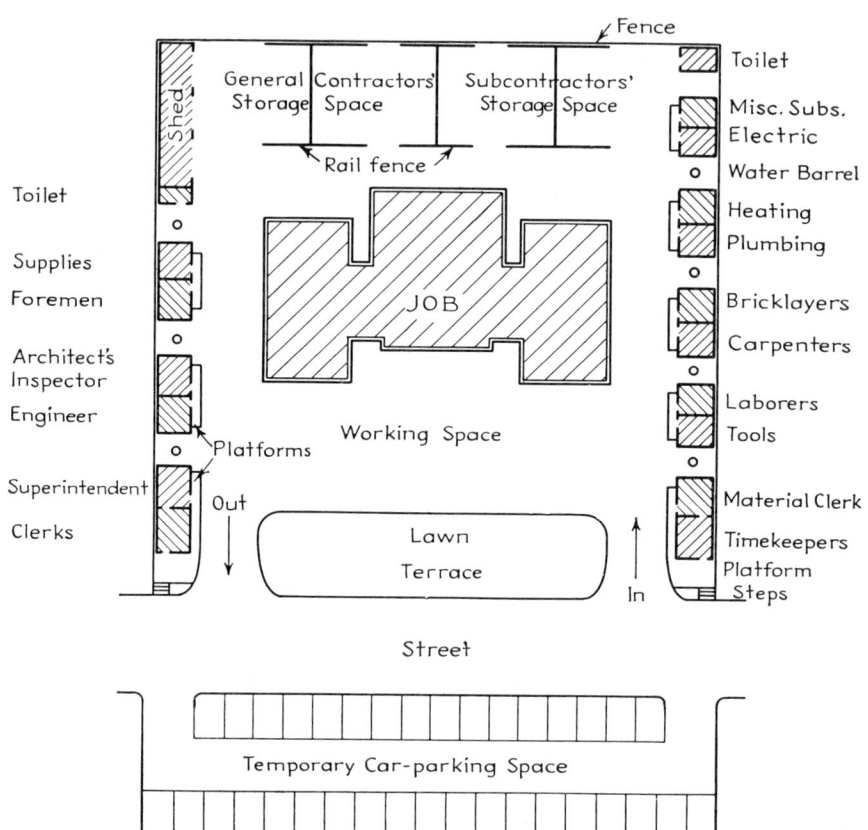

*Fig. 10-1* Layout of a construction job.

for the health and comfort of the men. Trade unions demand these and other things besides. Figure 10-1 shows the layout of the temporary construction and conveniences for a typical job. Fortunately, at this project an open plot across the street was available, where parking space was provided for the many cars in which the men came to work. Figure 10-2 shows the plan of the temporary job office building used on a large project. This shanty would cost as much to build as a good-sized house.

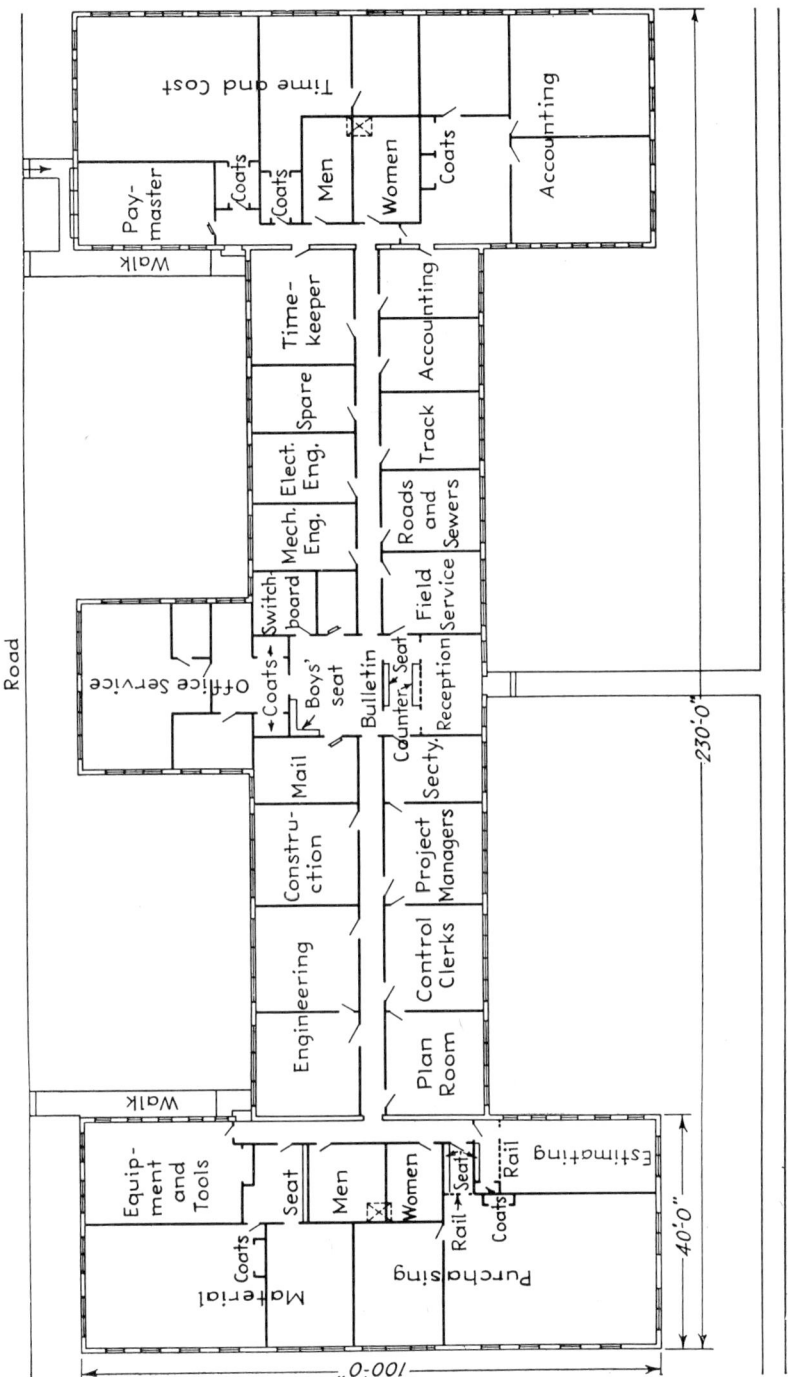

*Fig. 10-2* Job office on a large project.

Openings in the walls for doors and windows often have to be closed before the doors and the window glass have arrived at the job. The custom is to tack muslin in the openings that are not needed for access to the work. Some openings, however, must be provided with temporary doors. Enclosures are often required, especially during the winter months, to protect the men from the weather before the walls are advanced far enough so that merely covering the door and window openings is adequate.

If a structure over 40 ft high is to be erected or if one over 25 ft high is to be demolished, a sidewalk shed is usually required, unless the distance from the structure to the sidewalk is considerable. In large cities, catch platforms must be erected on the front of a building that is over a certain height, while it is being demolished. Furthermore, the law demands that fences be built along the front of jobs in certain locations, and civic associations do their best to have main thoroughfares kept in tidy condition, with neatly built front enclosures for construction work.

**Utilities.** Water and light are common requirements for building work; yet many estimators neglect to make provision enough in their estimates for these items. Water involves not only the water charges, but also the installation, maintenance, and removal of the temporary lines, and the changing about of these lines when that is necessary. Similarly, for temporary light and power there is the charge for current, the cost of installing and maintaining the lines, the electrical union's demand that union electricians be kept in attendance merely to turn switches on and off, and the cost of bulbs and fixtures.

To heat a construction job in winter is a costly undertaking. Some of the elements of this cost are the installation and first cost of the system or devices used for heating, the fuel cost, the additional temporary enclosure work required, the labor of attending to the heating, the repairing and rearranging of the equipment, the loss of time resulting from the men's working under these conditions, and the extra cost of supervising, etc.

There is often need for temporary stairways or temporary treads for the permanent steel stairs in large buildings. The law requires these to be safe and secure, and this means that substantial guardrails must be provided. Temporary toilet and locker rooms must be considered, too.

There are many small items in the job office that add up to a considerable sum on large operations. One of these, and one that should not be called very small, is the job telephone cost. A monthly bill of from $50 to $200 is common on a busy job of fair size. Office furnishings, stationery, and supplies often amount to $200 on a job.

Repairs to streets, walks, lawns, and other parts of the property and of adjacent property, as well as permits and deposits in connection with them, sometimes account for a large group of cost items and must be considered on all jobs. In fine building work, all the finished surfaces, corners and mouldings of stone, marble, tile, cabinetwork, bathroom fixtures, and equipment of all kinds have to be protected from damage that might result from exposure before the building is completed and from damage that might be caused by men or materials for other parts of the work. Repair costs for all these items and for other parts of the building, as well as for broken glass, broken lighting fixtures, etc., always make up a considerable sum on a building of even average size, especially if a series of accidents occurs on the job.

Specifications generally call for buildings to be left "broom clean" by the contractor. Sometimes, however, the specifications or some other part of the contract will require that every part of the building be left clean and polished and ready for occupancy, so that no further work of this kind will be necessary. In any case, a great amount of rubbish will have to be removed and considerable plain cleaning must be done. The work of gathering together a large assortment of miscellaneous waste material, crates, cans, broken glass, etc., and the removal of all this from the various floors of a building, besides the sweeping of the floors, account for one part of the cleaning cost. Another part covers the loading of the stuff on dump trucks and the hauling and disposal of it. A third part of this kind of expense may be necessary—that for washing all the window and door and other glass in the building. Then a fourth part that may have to be added is for cleaning the floors all over again and polishing them. Experienced estimators, knowing that a gang of men may be required for each of these operations for a number of days, take care to figure the cost accordingly.

**Special Services.** Surveys and engineering and other special services are frequently specified to be furnished by the contractor, and he may find these desirable, even though they are not specifically called for in his contract. If they are not included under some other division of the estimate, they should be placed under the Job Expense heading. Even the cost of accurately laying out the job and building and maintaining the batter boards must be considered. Special drafting work and the cost of blueprints, job signs, and progress photographs also come into this category. After the corners of the building have been accurately determined, the lines running over them are extended to the batter boards where the reference marks are maintained until the walls are built. Other reference lines provided for on the batter boards are the line of the footings and the excavation line. (Fig. 10-3).

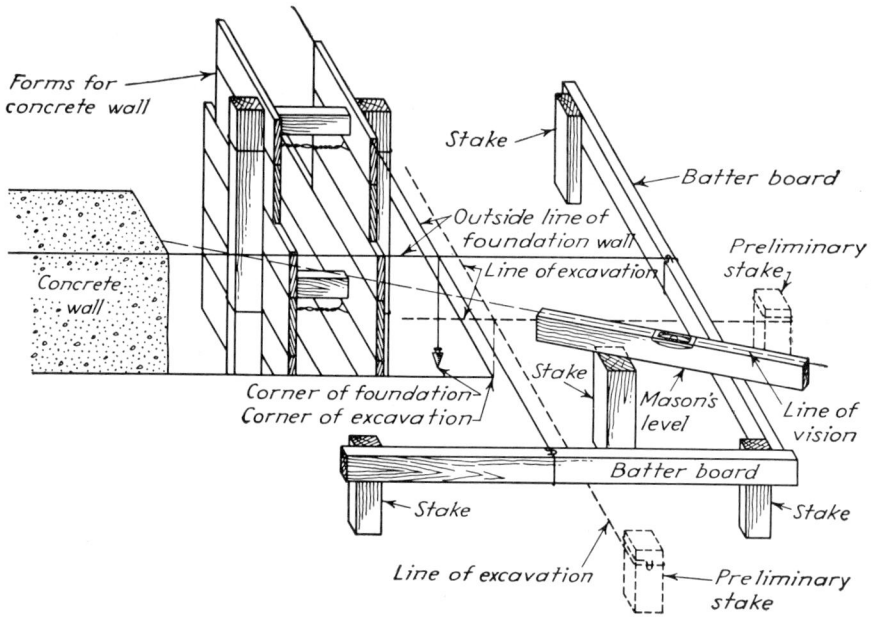

*Fig. 10-3  Batter boards and lines.*

**Bonds.** Surety bonds, guarantees, deposits, etc., are general items of job cost. The cost of a surety bond may run as high as 2 per cent of the total amount of the contract. The trouble of securing a bond and the time and expense of office work in connection with it all through the job and after the job is completed are elements of cost that some contractors seem to ignore completely; but these should be taken into account. Similarly, the trouble and expense incident to guaranteeing work may run for a year or more after the job is finally accepted. If any cash deposits are demanded, or if any money is to be held for a length of time, the wise estimator will make careful note of each such item on his estimate sheets and will charge the proper amount of interest as a part of the cost. He will also put in proper amounts for possible repairs, maintenance, etc., and service charges for the incidental trouble.

Contract bonds serve a twofold purpose. They evidence the qualifications of the contractor to perform the contract which is bonded, and they provide protection to the owner and the trade creditors against the consequences of a default (see Fig. 10-4).

Most surety companies use standard forms of application and of financial statement which are prepared expressly to obtain preliminary information for underwriting purposes. The underwriters determine the financial responsibility of a contractor, not only on his ability to meet job payrolls and to pay for materials, but on the indicated availability of sufficient current assets over current liabilities to provide reasonable assurance that increased cost of performance, or losses caused by defaulting subcontractors, by casualty, or by other liability will be paid without interfering with the progress of the work.

Financial statements must necessarily be judged on the liquidity and availability of working capital and the probable sufficiency of secondary assets to provide the reserve needed against contingencies. The conditions of the particular contract to be bonded naturally determine the qualifications a contractor must possess. The contract price is the primary standard of measurement of the work to be done. If the reasonable certainty of performance at the contract price, or under the conditions of the contract, is established, a bond may be obtained.

Subcontracted work does not automatically reduce the general contractor's liability, because he must, in effect, guarantee the subcontractor's responsibility and ability to perform, and he must pay the subcontractor to avoid a breach of the subcontract. The time of performance, weather conditions during this time, liquidated damages, whether terms of payment are sufficient to meet contract costs, liability for subsurface or other physical conditions, contractual liability for damage to property or injuries to persons, coordination of work with that of other independent contractors—all these are conditions which underwriters may assess to determine a contractor's eligibility for a bond.

**Insurance.** Insurance is a very considerable part of the cost of doing construction work. The proper place to put insurance costs that are based on labor is in each of the work divisions of the estimate and in

# PERFORMANCE BOND

KNOW ALL MEN BY THESE PRESENTS:

That ................................................................
(Here insert the name and address, or legal title, of the contractor)
as Principal, hereinafter called Contractor, and ...................... as Surety,
(Here insert the legal title of Surety)
hereinafter called Surety, are held and firmly bound unto ........................
(Here insert the name and address, or legal title, of the owner)
as Obligee, hereinafter called Owner in the amount of ..........................
................................................................ Dollars ($      )
for the payment whereof Contractor and Surety bind themselves, their heirs, executors, administrators, successors and assigns, jointly and severally, firmly by these presents.

WHEREAS, Contractor has by written agreement dated ................
entered into a contract with Owner for ...........................................
................................................................................
in accordance with drawings and specifications prepared by .....................
................................................................................
(Here insert full name and title)
which contract is by reference made a part hereof, and is hereinafter referred to as the CONTRACT.

NOW, THEREFORE, THE CONDITION OF THIS OBLIGATION is such that, if Contractor shall promptly and faithfully perform said CONTRACT, then this obligation shall be null and void; otherwise it shall remain in full force and effect.

Whenever Contractor shall be, and declared by Owner to be in default under the CONTRACT, the Owner having performed Owner's obligations thereunder, the Surety may promptly remedy the default, or shall promptly

    (1) Complete the CONTRACT in accordance with its terms and conditions, or

    (2) Obtain a bid or bids for submission to Owner for completing the CONTRACT in accordance with its terms and conditions, and upon determination by Owner and Surety of the lowest responsible bidder, arrange for a contract between such bidder and Owner and make available as work progresses (even though there should be a default or a succession of defaults under the contract or contracts of completion arranged under this paragraph) sufficient funds to pay the cost of completion less the balance of the contract price; but not exceeding, including other costs and damages for which the Surety may be liable hereunder, the amount set forth in the first paragraph hereof. The term "balance of the contract price," as used in this paragraph, shall mean the total amount payable by Owner to Contractor under the CONTRACT and any amendments thereto, less the amount properly paid by Owner to Contractor.

Any suit under this bond must be instituted before the expiration of two (2) years from the date on which final payment under the CONTRACT falls due.

No right of action shall accrue on this bond to or for the use of any person or corporation other than the Owner named herein or the heirs, executors, administrators or successors of Owner.

Signed and sealed this        day of        A.D. 195

................................................(Seal)
*Principal.*

IN THE PRESENCE OF:      ................................................(Seal)
*Surety.*

*Fig. 10-4  A performance bond.*

the job-expense division of the estimate, as well. In this way the various rates will best be taken into account. Some estimators, however, prefer to compute the total insurance requirements and then enter a lump sum of money, to cover it, in the job-expense sheet or in the summary of the estimate. All the workmen and supervisors and other employees on the job are subject to some form of insurance that has to be paid for by the employer. This includes state workmen's compensation and unemployment insurance, Federal social security and old-age pension insurance, public liability insurance, and others. The total cost for these runs between 11 and 20 per cent of the payroll amount for regular building work and much more for demolition, steel erection, and other specially hazardous lines of work. In addition to the insurance items that are based upon the payroll, other insurance is usually carried by contractors, such as contingent insurance on the subcontractors, fire insurance, and automobile insurance. Definite entries must be made on the estimate sheets to cover every kind of insurance carried.

## Overhead

*Overhead expense* is the name given to all those items that are necessary for the general maintenance of a business. This expense continues even when there is no contract work on hand. It does not belong exclusively to any particular part of the contract work and is therefore spread over all the work by prorating a percentage to each job, based on the cost of the job.

All expense that is not incorporated in other entries in the estimate must be considered as being included in the Overhead entry. The percentage allowance, therefore, must be large enough for this purpose. This is emphasized because many contractors do not fully realize the many items of expense involved in maintaining a business. The percentage is at best an approximation based on the contractor's own cost records and his judgment as to the requirements of each individual job. Some jobs may require a smaller allowance than others because the work to be done involves large subcontract items which require little more actual office effort than small items would require. Obviously, small jobs require more, in proportion, than do large jobs. All jobs must carry their share of the cost of maintaining the business while no work is on hand.

Tools, equipment, trucks, etc., used on the jobs should not be charged to overhead expense. The cost of permits, bonds, supervision, drafting, job gratuities, insurance and taxes in connection with these items, and other expense directly connected with the jobs should not be charged to overhead expense. All these items should be charged to the individual jobs. They should be included under the general job-expense heading and other headings in the estimate to suit the requirements for each job.

TABLE 10-2 OVERHEAD EXPENSE

| | | | |
|---|---:|---|---:|
| Employees: | | Rent | $16,800 |
| Manager | $20,000 | Furnishings (depreciation) | 800 |
| Manager's secretary | 4,420 | Stationery and supplies | 742 |
| Solicitor | 10,000 | Postage | 451 |
| Chief estimator | 15,000 | Electricity | 1,422 |
| Contract clerk | 7,020 | Telephones | 1,805 |
| Estimators | 23,400 | Legal expense | 1,000 |
| Payroll office | 15,600 | Advertising | 1,684 |
| Accountant | 8,000 | Trade reports | 206 |
| General superintendent | 15,000 | Use of autos | 3,817 |
| Material clerk | 7,280 | Car and taxi fares | 1,051 |
| Stenographers | 7,280 | Travel expense | 1,583 |
| Receptionist | 3,380 | Entertaining | 975 |
| Plan clerk | 3,380 | Gratuities and contributions | 210 |
| Vacations (included above) | | City business taxes | 10,822 |
| Sick leave (included above) | | Miscellaneous | 2,746 |
| Christmas bonuses | 14,200 | | |
| Social security taxes | 1,152 | For year 1958 | $208,828 |
| Unemployment taxes | 1,620 | Job cost for year, $2,783,400 | |
| Insurances | 5,982 | Overhead expense, 7½% | |

Although not always so handled, the term *overhead* should be reserved for the cost of running the main office. If we think of all the expense necessary to maintain a contracting business which has no work on hand, then we will have a picture of the items concerned. Table 10-2 gives a list of such items as applied to the construction company shown in Fig. 4-6 in Chap. 4. This does not include any engineering or drafting expense, as these items would be directly charged to the jobs requiring them. A smaller business would cost less to maintain but, in any case, the expense must be spread among the jobs that are actually performed. By keeping track of the overhead expense, a contractor will know after a few years the exact relation it bears to the job cost of the work actually done.

**Working Conditions.** State labor laws cover the design, construction, and operation of factories, manufacturing plants, and mercantile establishments and include the work of constructing buildings. In New York the labor law is administered by the State Industrial Board and is enforced through the State Commissioner of Labor and the usual channels of law enforcement. This commissioner also administers the Workmen's Compensation Law.

In the supervision of places where people work, the proper sanitation of these places is included, and they are controlled for guarding against and minimizing fire hazards, personal injuries, and diseases. It is the policy and intent of the law that all places to which it applies shall be so constructed, equipped, arranged, operated, and conducted in all respects as to provide reasonable and adequate protection to the lives, health, and safety of all persons employed in them.

On building-construction work, inspectors from the Labor Department make careful note of the scaffolding, ropes, planks, ladders, hoistways, temporary protection, and equipment; and it is against the law for any contractor, superintendent, foreman, or anybody else who directs work in any way to allow workmen to be endangered.

In steel-frame construction, the entire floor framing must be planked over on the floor where the steel erectors are working. Floor arches of concrete or some other material must be installed as the building progresses and, in wood-floor construction, the underflooring must be likewise installed as the building progresses. If double wood floors are not used, the floor below the one being worked on must be planked over. Elevator shafts and hoistways must be fenced in at each floor. All persons are prohibited from riding on any platform hoist or on any elevator car that is not equipped and operated for carrying passengers; and all persons are forbidden to ride on any concrete bucket, derrick, or other hoisting apparatus or on loads being hoisted.

All factories must be registered and are inspected periodically. They must have fire drills. All halls, stairs, elevators, etc., must be kept lighted during working hours; proper safeguards must be maintained at all elevators, hoistways, etc.; and toilets must be kept clean and in good working order. Estimators should realize that these requirements, with the exception of the fire drills, also apply to construction jobs and that the cost involved has to be provided for in their estimates.

## Changes

Few contractors and almost no architects appreciate the amount of indirect expense involved in changes that are ordered while a job is in progress. Sometimes this work should have been included in the specifications, and the extra work thus made necessary is due to the oversight or lack of experience of the architect. When such items come up, the architect may have an impulse to cover them up rather than to make a straightforward explanation to the owner. In that case, work is done for the owner without his paying anything for it.

Some architects will tell the owner, and really believe it to be true, that their specifications and plans are so definite and binding that all bids may be judged by the figures alone. This is a fallacy, in spite of the assumption that none but first-class contractors are allowed to figure the job.

Certain speculative operators have been known to swindle their contractors and subcontractors by having them estimate on incomplete specifications, certain pages being temporarily removed. Happily, however, this is not at all a common occurrence.

Contractors should be asked to estimate only when a full set of properly prepared plans and properly written specifications are put at their disposal for an ample length of time to serve this purpose. Furthermore, no alternate prices or unit prices should be asked for, nor should any addenda be issued, until after the bids have been submitted and at least half the bidders have been eliminated.

## Summary

A summary enables the estimator and the contractor to see quickly the relation and the proportions of the various branches of the work. This is valuable as a general check on the amounts extended, and it also serves as a basis for determining the office overhead and the profit. One danger in summarizing an estimate, however, is that some of the totals may not be carried over from the regular make-up sheets or that last-minute changes may be made in some of the items and the corresponding amounts may not be changed in the summary. For this reason, a check should be made by adding up all the individual sheets of the estimate separately from the summary, to see that this total agrees with that on the summary.

Many of the items in building work are sublet. Some estimators make up approximate estimates of their own for these items, as a check upon the prices received from subcontractors. These approximations should always be used only as a check, however, unless the estimator is thoroughly familiar with the work and with the pricing of the items thus figured. The custom is for contractors to send out to their regular group of subcontractors, postal-card requests for estimates, when the plans are first received. It is customary also for the successful contractor to favor those subcontractors whose figures were received before the contractor's bid was submitted and were used for arriving at the amounts entered in the summary of the bid. This is only fair, and subcontractors soon learn whether it is worth their while to continue to estimate for a particular contractor. The ideal condition—one that ethics demands—is that the lowest responsible bidders in all lines shall be given the work in their line without further question and without any changes in their prices.

Figure 10-5 shows a typical summary sheet of an estimate. The summary will also serve to illustrate a typical list of divisions, with the headings given in the usual order.

Figure 10-6 gives a typical general job-expense sheet, from which the amount for the first item on the summary is obtained.

On the specimen estimate sheets, note the method of treating the main heading, the headings for the main divisions, and the subdivision, or key, headings. The main heading is placed on the summary and on the first sheet of the estimate only. Follow the neatness and general arrangement and the consistent order of the figures shown on all the sheets.

### ■ EXERCISES

1. Name 10 items of expense that should be considered under the heading Job Expense.
2. What cleaning and rubbish-removal work are done on a building job and therefore require consideration as elements of cost?
3. What does an estimator need to know about insurance?
4. Name 10 items of expense that should be considered as included in the main-office, or overhead, expense.
5. Make a general job-expense sheet for the house estimate.
6. Make a summary sheet for the house estimate.

## SUMMARY OF ESTIMATE

STORE AND SHOWROOM BUILDING.            Submitted
Webster Ave. & 188th St.                 April 1, 1959.
Owner: Mayfair Co.              Wells & Brown, Archts.

| | |
|---|---:|
| General Job Expense | $ 3,480 |
| Excavating | 5,478 |
| Concrete Work | 7,120 |
| Mason Work | 12,654 |
| Granite | 300 |
| Cast Stone | 1,050 |
| Structural Steel | 12,200 |
| Miscellaneous Iron | 675 |
| Concrete Floor & Roof Arches | 16,105 |
| Cement Finish Work | 3,555 |
| Roofing & Sheet Metal | 2,424 |
| Dampproofing | 265 |
| Caulking | 115 |
| Plastering | 4,937 |
| Carpentry | 2,681 |
| Kalamein Work | 452 |
| Rolling Steel Doors | 400 |
| Peelle Shaft Doors | 600 |
| Overhead Garage Door | 140 |
| Tile Work | 372 |
| Rubber Flooring | 1,347 |
| Metal Toilet Partitions | 110 |
| Finish Hardware | 300 |
| Glass & Glazing | 227 |
| Painting | 1,484 |
| Plumbing | 2,750 |
| Heating | 2,552 |
| Oil Burner | 800 |
| Electric | 1,670 |
| Elevator | 3,000 |
| Job Cost | 89,243 |
| Overhead 7% | 6,247 |
| Total Cost | 95,490 |
| Profit 4% | 3,820 |
| BID | $99,310 |

*Fig. 10-5*   *Summary of estimate.*

## J. E. EKEBLAD CONSTRUCTION CO. Inc.
### BUILDERS
389 LEXINGTON AVENUE
NEW YORK CITY

Job *Fordley Bldg.*
*Syracuse, N.Y.*

Date *March 10, 1959*
Est. No. *481*

### JOB EXPENSE

| | Description | | | Amount |
|---|---|---|---|---|
| | Superintendent-Asst. Supt.-Job Runner | 15 wks. | 189. | 2835 |
| | Engineers-Rodmen | 15 wks. | 102. | 1530 |
| | Timekeepers-Checkers-Bookkeepers-Stenographer-Expeditor | 15 wks. | 102. | 1530 |
| | Watchmen-Waterboys | 12 wks. | 80. | 960 |
| | Temp. Office & Equipment  8' × 20' | | | 450 |
| | " Sheds-Cement Shed-Warehouse  50. | | | 50 |
| | " Toilets | 2 | 20. | 40 |
| | " Roads | | | — |
| | " Light & Power | 13 wks. | 22. | 286 |
| | " Water (Conn.-Piping-Protect-Water) | | | 75 |
| | " Lights-Guards-Barricades-Ladders-Stairs-Protection  20.  40.  40. | | | 100 |
| | " Doors-Windows | | | 25 |
| | " Fences | | | — |
| | " Bridge-Lighting | | | — |
| | " Heat | | | — |
| | Travel & Transportation | | | 30 |
| | Trucking | | | 200 |
| | Telephone & Telegraph-Archts. | 15 wks. | 6.- | 90 |
| | Small Tools-Tarpaulins | | | 50 |
| | Plant & Equipment | | | 300 |
| | Plank & Horses | | | 200 |
| | Pumps & Pumping-Fuel-Attendance | 4 days | 60. | 240 |
| | Surveys-Laying Out-Batter Boards | | | 40 |
| | Permits (Street-Building-Vault)  20. | | | 20 |
| | Drafting-Shop Drawings-Blueprints | | | 400 |
| | Rubbish Removal-Cleaning-Final Cleaning | 4800 SF | .03 | 144 |
| | Clean Glass-Glass Breakage | | | 60 |
| | Protect Finish Floors-Stairs | | | 50 |
| | Cut & Patch | | | 250 |
| | Photographs-Samples-Tests-Guarantees  60  120.  400. | | | 580 |
| | Winter Conditions-Snow Removal | | | — |
| | Hoist & Hoisting (Engine-Installed-Remove-Service-Fuel Tower-Platforms-Head-Cable) | 8 wks. | 420. | 3360 |
| | Project Signs | | | — |
| | Project Office (Light-Heat-Janitor-Equipment) | | | — |
| | Contractors Reports | | | — |
| | | | | 13,895 |
| | (Include insurance and taxes in each item) | | | |

*Fig.* 10-6  A job-expense sheet.

CHAPTER 11

# *excavating*

Excavating for building work generally consists of digging a big hole, called the *main excavation* or *general excavation* and digging other parts, called *pit excavating* or *trench excavating,* etc. The work is measured in cubic yards. The three dimensions of each part are taken in feet and these are multiplied together to get cubic feet. This result is divided by 27 to get cubic yards, since there are 27 cubic feet in one cubic yard.

Plans never show the outline of the excavation, nor do they generally indicate the ground level, or original grade, of the site. It is necessary for the estimator to visit the place and determine the conditions that will affect the size of the excavation required. The outline of the banks of the excavation should then be drawn on the cellar plan and a profile of the excavation drawn on one of the cross sections. By this means the estimator is better enabled to visualize the actual excavating work. If a plot plan or survey is furnished with the plans, this should be referred to for any information it may contain relative to the ground, the location of the building, trees that may have to be protected or removed, and other data that will affect the estimate.

**Space Allowance.** Space must be allowed all around the outside of the structure for proper working room for installing the temporary forms for the footings and the walls. The projection of the footings beyond the outer face of the walls must be taken into account, as well as the natural sloping of the banks of the excavation.

The depth of the excavation is measured from the present, or original, grade and not from the finished condition that may be indicated on the plans and sections. The deeper the excavation is to be made, the more thorough should be the study of the profile that the banks around the excavation will take—not only at the time the excavation is being dug but afterward, also. If a deep excavation is to be left open for several weeks or months, it will be necessary to provide bracing for the banks or else to make the excavation wide enough at the top so that earth washed down by rain or because of continual disturbance by working conditions will not hinder the proper construction of the foundations.

Pits and other handwork must be kept separate in the estimates, so that they may be figured at different prices from those for the main excavation. The main excavation work is usually done with a power shovel or some other large piece of apparatus, unless the amount to be removed is small or, for some other reason, the job does not warrant the use of large equipment.

All items of expense directly due to the excavating work are provided for under this heading. Temporary guardrails, ramps, and other provisions for doing the work safely and economically are noted in the estimate and considered in the unit cost per cubic yard for completely removing the excavated material, or else they are listed and priced separately.

Text and reference books refer to the "angle of repose" in connection with excavating. This is largely theoretical and has little practical value for builders. The angle is that which the bank of earth or other material makes with the horizontal without being held in place. Tables are given for reference, but a glance at the wide range of the figures in them will convince anybody of their worthlessness in accurate estimating. The only safe procedure is to know the locality well enough to be fairly sure of the probable ground conditions.

When soft banks are to be left standing, it is necessary to brace them or to drive sheeting for their support. This is called *skeleton sheeting* if only light bracing will suffice, or *sheet piling* if a continuous sheeting is required. Both need heavy wales and bracing to hold the sheeting in place. Ordinary wooden sheet piling consists of a continuous line of vertical planks held against the sides of the excavation by these horizontal timbers, called *wales* or *breast timbers*, which are in turn supported by shores or by cross braces reaching to the opposite side of the excavation. Planks 2 in. thick are generally used for depths up to 12 ft and 3-in. planks for depths up to 18 ft.

Steel sheet piling is used for deep excavations where great pressures must be supported or in ground that contains water.

Isolated buildings of regular type, where only plain dirt excavation is to be considered, present no particular problem in estimating the costs under the excavating heading. By allowing about 18 in. of clear space all around the exterior walls, the matter resolves itself into a case of care in measuring the plans and care in determining the present ground levels. The work is computed in cubic yards, as it is—in the ground—before being removed. If the work were priced loose or by the truckload, the natural expansion of the material would have to be considered. This expansion amounts to an increase of about 20 per cent for earth and up to 50 per cent for gravel or rock.

*Ordinary excavating* denotes the kind that can be done with shovels by hand, if this method is otherwise desirable. *Heavy earth* denotes clay or any other dense formation for which picks would be required before shoveling could be done. *Hardpan* denotes a tough formation of gravel that is held together with a natural cementing material and that cannot practicably be loosened with an ordinary pick. *Rock excavating* is excavation of either solid rock or a formation of large boulders, which generally require blasting to break them up.

Not enough care is given to estimating excavation work, especially where conditions are complex. Frequently it is necessary to install considerable temporary construction work before the excavating work may be completed. Rock excavation, ground water, adjoining buildings, and streets often make the work so complicated that the excavating becomes an engineering problem, requiring exhaustive research and planning. In the case of large buildings in New York City, there are many additional plan difficulties and operating handicaps, and only the most highly organized specialists in this field are capable of properly estimating the work and doing it. The excavating of rock under these conditions and the removal of dirt and mud incidental to caisson work and pile driving are not ordinary estimating items.

In estimating for ordinary work, proper study should be made of the nature of the material to be removed, the general method of removal, special handling that may be necessary for topsoil or plants, temporary storage of material for later backfilling, method of disposal of the excess material, distance to the dump, cost of dumping, equipment and trucking costs, permits, protection of sidewalks and pavements, temporary roads and runways, ramps, safeguards, watchmen, foremen, lights, labor, compensation insurance, public liability insurance, employment and old-age pension insurance, general and overhead expense, risks, weather conditions, breakdowns, cave-ins, and rehandling.

In estimating for complex excavations, engineering skill is necessary and should be provided. Most excavations should be considered to be complex if they have rock or water to be cared for or if they are more than 10 ft deep or if there are adjoining buildings in which the foundations do not extend down as far as the bottom level of the new excavation.

Excavations are always made deep enough to enable the foundations to be extended at least 4 ft below the finish grade, in order to escape the action of frost. At lower levels, also, soils are more condensed and less moist and will bear a greater load. Footings must be made wide enough not to exceed the bearing capacity of the ground. Rock, of course, in its native condition, will carry any load put upon it. In preparing a rock bottom, it is necessary to trim the rock to a level surface under all footings.

In soft ground, piles are sometimes driven on which to rest the foundations. Wood piles are used only where they will always be below the ground-water level; otherwise, they would soon rot away. Concrete piles are driven or are cast in place after a steel shell has first been driven.

**Site.** The site should be visited by the estimator so that he may ascertain the type of ground, whether it is hard or soft, wet or dry, etc. The labor cost varies according to these conditions. The general method of doing the work must also be given consideration at the site—whether it shall be done by hand with picks and shovels, or by drag scrapers, bulldozer or power shovel, or some other means. Large equipment is used only when large amounts of work are to be done or when time is an essential element and high costs may be sustained in order to advance the date when actual construction can be started.

**Methods.** A bulldozer is used to clear the ground, remove trees and shrubs, and scrape off the topsoil. Shallow excavating, on jobs where the material is not to be hauled away but merely taken out and then spread over the balance of the plot, is also done with a bulldozer. In such cases an ordinary house excavation requires the use of a bulldozer for only 1 day.

A power shovel, usually operated by a diesel engine, is required when the excavated material is to be loaded into trucks. A *loading machine* is a variation of a power shovel, and the larger types are as effective as power shovels of ordinary size.

Back hoes and ditching machines are used for digging pier holes and trenches and for other excavating of this type.

The cost of maintaining a shovel involves expense for fuel, repairs, depreciation, and the time required to move it to and from the job, as well as the payroll and incidental costs of the crew. A ¾-cu-yd shovel can dig and load about 250 cu yd of ordinary earth per day into trucks if there is free room for it to operate properly. In cramped quarters it will do much less.

Motor trucks usually carry from 4 to 12 cu yd per load, depending upon the capacity of the individual truck. Five cubic yards is the capacity of most of the trucks in general use. On very large operations, other equipment is often used, such as enormous *carryalls* which cut into the ground and carry away 15 to 30 cu yd at a time. Obviously, a job has to involve thousands of cubic yards in order that such large pieces of equipment can be operated economically.

The last material removed in an excavation is generally used for backfilling behind the foundation walls and elsewhere in those spaces that were excavated for working room in excess of the exact requirements for the structure itself. Sometimes this material may be taken directly from its original location and deposited where it is needed as backfill. This eliminates all extra handling, as well as the trucking costs and dump charges. Each job should be analyzed separately in order that this line of reasoning may be developed as it applies to the particular case.

The kind of material to be removed will greatly affect the cost per cubic yard. Obviously, therefore, it is necessary to determine and note in the estimate whether the excavation is to be plain dirt, filled ground, sand, hard soil, or whatever the material to be removed may be. Often several kinds are found in one job and the computation of volumes and the pricing must then be adjusted accordingly.

The ground surface may be sloping or at various levels. In cases of this kind or cases where several materials are to be handled, it is necessary to divide the plot into as many small portions as will be necessary for computing the volumes and for grouping them so that the right unit prices may be applied.

Figure 11-1 shows a simple estimating sheet such as is used by some estimators. This is on a standard journal-ruled pad obtainable in practically all stationery stores.

In the hands of an experienced estimator this form is quite satisfactory for simple jobs. Men with little experience, however, should use regular estimating forms like those in the chapters which follow. These provide more and better guidelines and thus help to eliminate errors. Furthermore, they do not require the numerous ×'s between figures, as guidelines take the place of these.

In Fig. 11-1, the estimator has good headings and subheadings, the use of which is always to be recommended. The statement "plain dirt" clarifies the work at the outset as to the kind of excavating that is involved. The estimate is then divided into two parts, power-shovel work and hand work. The estimator has noted the fact that the excavating done by the power shovel on this job will have to be hauled to the dump. Short notes like this should be inserted all through the estimate so that anyone reading it may quickly get a good mental picture of the work and conditions pertaining to it. Further down is another good note stating that certain material will be laid aside and used later for backfilling.

This is an illustration of an estimate sheet that is well arranged and on which the items are named and are grouped so that various unit cost prices may be applied. These prices, of course, are for this particular job only and must not be thought of as standard reference prices suitable for other jobs.

The main excavation on this job consisted of the cellar excavation and an extension of this to provide space in which to construct a freight-elevator shaft and an adjoining book-storage vault. The excavating involved is all of the same character as to the depth and the manner of doing the work, and therefore the two items shown are grouped together so that one unit cost price may be applied.

There are 20 column footings on the job being considered, and they will require that pits, or pier holes, of the sizes noted, will have to be dug. These are below the bottom of the main excavation.

The wall footings are actually 12" deep. However, the bottom of the main excavation on this job is 6" below the top of these footings,

140  BUILDING CONSTRUCTION ESTIMATING

| | | | | | | | | |
|---|---|---|---|---|---|---|---|---|
| | | 1239 Webster Ave. | | | | Mar. 20, '59 | | |
| | | Excavating (plain dirt) | | | | | | |
| | | Power Shovel | | | | | | |
| | | Haul to dump | | | | | | |
| | | General | | | | | | |
| | | 90 x 88 x 9-3 | 72760 | | | | | |
| | | Elev. shaft, etc. | | | | | | |
| | | 19 x 30 x 9-3 | 5273 | | | | | |
| | | 27)78533 | | | | | | |
| | | | 2909 CY | 1.25 | 3636 | | | |
| | | Hand Work | | | | | | |
| | | use for backfill | | | | | | |
| | | Col. ftgs. | | | | | | |
| 3 | | 5 x 5 x 1-6 | 112 | | | | | |
| 8 | | 8 x 5 x 1-6 | 480 | | | | | |
| 7 | | 7-6 x 7-6 x 1-6 | 590 | | | | | |
| 1 | | 8-6 x 8-6 x 9 | 646 | | | | | |
| 1 | | 8 x 8 x 4 | 256 | | | | | |
| | | 27)2084 | | | | | | |
| | | | 77 CY | 1.80 | 139 | | | |
| | | Wall ftgs. | | | | | | |
| N | | 64 x 2-6 x 0-6 | 85 | | | | | |
| S | | 64 x 3 x | 96 | | | | | |
| E | | 54 x 3 x | 81 | | | | | |
| W | | 55 x 3 x | 83 | | | | | |
| | | 27)345 | | | | | | |
| | | | 13 CY | 1.50 | 20 | | | |
| | | | | Fwd | 3795 | | | |

*Fig. 11-1* An excavating estimate.

and therefore it is only necessary that the footing trenches be excavated an additional 6″ deep.

**Costs.** A simple excavation, as for a suburban house, costs the contractor between 75 cents and $1 per cu yd. This is for simply loosening ordinary dirt and spreading it over the balance of the plot with the aid of a bulldozer. Sometimes the rich topsoil is taken off and piled up, to be spread carefully after the rough grading has been done.

An excavation for a building in the heart of the business section of a big city may cost the contractor between $2.50 and $5 per cu yd. This is for digging, loading, hauling, and disposing of the excavated dirt, together with the incidental expense of dumping charges, lost time, temporary protection, watchman's services, etc.

If there is rock or if the banks require bracing, the extra cost of such items will have to be computed separately. If they are considered in the cubic yard price, this may well amount to between $10 and $20 per cu yd.

**Building Codes.** Any person causing an excavation to be made must provide such sheet piling and bracing as may be necessary to prevent the earth of adjoining property from caving in before permanent supports have been provided for the sides of the excavation. Also, whenever provisions are lacking for the permanent support of the sides of an excavation, the person causing such excavation to be made must build a retaining wall at his own expense and on his own land. This retaining wall must be carried to such a height as to retain the adjoining earth. Often it is required to be properly coped and provided with a substantial guardrail or fence. Some codes also provide that whenever an excavation is carried to a depth of more than 10 ft below the curb level, the person causing it must preserve and protect the adjoining property and structures and provide shoring, underpinning, etc., if necessary, regardless of how far down the adjoining foundations may extend.

■ EXERCISES

1. Why is the depth of an excavation measured from the original grades, instead of from the finish grade that can be seen on the architect's elevations?
2. About how much extra space is required around the outside of the walls below grade for the proper construction of them? Show by a sketch.
3. Compute the amount of excavating in the following portions of an excavation:
    a. 109'0" × 86'0" × 12'0"
    b. 22'6" × 14'8" × 10'9"
    c. 146'0" × 4'0" × 3'2"
4. Write at least 300 words of notes on excavating work, taken from one or two of the books listed in the Bibliography. Name the books used.
5. State the meaning and use of the following terms, and give sketches if possible: original grade, topsoil, hardpan, backfill, sheet piling.

6. Make a complete estimate of the excavating work required for the house shown in Chap. 6. Assume that the material is ordinary earth. The specifications are as follows:

### EXCAVATING

Remove 9 in. of topsoil from the entire width of the property and for a distance of 60 ft back from the sidewalk, pile this at the rear of the property, and spread it later over the front lawn and terrace and elsewhere as directed by the architect. None of this material is to be removed from the premises.

Excavate for all footings, foundation walls, areas, steps, walks, and driveway, first removing and stacking the topsoil as noted above. Excavate for a sufficient distance from the foundation walls to allow for inspection and to permit the various trades to install their work.

Remove excess subsoil from the premises, and grade and backfill to the finish levels shown on the drawings.

The seeding and landscaping will be done under a separate contract.

CHAPTER 12

# concrete foundations

Concrete and cement work is usually divided into several parts for convenience in estimating. Sometimes all the work is placed under one heading, but portions are often separately listed under such headings as Concrete Arches and Cement Finish.

The general method is to list the different items in each division approximately in the same order as that in which they will be constructed. Take the footings in the present division and complete them before proceeding to the next item to be listed under Concrete Foundations.

Include all the forms, reinforcing steel, and concrete in each item—if the item requires all three parts.

Adopt a fairly uniform procedure when listing quantities for any kind of work. Follow a regular system also when putting down the dimensions: length first, width or thickness next, and then height. This is a custom, and it makes the figures easier to identify—at least, to some extent. Separate all the work into small items or groups for ease in checking and pricing.

Put under this heading, Concrete Foundations, all the wall and column footings, cellar or basement walls, area walls, pits, boiler and tank foundations, and all other items of bulk concrete below the first floor. Omit the cellar or basement floors and all other pavements, however, as these would undoubtedly be installed much later than the other work mentioned and should, therefore, be listed with work done at the later period. It is customary to list floors, walks, and other pavements that are laid directly on earth under the heading Cement Finish.

Some estimators measure all the concrete shown on the plans, then all the formwork and all the reinforcing, separately. This method may seem concise and systematic because it keeps each of these three parts by itself. A better method, however, is to complete each item shown on the plan before going to the next. Start with the footings, as first stated, and list all the forms, reinforcing, and concrete for one kind of footing, such as the column footings; then take another kind of footing, such as the main-wall footings, and list all the forms, reinforcing, and concrete in these. Put a minor heading before each item of work thus handled and proceed in the same manner with the main walls, inside walls, area walls, etc., completing each one before considering the next.

This method is better because it has been found that fewer errors of omission and computation are made when it is used. It is better, too, because it enables the estimator to keep the work in groups that are complete in themselves. This is important in the later analysis of the construction details, in the analysis of the unit costs, and in making adjustments in the estimate.

All the work is estimated complete, in place, and the price thus includes all the material, labor, use of equipment, and incidental direct expense of every description. Each of the three parts of each item of work is listed and priced separately: the forms, the reinforcing steel, and the concrete. If no reinforcing steel is called for, this part is omitted or a note is inserted to the effect that none is required.

*The formwork* is priced per square foot of contact of the forms against the concrete.

*The reinforcing steel* is priced per pound of steel. The number of linear feet of each size of bar called for is listed, and then, from this information, the weight is computed.

*The concrete* is priced per cubic yard. Each different mixture is kept separate in the estimate, so that different prices may be applied.

Each portion of the concrete shown on the plan is, then, first given a minor heading in the estimate. It is next subdivided under the headings, Forms, Reinforcing (when the item calls for reinforcing to be provided), and Concrete.

The forms are totaled in square feet for each of these minor headings and are later priced at the contractor's cost per square foot.

Where reinforcing bars are called for, they are totaled in linear feet of each size, in each minor heading, and these are turned into pounds ready to be priced at the contractor's cost, per pound, complete in place. The concrete is likewise totaled, first in cubic feet and then in cubic yards, ready to be priced at the contractor's cost per cubic yard.

Figure 12-1 shows the method of entering concrete-work items in the estimate. The job used here called for crossbars to act as ties. As these are similar to reinforcing bars, they are listed along with the reinforcing bars. This form is a good one for estimating general building work and is better than the one shown in Fig. 11-1. The unit-price column in this form is used to show the estimated contractor's total direct cost per unit of work, including the cost of material, labor, taxes, insurances, and all incidental direct expense.

**Formwork.** Forms must be strong and rigid, and they must be well braced to prevent bulging. Wall forms have to resist a pressure of about 145 lb per sq ft for each foot of height, to hold wet concrete. At the same time, however, forms must be so arranged that they may be rapidly put in place and readily taken apart for re-use.

Temporary forms are used to hold the wet concrete until it has set sufficiently to allow removal of the forms safely. For ordinary walls forms may be removed in about 2 days, but for beams and floor slabs a week or more is required. The forms are used over again for other parts of the construction, where possible. Great care is taken in the construction of the forms so that the exact shape of the concrete will be formed and so that they will safely hold the great load entailed by wet concrete and the shocks of working conditions.

Forms should always be used for footings, except in extreme cases where it can be plainly seen that they may profitably be omitted. Most specifications call for forms to be used for footings. When the excavating is very hard and the banks will stand without any caving in, even after a rainfall, then the excavation for the footings may be cut neat to the line of the concrete footings (if permitted by the specifications) and the trench thus formed filled to make a footing. If it is necessary to excavate within the building after the footings are thus formed, in order to place the floor or the base under the floor, it will probably not be worthwhile to consider omitting the footing forms. Footing forms are quite simple. If they are omitted, considerable extra time is consumed in trimming the trench sides very neat, and some of the concrete will undoubtedly be wasted. Furthermore, if space is needed

146  BUILDING CONSTRUCTION ESTIMATING

| | EST. 481 | | Concrete | | | | | | | | | SHEET 5 | | |
|---|---|---|---|---|---|---|---|---|---|---|---|---|---|---|
| | | Wall Ftgs | | | | | | | | | | | | |
| | | Forms | | | | | | | | | | | | |
| W | 2 | 59'-0" | 1'-0" | | | | 1 | 1 | 8 | | | | | |
| N | 2 | 84-0 | 1-0 | | | | 1 | 6 | 8 | | | | | |
| S | 2 | 82-0 | 1-0 | | | | 1 | 6 | 4 | | | | | |
| E | 2 | 59-0 | 1-0 | | | | 1 | 1 | 8 | | | | | |
| E | 2 | 79-0 | 1-4 | | | | 2 | 1 | 1 | | | | | |
| | | | | | | | 7 | 7 | 9 | SF | .28 | 2 | 1 | 8 |
| | | Reinfg. #5 | | | | | | | | | | | | |
| W | 7 | 59' | | | | | 4 | 1 | 3 | | | | | |
| N | 5 | 84 | | | | | 4 | 2 | 0 | | | | | |
| S | 7 | 82 | | | | | 5 | 7 | 4 | | | | | |
| E | 7 | 59 | | | | | 4 | 1 | 3 | | | | | |
| E | 5 | 79 | | | | | 3 | 9 | 5 | | | | | |
| | | | | | | | 2 | 2 | 1 | 5 LF | | | | |
| | | | | | | | x | 1. | 0 | 4 | | | | |
| | | | | | | | 2 | 3 | 0 | 4 LB | .12 | 2 | 7 | 6 |
| | | Ties #3 | | | | | | | | | | | | |
| | 45 | 2'-4" | | | | | 1 | 0 | 5 | | | | | |
| | 63 | 1-10 | | | | | 1 | 1 | 3 | | | | | |
| | 62 | 2-4 | | | | | 1 | 4 | 5 | | | | | |
| | 45 | 2-4 | | | | | 1 | 0 | 5 | | | | | |
| | 60 | 1-10 | | | | | 1 | 0 | 8 | | | | | |
| | | | | | | | 5 | 7 | 6 | LF | | | | |
| | | | | | | | x | .3 | 8 | | | | | |
| | | | | | | | 2 | 1 | 9 | LB | .30 | | 6 | 6 |
| | | Concr. 1:2½:5 | stone | | | | | | | | | | | |
| | | 59' | 2'-4" | 1'-0" | | | 1 | 3 | 8 | | | | | |
| | | 84 | 1-10 | 1-0 | | | 1 | 5 | 4 | | | | | |
| | | 82 | 2-4 | 1-0 | | | 1 | 9 | 1 | | | | | |
| | | 59 | 2-4 | 1-0 | | | 1 | 3 | 8 | | | | | |
| | | 79 | 1-10 | 1-4 | | | 1 | 9 | 3 | | | | | |
| | | | | | | 27 )| 8 | 1 | 4 | CF | | | | |
| | | | | | | | | 3 | 0 | CY | 21.50 | 6 | 4 | 5 |
| | | | | | | | | | | | Fwd. | 1 | 2 | 0 5 |

*Fig. 12-1* A concrete estimate.

outside the walls anyway, for easier installation of the wall forms or for waterproofing the walls or installing outside drains, etc., then surely the method of omitting the footing forms would seem uneconomical.

It is obvious that forms for walls and wall footings will not cost the same price per square foot of contact as forms for column footings and other isolated items. There is no direct relation between the quantity of forms and the quantity of concrete, either, Every difference in the size and shape of the concrete to be constructed will make a change in the quantity or type of forms required. This is another reason for adopting the method recommended for estimating concrete work, so that the various kinds of formwork may be kept properly separated and be priced differently where the character of the work in the items calls for different prices.

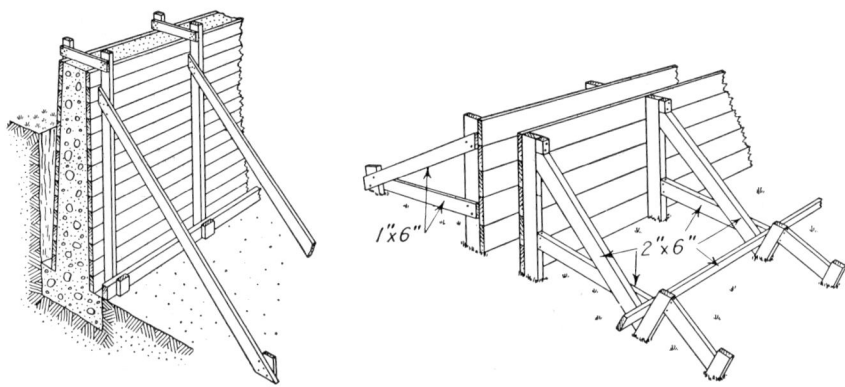

*Fig. 12-2  Wall forms.*

Wall forms are removed within 2 or 3 days after the pouring of the concrete is completed in them, but thin walls in winter are allowed at least 5 days. Ordinary floor- and roof-slab forms are stripped in 6 or 7 days in summer and 2 weeks in winter. Beams and girders require about 2 weeks in summer and 4 weeks in winter. Columns require 2 days in summer and 4 days in winter. Girders and slabs should be shored as the forms are being removed and left supported for several days longer (see Fig. 12-2).

Several factors enter into the cost of formwork, not the least of these being the possible re-use of the lumber or of made-up panels, or, as in the case of column footings, of the entire form box. Besides, it may be that not all the wall forms on a job will take the same unit price, because one series of walls may require more bracing than others or the lumber in one may not be available for further use afterward.

The square feet of contact area of the forms against the concrete is found by measuring, on the plans, the concrete surfaces that will require this temporary support. The dimensions of each item are entered in some suitable order on the estimate sheets in feet and inches, and the total square feet extended for each part, or for each group, when a group can readily be made.

**Form Material.** Form boards are generally $\frac{7}{8}''$ thick, and widths of 6" to 10" are used. Bracing lumber is $2'' \times 4''$, $3'' \times 4''$, $4'' \times 4''$, $2'' \times 6''$, etc. The amount of lumber required for formwork, when the material is used but once, varies from $1\frac{1}{2}$ to $3\frac{1}{2}$ board feet per square foot of form contact. Nails used are 6d, 8d, and 10d mostly, and about 8 lb are required for 100 sq ft of forms. Tie wire used is No. 8, No. 10, and No. 12 black annealed wire, and about 5 lb are required for 100 sq ft of forms. This wire comes in bundles of 100 lb. The three sizes equal about 14, 20, and 33 ft, respectively, per pound. Nails come in 100-lb kegs. Metal and plywood forms are also used.

**Reinforcing Steel.** The amount of reinforcing steel bars is obtained by listing the number of bars of each size and length required. Reference to the specifications and to the notes on the plans may be necessary in order to find some of this information. The total length in linear feet of each size is computed from the entries thus made in the estimate, and this figure is multiplied by the number of pounds that the bar weighs per linear foot. This unit weight is sometimes given on the plans or in the specifications, but more often it is not and must be looked up in a steel handbook or in one of the standard builders' reference books. The total weight of steel is extended for each item or for each group of similar items.

Wire mesh, square or triangular, is made of wire of several gauges, sometimes two gauges being used in one make of mesh. The intersections are generally welded. Mesh is measured by finding the number of square feet of area to be covered and adding 10 per cent for the laps. The gauge and spacing of the wires or the weight per square foot must be stated in the estimate.

The next chapter, dealing with reinforced-concrete construction, contains numerous illustrations of formwork, reinforcing, and methods of construction that are applicable also to concrete-foundation work.

A specification may call for the reinforcing steel bars to be tied at every intersection. This is somewhat different from one requiring merely that enough intersections shall be tied to keep the bars in proper position. The tie wire has no value as reinforcing. If bars are placed a few inches apart, two ways, there will be a great many intersections, and the time required for tying may be four or five times that for plain tying. An experienced estimator notices this in specifications and gauges his pricing accordingly. He may not take written notes of such deviations from the usual run of work, but they all tend to set his mind for the pricing range he is going to use—a low range for jobs that have liberal specifications and higher ranges for those of more rigid requirements. A wise estimator will not assume too much liberalization of vaguely written specifications, as he may be surprised later to learn that an inspector on the job for the architect is of the type accustomed to rigid specifications, and it may be embarrassing to keep differing from him regarding the interpretation of the requirements.

**Reinforcing Bars.** Bars are referred to according to their nominal diameter or by their size number. These numbers and the weights of the bars, per linear foot, are given in Table 12-1.

TABLE 12-1

| Bar No. | Pounds per foot |
|---|---|
| 2 | 0.17 |
| 3 | 0.38 |
| 4 | 0.67 |
| 5 | 1.04 |
| 6 | 1.5 |
| 7 | 2.04 |
| 8 | 2.67 |
| 9 | 3.4 |
| 10 | 4.3 |
| 11 | 5.31 |

The No. 2 bars are plain round bars. The No. 9, 10, and 11 bars are deformed round bars, equivalent respectively to square bars 1", 1⅛", and 1¼" thick. The other bars are deformed round bars and the numbers are based on the number of eighths of an inch in the diameter. A No. 5 bar is ⅝", a No. 4 is ½", etc.

Figure 12-3 shows the reinforcing bars in a typical column footing.

**Concrete.** The number of cubic yards of concrete is obtained by measuring the concrete indicated on the plans. The dimensions for this are carefully entered and computed to get cubic feet for each entry and then totaled in cubic feet for the entire item being considered. Before this is extended into the final column, it is divided by 27 to turn it into cubic yards, and this amount is extended for pricing.

Many estimators leave their concrete items in cubic feet instead of turning them into cubic yards as was formerly the universal custom. Even these, however, usually turn the figures into cubic yards for figuring such items as runways and for progress-schedule entries, etc. So little saving in written work and mental effort is involved that it seems hardly worthwhile to vary from the custom used by most contractors, that of considering concrete in cubic yards for most purposes.

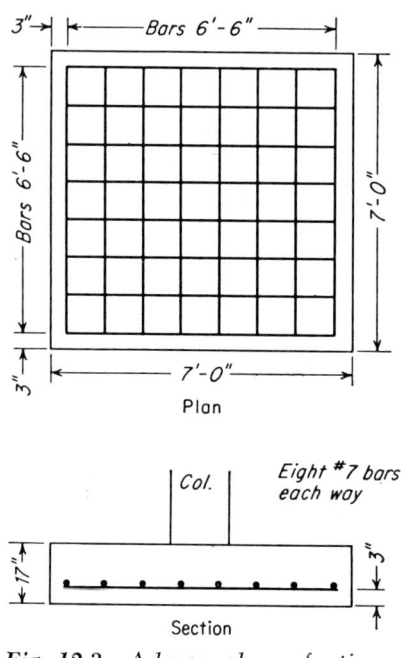

Fig. 12-3  *A large column footing.*

Freshly poured concrete should be worked around in the forms so as to compact the mass before it begins to set and to remove any air pockets that may otherwise form in the concrete. This is done by pushing iron or wood rods down into the concrete and churning it about, or by using special tools or mechanical agitators that are available for this purpose. Specifications for good work require that this shall be done and also call for the tamping or ramming of concrete that is poured on the ground to form pavements, etc. The unit price for concrete work should be increased to take care of this expense, if it is called for, or a separate entry should be made in the estimate to cover it. Concrete begins to set about 30 minutes after being poured and should not be disturbed in any way after that time until it is thoroughly set. In fact, it should be protected against disturbance.

Concrete is made by mixing cement and sand with a coarser material, called the *coarse aggregate*. Sand is called the *fine aggregate*. The coarse aggregate is usually either coarse gravel or crushed stone, except in lightweight floor and roof arches, where cinders are used instead. Steel bars or wire mesh are embedded in the concrete when tensile strength is necessary as in beams, girders, floor arches, etc. When it has any form of reinforcing, concrete is called *reinforced concrete*. Without reinforcing, it is called *plain concrete*.

In the mixture of concrete, as the student should realize, the sand, cement, and water fill in the spaces between the pieces of broken stone or gravel or other material that forms the coarse aggregate. More than one cubic yard of materials, therefore, is required to produce a cubic yard of mixed concrete. For example, in a 1:2:4 mix, six bags (equaling 6 cu ft) of cement, 12 cu ft of sand, and 24 cu ft of coarse aggregate are required, as well as about 6 cu ft of water. Separately, these total 48 cu ft, but when mixed together they fill a space of only 27 cu ft, or 1 cu yd.

The expression 1:2:4 means 1 part of cement to 2 parts of sand to 4 parts of the coarse aggregate, measured separately and dry. One bag of cement measures about a cubic foot and is always taken as one cubic foot. Note that the coarse aggregate measures almost a cubic yard, by itself, for each cubic yard of concrete produced. It is the spaces between the particles that furnish room for the other ingredients. However, concrete is spoken of and ordered as so many cubic yards of concrete in the structure or member of the structure under consideration.

For ordinary work the concrete mix is generally proportioned by specifying the volume of each of the ingredients to be used. The coarse aggregate is mentioned by name. For work of any importance, it is now the practice to specify the strength of concrete desired, instead of the proportions of the mix: for example, 2,500-lb concrete, 3,750-lb concrete. This means that the concrete must have the strength specified, per square inch, after 28 days—the period generally accepted as the complete curing period for concrete. Most concrete is bought ready-mixed and the companies supplying it have accurate measuring devices at their plants to produce the concrete desired.

The mixtures of concrete that are generally used form concrete with a compressive strength of between 2,500 and 4,000 lb per sq in. The strength depends upon the proportion of cement, mainly, and on the amount of water included, the quality of the workmanship, and other factors. Important work must be given constant supervision by the superintendent, the concrete foreman, the carpenter foreman (as to the formwork), the reinforcing foreman, and the job engineer. This means that important concrete will cost much more for the general job expense as well as for the actual higher cost of the concrete ingredients, if a richer mix is called for, and for more lumber and labor required for forms that are better made than the ordinary. On a job of reinforced concrete with very rigid requirements, the complete structure may cost 50 per cent more than on one of ordinary type.

In concrete, between 6 and 7 gal of water is required for each bag of cement used. Thus, if six bags of cement are used per cubic yard of concrete, between 36 and 42 gal of water is needed for the cubic yard of concrete. This is a considerable amount of water and, if the concrete is mixed at the job, adequate provision must be made for obtaining the water and supplying it at the point of mixing. In some localities a meter must be installed, and the water must be paid for. These are items of expense that the estimator must provide for in the estimate, wherever they are required. If long lines of temporary water piping have to be installed and maintained, especially in winter weather, the expense will be considerable.

The chemical action between the cement and water used in concrete is what causes the hardening, or setting, of concrete. This is a slow process, which continues for months. If the concrete is allowed to dry out too quickly, it does not gain its full strength; therefore, specifications frequently ask that special attention be given to slowing up the drying process, or "curing" the concrete. This is done particularly in the case of concrete floors. Material to mix with the concrete or to apply on it after it is poured is sometimes called for. Whatever the case, it is the concern of the estimator to figure the cost involved and to provide for it in his estimate.

**Joints.** Because it is usually not known just where each pouring of concrete will end, to figure the number and type of construction joints that will be required is impossible. However, in every concrete estimate an item should be entered for construction joints, and an approximation should be made as to the number of them and the average cost of each one or the cost per linear foot shown. Also, if dowels or special treatment of the joints are specified or required for proper bonding of the new concrete to the old work, these must, of course, be taken into account, either by figuring in detail the work involved or at least by entering an approximation of the total cost expected to be incurred.

Expansion joints are often specified or shown on the plans. These require study as to their requirements of material and labor. They are usually priced on the basis of the linear foot of joint for regular joints, or by the price per joint for special or elaborate joints.

CONCRETE FOUNDATIONS 153

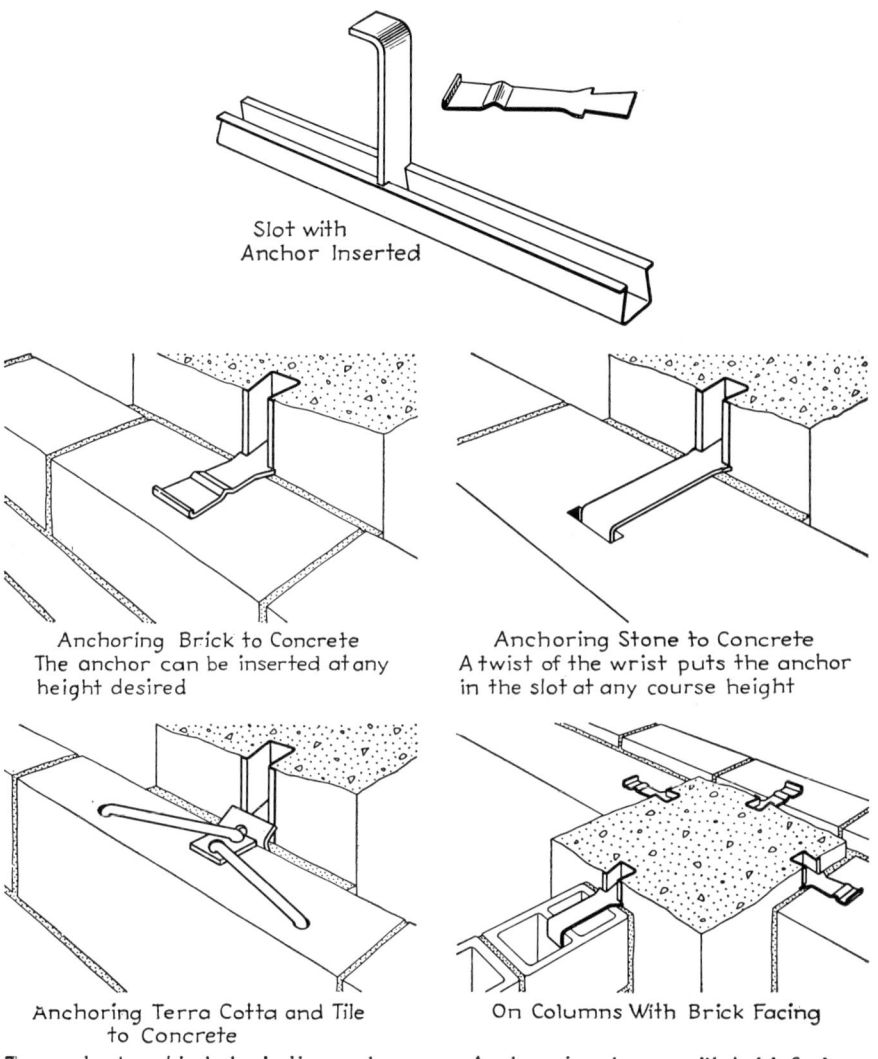

Slot with Anchor Inserted

Anchoring Brick to Concrete
The anchor can be inserted at any height desired

Anchoring Stone to Concrete
A twist of the wrist puts the anchor in the slot at any course height

Anchoring Terra Cotta and Tile to Concrete

Two rods placed in holes in the anchor which may be inserted at any height hold the terra cotta or tile tightly to the concrete

On Columns With Brick Facing

Anchors in columns with brick facing and tile partition

*Fig. 12-4  Masonry anchors and slots.*

**Inserts.** Piping for plumbing, heating, and electric lines running through concrete walls and floors requires some work on the part of the concrete gang or of the carpenters installing the forms. Anchors for structural steel and for securing several kinds of materials used by other trades must be installed in the concrete, and inserts of various kinds that are employed in building work are built into the concrete. These should all be counted by the estimator and priced properly under the heading Concrete Work for the labor of installing and bracing them and, if required, for the cost of the item or material that is so built in. Anchors, especially, must be given care in order that they may project at just the right location and at the right distance from the face of the concrete. Sometimes two carpenters will spend an hour properly placing a single anchor in the forms. This would mean a cost of about $9 for each such anchor, to cover the labor, or about $12 each as the total charge that would be made to the owner if the general expense, overhead, and profit items are considered (as, of course, they should be if one is considering the effect an item has on the contract amount). Thus, if there were 50 such anchors required and they were forgotten in the estimate, the bid for the job would be about $600 lower than it should be, on this one item alone (Fig. 12-4).

**Cement.** Cement comes in bags of 94 lb each. A bag counts as 1 cubic foot of volume in mixing. Four bags make a barrel, although there is no barrel actually concerned.

**Aggregate.** Sand, called *fine aggregate*, weighs about 2,700 lb per cu yd. Gravel also weighs about 2,700 lb and broken stone about 2,600 lb per cu yd. Gravel and stone are graded as to the size of a ring that each piece will pass through. Thus, ¾" gravel is the size that will pass through a ring ¾" in diameter. All coarse aggregate is more than ¼" in size; sand is ¼" or less in size. Clean cinders from hard coal, which are used in cinder concrete, are sold by the cubic yard.

**Costs.** Formwork usually costs the contractor between 25 and 50 cents per square foot of contact. The cost will depend upon the quantity of work involved, the type of structure, the re-use of the form material, etc.

Reinforcing steel costs 8 to 20 cents per lb in the regular types of footings and walls, the cost depending upon the amount of work, the bending required, the number of intersections to be tied, etc.

Concrete costs between $20 and $30 per cu yd, depending especially upon the mix required, as well as on the quantity of work and other factors.

Material quotations must be examined carefully before being used, especially those that are much lower than others for the same kind of material. Often the firms quoting do not know exactly what is called for in the specifications and will submit prices anyway, in order to get their names before the bidders on a particular job. In these cases the quotations usually state exactly what the vendor proposes to furnish, and it is a simple matter to check with the specifications to make sure that the requirements are met. There may be a difference in the size of the material, in the cases of the low quotation and those that are higher. This may cause the contractor who uses the low prices more expense in handling the materials, or the sizes proposed may not satisfy the architect. There may be a difference in the manner of shipment, the low quotation perhaps providing for shipments in large lots or by freight instead of directly to the job by truck.

Temporary runways are usually required for the concrete work. Total the number of cubic yards of concrete that are to be poured and place a price per cubic yard to take care of these runways, unless the unit price for the concrete has been raised enough to include the runways. A price of 40 to 60 cents per cu yd will take care of the expense for the material and for installing, maintaining, shifting, and removing the runways.

Cement stored at the job for making concrete or mortar must be protected against dampness. On a job of any considerable size, a weather-tight shanty is usually built, to be used exclusively for this purpose. On smaller jobs or where the cement is to be stored for only a few days at a time, tight covering with tarpaulins will suffice. Even this requires material and labor, however, and it is a question whether it would not be better to construct a shanty in the first place rather than to be placing and removing covers every day and perhaps ruining the material used for this purpose. In addition, storms may reach cement that is protected by covers not properly tied down, and thus a whole pile of cement may be ruined.

When items of temporary protection, such as cement and some other materials entail, are left for consideration under the general job-expense heading, they frequently are not given sufficient study. It is well, therefore, to list all these items on the various work-division sheets for each trade, as well as on the general-expense sheet. In this way, the items will probably be given the consideration they should have.

Foremen and other supervisory employees should be listed on all the work sheets. Even if they are not priced, the expenses thus noted will be brought out for consideration; and those contractors who price their work complete (including the foremen's time in the unit cost) may reconsider some of the unit prices or enter an amount to take care of additional expense in this regard, which they might otherwise overlook. Similarly, the unloading and stacking and protection of materials should be listed. If large lots of material are received and must be stored for a length of time, considerable expense is incurred over and above the cost of materials that are brought directly into the work and used without further handling.

**Building Codes.** Building codes contain provisions for concrete construction. Forms must conform to the shape, lines, and dimensions of the members as called for on the plans. They must be substantial and sufficiently tight to prevent leakage. They must be properly braced or tied together, so as to maintain position and shape and ensure safety to workmen and passers-by.

The removal of forms must be carried out in such a way as to ensure the complete safety of the structure. Where the structure as a whole is supported on shores, the vertical forms may be removed within 24 hours. The other forms and the shoring must be kept in place until the concrete has acquired sufficient strength to support its weight and the load upon it.

Metal reinforcement must be accurately placed and secured and must be supported by concrete or metal chairs or spacers or by metal hangers. It must be protected by at least ¾" concrete in walls and slabs, 1½" in beams and girders, and 2" in columns. In footings the reinforcing is usually required to have a minimum covering of concrete 3" thick. Splices in reinforcing rods or bars must usually be 80 times the diameter in length, and the ends of rods and stirrups must be well anchored.

Greater care is now demanded by architects in regard to the manufacture and placing of concrete. The concrete must be properly proportioned and properly poured; forms must be substantial. In general, concrete is receiving more attention than it formerly did. All this tends to slow up the progress of the job from the contractor's standpoint, and the rigidity of the inspection must, therefore, be considered by the estimator.

Concrete work requires equipment of good quality. As it is heavy work, the wheelbarrows, runways, chutes, bar benders, tampers, etc., must be extra strong in order to withstand the rough use they receive.

### ■ EXERCISES

1. Name the materials used in concrete work.
2. What parts of a building are generally estimated under the heading Concrete Foundations?
3. By what units of measure are formwork, reinforcing bars, and concrete priced?
4. Compute the amount of concrete in the following walls:
    a.  132'0" × 12'6" × 1'4"
    b.  47'8" × 13'9" × 2'2"
    c.  22'9" × 6'2" × 1'0"
    d.  34'3" × 10'4" × 1'3"
5. What is meant by 1:2:4 concrete? 3,750-lb concrete?
6. State average cost prices for the following kinds of work, in place:
    a. Forms for foundation walls
    b. Concrete in foundation walls
    c. Reinforcing bars
7. Write at least 300 words of notes on concrete foundations, from two or three of the books listed in the Bibliography. Name the books used.
8. State the meaning and use of the following terms, and give sketches where possible: footings, wall forms, reinforcing bars, coarse aggregate.
9. Make a complete estimate of the concrete footings for the house shown in Chap. 6. The specifications state as follows:

### CONCRETE FOOTINGS

All footings shall be 1:2:4 crushed rock or gravel concrete, poured into substantial forms. All wall footings shall contain three continuous lines of ⅝" diameter bars held in alignment by ⅜" diameter bars every 18". The column footings shall each contain four ⅝" diameter bars in each direction, a total of eight bars in each footing. All bars shall be spaced as directed by architect.

CHAPTER 13

# concrete floors and roofs

Some estimators set up two separate headings, one for the concrete floor and roof construction and one for the cement work. On small jobs this is not necessary, but on larger jobs it offers a better opportunity for analyzing the cost of the items included in these two lines.

## Arches

Concrete arches is the term used to denote the type of floor and roof construction commonly found in steel-frame buildings (Fig. 13-1).

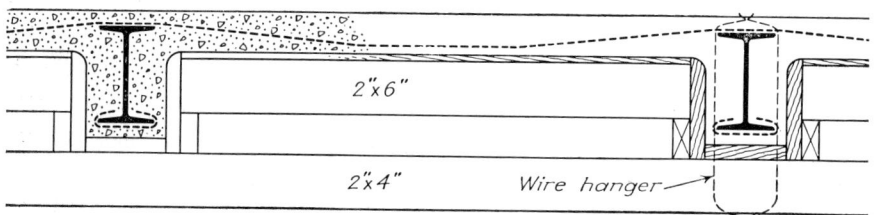

Fig. 13-1   Concrete arches.

The construction usually consists of cinder-concrete slabs, reinforced with heavy wire mesh, which are supported by the steel beams and girders. It is this heavy woven-steel cloth that carries the load—much as the steel cables carry the load on a suspension bridge. The steel beams and girders are encased in the cinder concrete at the time the slabs are poured. This bonds the whole steel frame and the concrete floors and roof together. The concrete also protects the steel from fire, for as one should know, steel will fail if exposed to extreme heat. Special wire wrappings, called *soffit clips*, are first placed around the bottom flanges of the beams and girders in order to bond the concrete to the steel.

All the temporary formwork, or centering, and the shores or braces that support the forms from below, or the hangers that support the forms from the steel beams over them, are included under this heading. The method of support is optional with the contractor. Generally speaking, hanging the forms from the steel beams and girders is found to be economical. In other types of concrete floor and roof construction, however, it is more desirable to jack up the formwork from below. Lumber used in formwork of this kind is always used many times over.

The fill that is placed on the slabs, between the slab (arches) and the finish flooring, is included by most estimators under the heading Concrete Arches. Sometimes it is put under Cement Finish Work, instead. Roof fill (crickets), to give slope to the top surface for drainage of rain water, is handled with the floor fill. This fill material usually consists of a lean mixture of cinder concrete. In the case of sleeper fill for wood-floor finish, however, dry cinders are used.

A quick and fairly common method of estimating floor and roof arches is to compute the number of square feet of each and then apply a price per square foot. The cost per square foot is built up to suit the thickness of the slab shown on the plans and the kind of wire reinforcement. It includes the temporary forms, the wire mesh, the cinder concrete, and all incidental expense. When this method of estimating is used, everything that affects the cost, such as the amount of work involved, the possible re-use of the forms, the economy of plant layout, and hoisting expense, is considered in building up the unit cost price.

## *Beams and girders*

The concrete portions of the beams and girders projecting below the slabs are listed separately. These are totaled in linear feet of beam and girder. The price is built up to suit the cross-section size of concrete. It is sufficient to provide for only a few different sizes for the encasing of steel beams and girders, as the cost per linear foot for the forms and the steel soffit clips (Fig. 13-2) will be practically the same for several sizes that are of approximately the same cross-section dimensions.

Concrete floor and roof slabs, complete with all formwork, reinforcing mesh, and cinder concrete, cost the contractor between $1.10 and $1.40 per sq ft, depending upon the amount of work involved

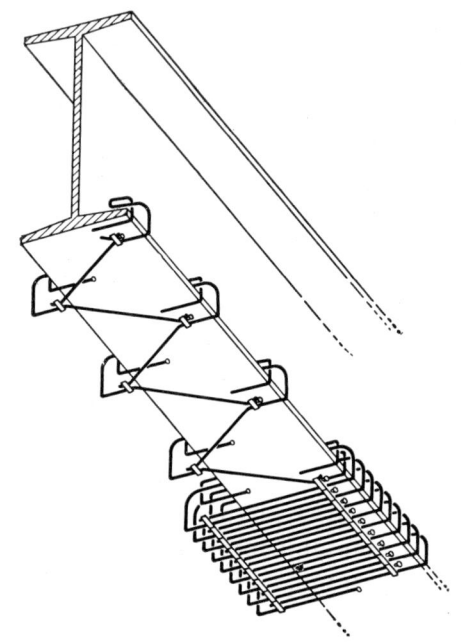

*Fig. 13-2   Soffit clips.*

and the composition and thickness of the slabs. Beam and girder covering costs between $1.20 and $2 per lin ft. Specially large girders require special analysis and pricing to suit the size.

Each portion of the building and each kind of work should be listed separately, in order that the proper unit prices may be applied. Floor and roof fill is best measured in cubic yards, although some estimators continue to use square feet, as in the listing of the floor slabs. A good method is to show the total quantity both in cubic yards and in square feet and then to check the pricing both ways.

Estimators should use all the plans for reference when estimating any one division and not depend upon only those drawings that are primarily concerned. The other drawings will often help toward a

CONCRETE FLOORS AND ROOFS 161

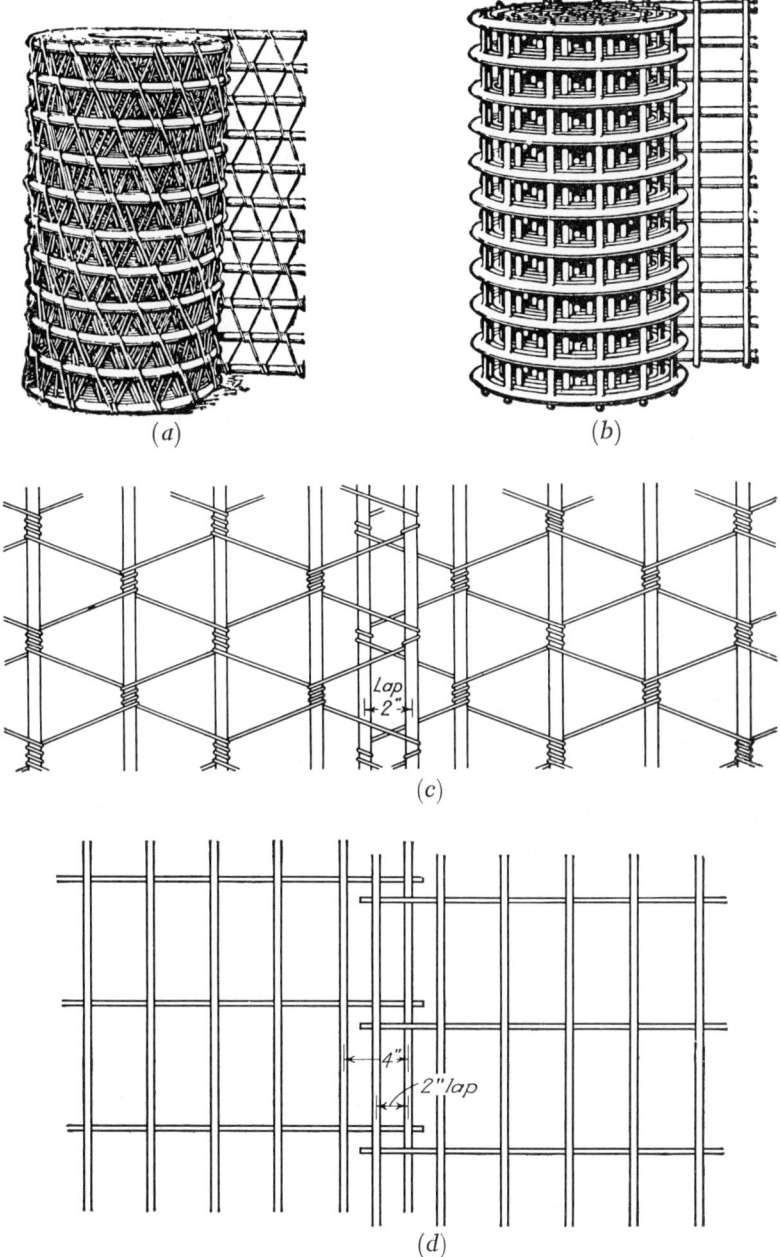

**Fig. 13-3** Wire mesh. (a) Triangular-woven mesh. (b) Rectangular-welded mesh. (c) Triangular mesh with 2″ side lap. (d) Rectangular mesh with 2″ side lap.

clearer understanding of the detail requirements. In the case of concrete floors and roofs with steel beams, the steel plans are referred to for the spacing of the beams and girders that are to support the construction. The steel itself is generally covered with concrete at the same time that the floor and roof construction is being installed.

Cinder-concrete slabs are usually of 1:2:5 mix, and this is used for the covering of the steel, as well. The reinforcing mesh is made of

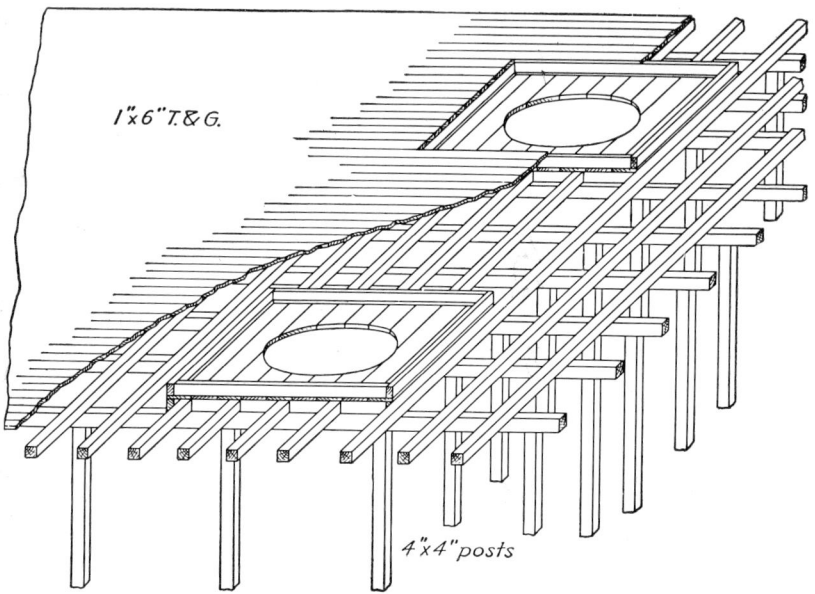

*Fig. 13-4  Formwork for flat-slab construction.*

cold-drawn steel wire, in rectangular or triangular mesh (Fig. 13-3). The forms are supported by heavy wire hangers from the steel beams and girders. Forms for this work should not be removed for a week after the concrete is poured.

**Hoisting.** Hoisting expense must always be considered. The method of hoisting is studied and proper figures are entered, either under the division of work requiring hoisting or in the job-expense sheet. If it is not listed on the sheets concerned with the divisions of work, the hoisting item should at least be noted on them in order to form a check toward making sure that this item is properly entered on the general-expense sheet.

**Other Types.** In concrete construction, such as the flat-slab, beam-and-girder, metal-pan, and hollow-block systems, the aim is to find the quantity of concrete in cubic yards, by taking the gross area and thickness of the floor or roof, and deducting the openings, and—in the pan and block systems—the volume of the blocks or of the voids formed

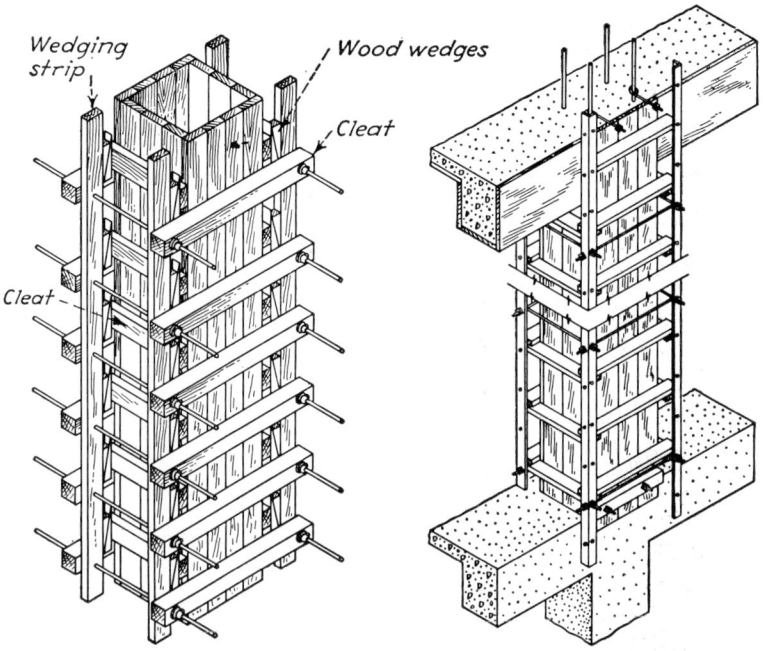

*Fig. 13-5  Column forms and clamps.*

by the pans. The number of linear feet of pans or blocks is separately listed. Figures 13-4 to 13-22 show types of concrete construction.

The total area of the floor or roof is taken for the supporting formwork. The steel bars are computed by listing the number and length of the bars of each size, and from this determining the total weight of steel. The top finish, if it is of cement, is expressed in square feet and is entered under the heading Cement Finish Work.

Concrete-joist construction is becoming more and more popular for floors and roofs. This consists of concrete girders and beams, with concrete joists between the beams, reinforced with steel bars. Note the use of the term *joist* here. It is needed when three members—

girders, beams, and joists—make up the construction. The joists are spaced quite close together and the spaces between them formed by metal pans or block fillers, as has been noted above.

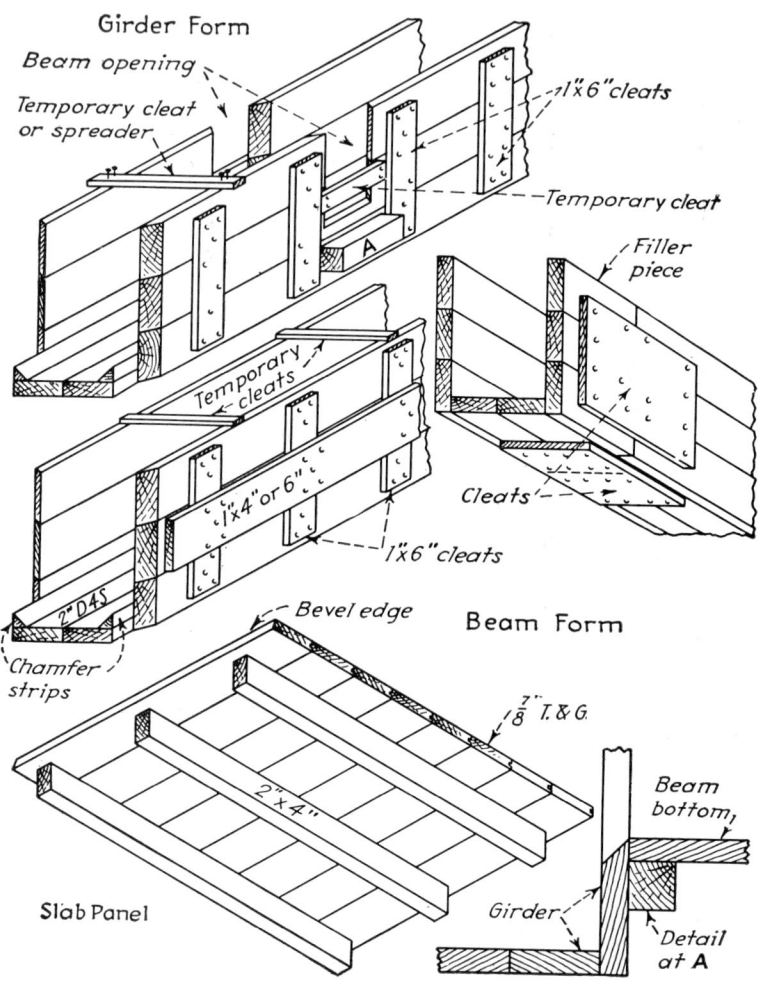

*Fig. 13-6  Girder and beam forms.*

Pans of lightweight metal are used when they are to be left permanently in place, and heavier pans when they are to be removed and reused. If clay or gypsum blocks are used, these are left in place. Each system has its own merits; and the architect, having selected the one best adapted to the requirements of the particular job, specifies or

CONCRETE FLOORS AND ROOFS   165

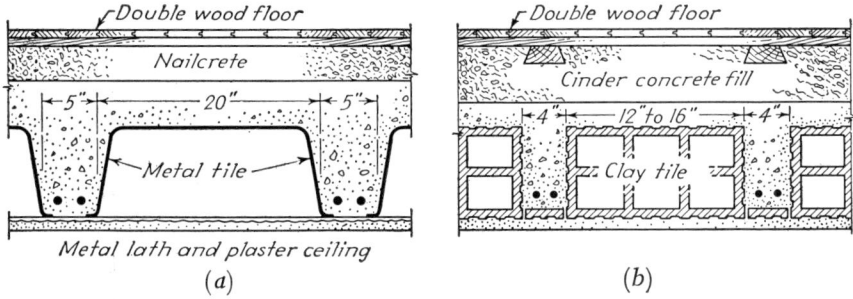

Fig. 13-7  (a) Pan and (b) block floor systems.

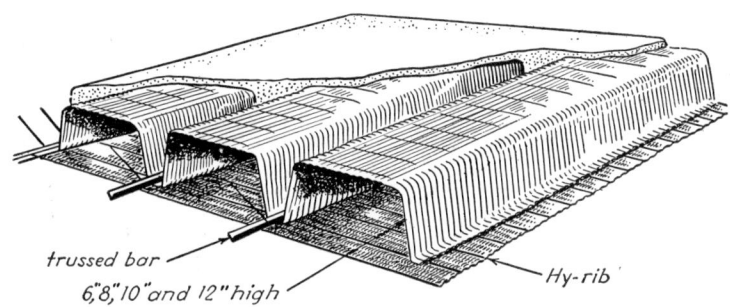

Fig. 13-8  One-way pan system.

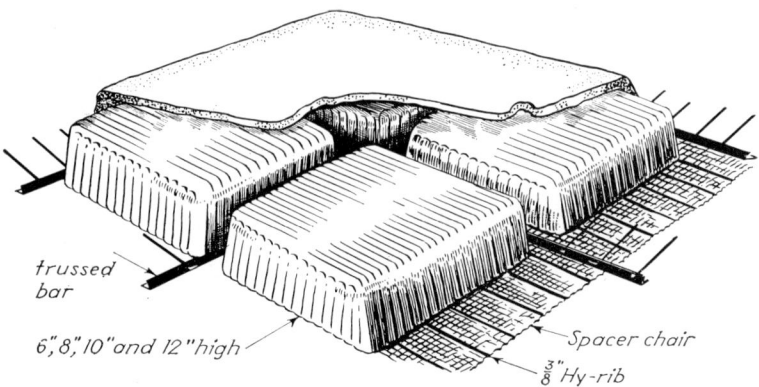

Fig. 13-9  Two-way pan system.

shows the system desired. If no ceiling is required on the floor below, the pan system is sometimes favored rather than the block system. The only form supports required for either of these systems are those under the girders, beams, and joists, as the pans or tile blocks form the sides of the girders, beams, and joists. For girders that are deeper than the joists, however, side forms are required for those portions that extend below. These side forms must be separately measured and totaled in square feet. Blocks and pans are both obtainable with closed

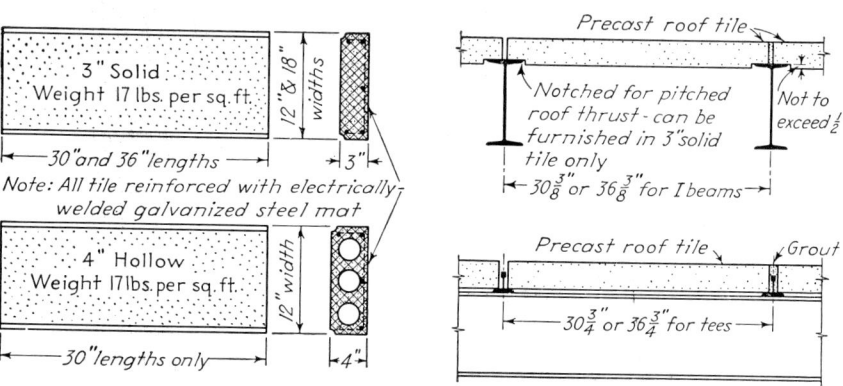

*Fig. 13-10* Short-span roof tile.

ends for use against the beams and girders and for the two-way system in which the joists run at right angles instead of parallel to one another.

Gypsum-tile blocks made for fillers are 19″ × 18″ and are 6″, 8″, 10″, or 12″ in depth. The concrete ribs or joists between the blocks are usually made 4″ or 5″ wide. The top slab is 2″ to 2½″ thick.

Metal pans of the removable type are No. 14 or No. 16 gauge, and those intended to be left in place are No. 26 gauge. The metal pans are laid end to end and lapped one corrugation. They are 20″ wide and from 30″ to 48″ long, for the one-way system.

**Precast Slabs.** Precast floor and roof slabs are also made, some of the types being patented. These run from 2″ to 6″ in thickness and are reinforced with wire mesh or rods. One type in fairly common use is 2½″ thick, 24″ wide, and 30″ long, and is designed for use in floors. Roof slabs are made 12″, 18″, and 24″ wide and in lengths of up to 7′0″. Concrete planks are made 2″ and 2¾″ thick and 16″ to 24″ wide.

CONCRETE FLOORS AND ROOFS 167

**Flat-slab System.** The *flat-slab* system of concrete construction consists of columns with flared caps and with thick portions of slabs over the caps, called *drops*. The balance of the floor or roof is a slab of uniform thickness. The reinforcing bars are run across the tops of

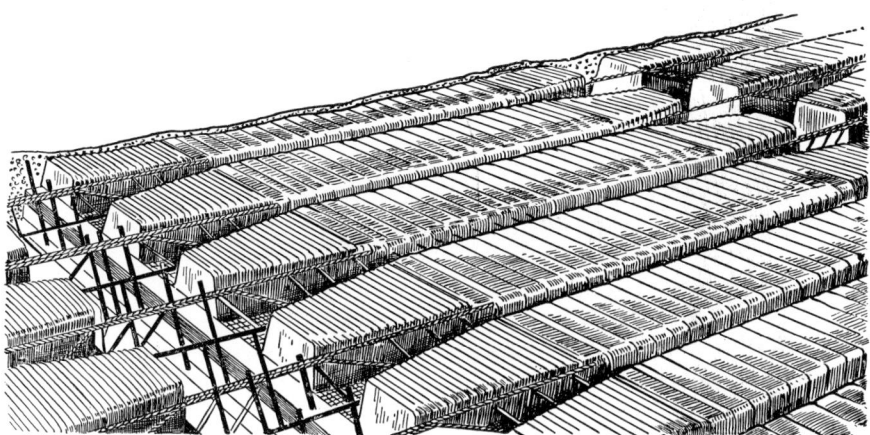

*Fig. 13-11   Cantilever pan system.*

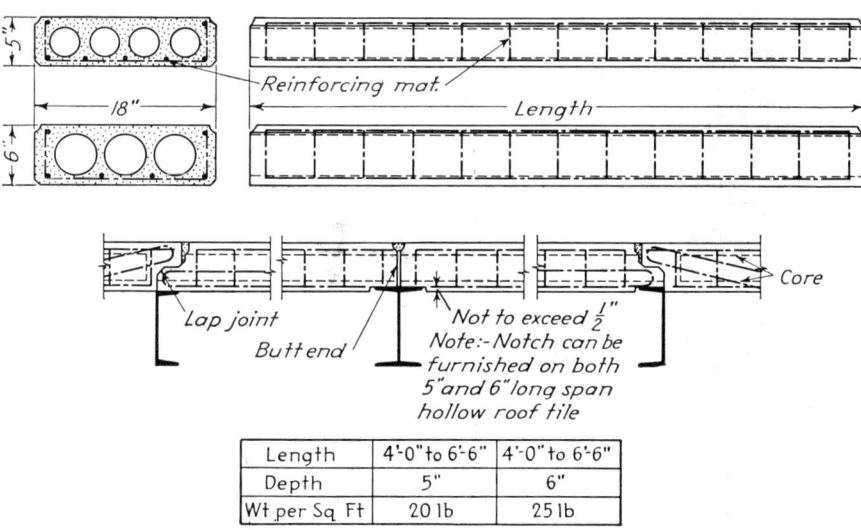

| Length | 4'-0" to 6'-6" | 4'-0" to 6'-6" |
|---|---|---|
| Depth | 5" | 6" |
| Wt per Sq Ft | 20 lb | 25 lb |

*Fig. 13-12   Long-span roof slab.*

the columns over the entire slab area, running either two ways or four ways. In the four-way system the bars run across the columns both ways and also diagonally across the columns.

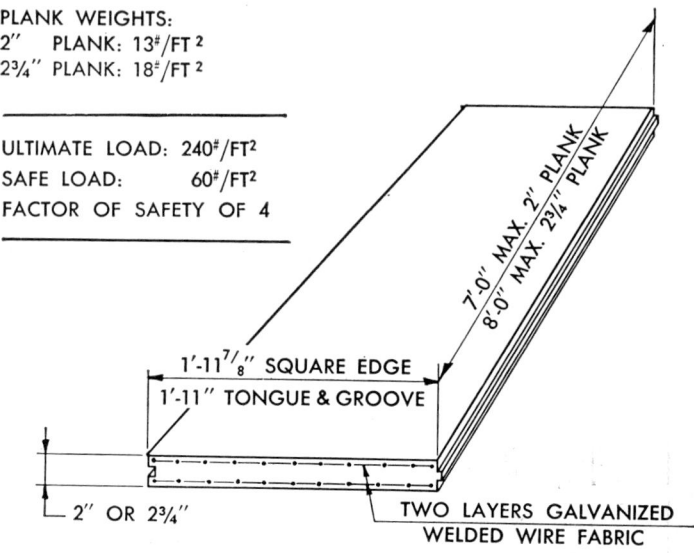

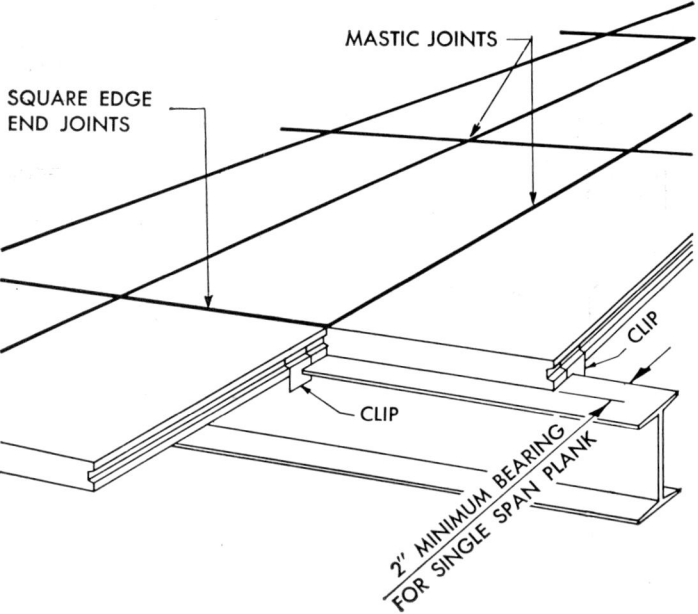

*Fig. 13-13* Concrete planks.

## Building code

The following are definitions of terms as used in the New York City Building Code:

**concrete.** A mixture of cement, fine aggregate, coarse aggregate and water.
**plain concrete.** Concrete without metal reinforcement.

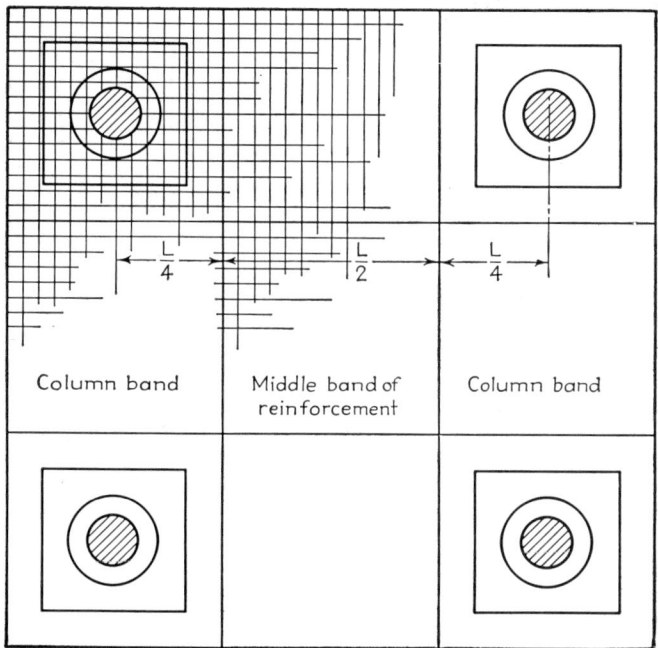

*Fig. 13-14   Two-way flat slab.*

**reinforced concrete.** Concrete in which metal is embedded in such a manner that the two materials act together in resisting stresses.
**controlled concrete.** Concrete where preliminary tests of the materials are made, and the concrete work is inspected in accordance with the code.
**average concrete.** Concrete where preliminary tests of the materials are omitted and where the concrete work lacks the inspection required for controlled concrete.
**aggregate.** Inert material which is mixed with cement and water to produce concrete, consisting in general of sand, pebbles, gravel, cinders, crushed stones, blast furnace slag, burnt shale or clay, or similar materials.

**crushed stone.** Bedded rock or boulders, broken by mechanical means into fragments of varying shapes and sizes.

**gravel.** Rounded particles, larger than sand grains, resulting from the natural disintegration of rocks.

**sand.** Small grains one-quarter of an inch or less in size resulting from the natural disintegration of rocks.

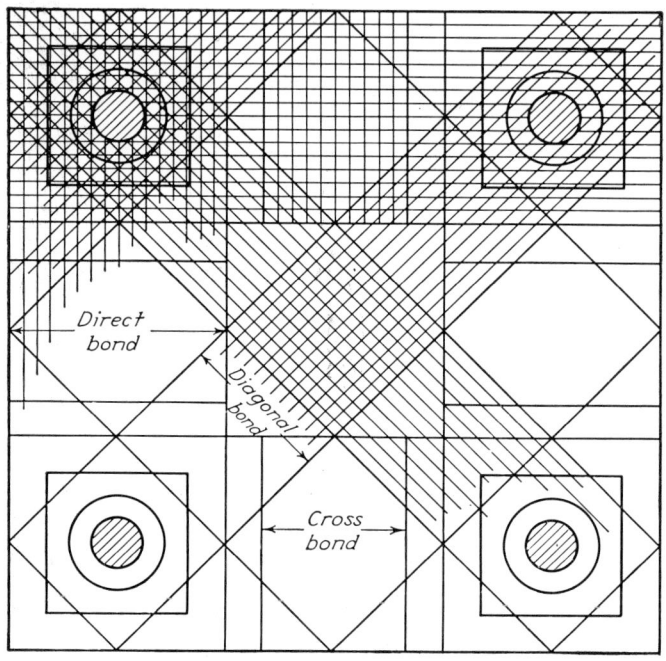

*Fig. 13-15   Four-way flat slab.*

**deformed bar.** A reinforcement bar with closely spaced shoulders, lugs or projections formed integrally with the bar during rolling. Wire mesh with welded intersections twelve inches or less apart in the direction of the principal reinforcing and with cross wires at least No. 10 gauge, may be rated as a deformed bar.

**flat slab.** A reinforced concrete slab generally without beams or girders designed to transfer the loads to supporting members.

**column capital.** An enlargement of the upper end of a column designed and built to act as a unit with the column and flat slab.

**column strip.** A portion of a flat slab panel one-half panel in width occupying the two quarter-panel areas outside of the middle strip.

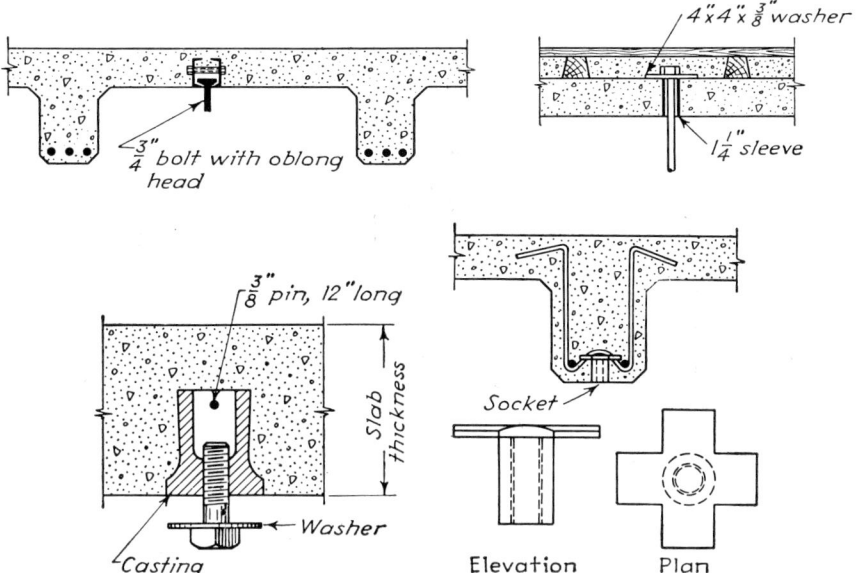

Fig. 13-16  Hanger inserts.

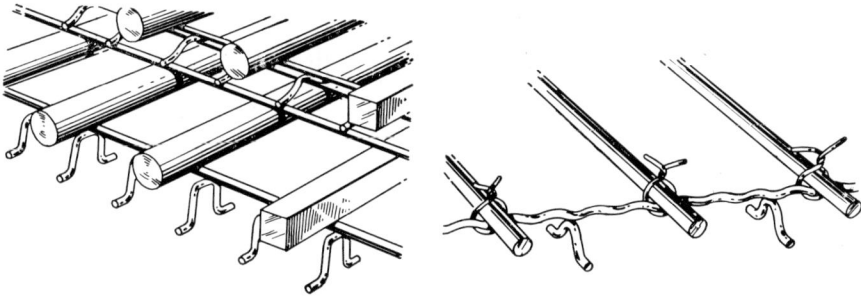

Fig. 13-17  Bar bolsters.

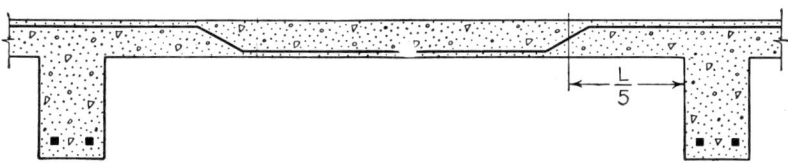

Fig. 13-18  Slab and beams.

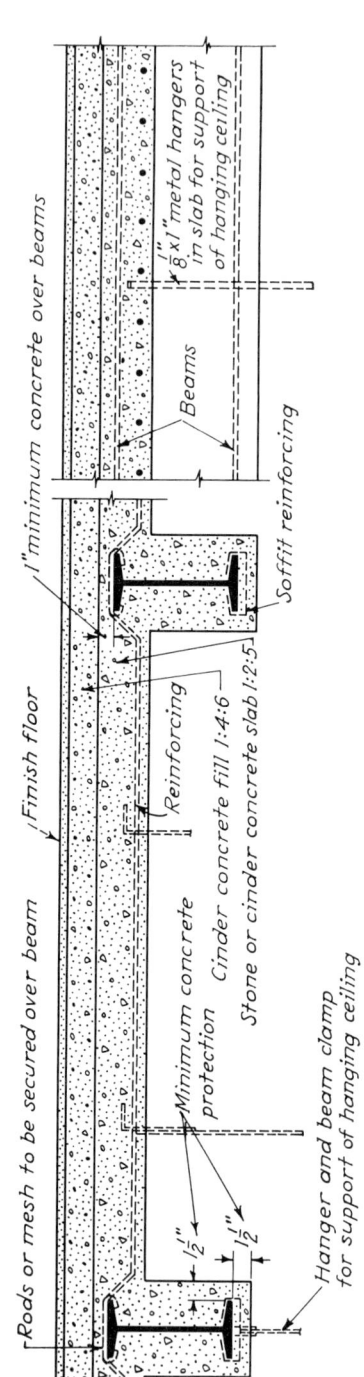

Fig. 13-19 Steel beams and flat arch.

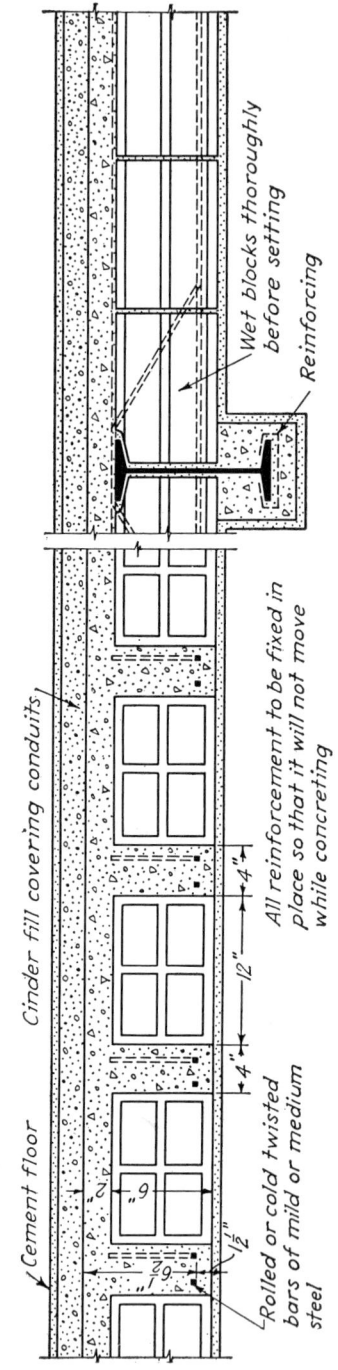

Fig. 13-20 One-way joist and block floor.

**middle strip.** That portion of a flat slab panel one-half panel in width, symmetrical with respect to the panel center line and extending through the panel in the direction in which moments are being considered.

**dropped panel.** The structural portion of a flat slab which is thickened throughout an area surrounding the column capital.

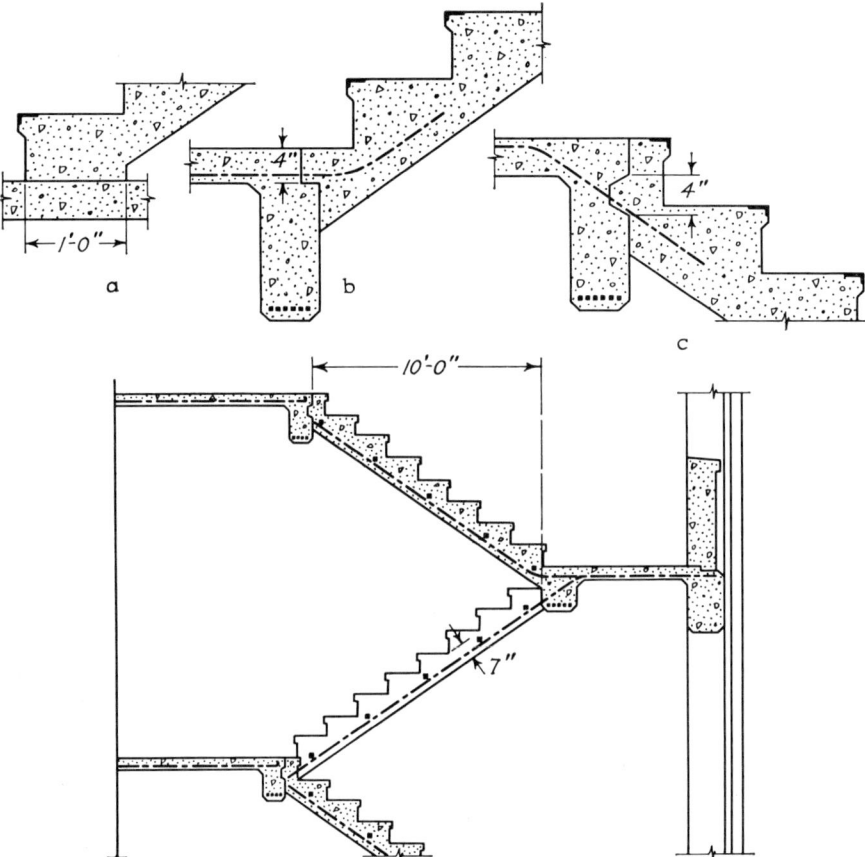

*Fig. 13-21  Concrete stairs.*

**combination column.** A column in which a structural steel section, designed to carry the principal part of the load, is wrapped with wire and encased in concrete of such quality that some additional load may be allowed.

**direct band.** A group of bars in a four-way flat slab system covering a width approximately four-tenths of the span, symmetrical with respect to the lines of centers of supporting columns.

**diagonal band.** A group of bars in a four-way flat slab system covering a width approximately four-tenths of the average span, symmetrical with respect to the diagonal running from corner to corner of the panel.

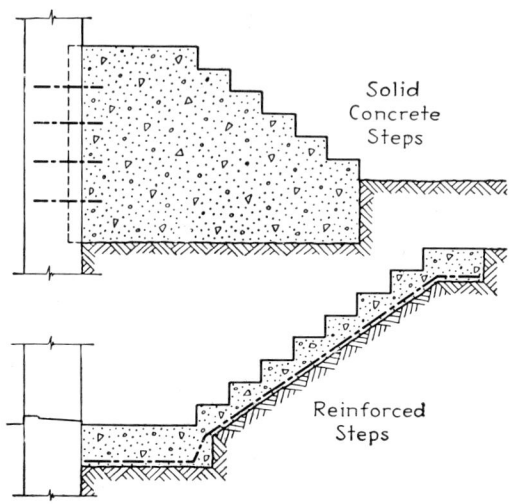

*Fig. 13-22  Concrete steps.*

● EXERCISES

1. Make sketches showing three different types of floor construction with arches and finish.
2. Write at least 300 words of notes pertaining to concrete floors and roofs, taken from one or two of the books listed in the Bibliography. Name the books used.
3. State the meaning and use of the following terms, giving sketches where possible: steel-beam fireproofing, floor fill, roof cricket, soffit clips.
4. Make a complete estimate of the concrete floor and roof arches and steel fireproofing based on the plans and specifications in Chap. 29.

CHAPTER 14

# *masonwork*

The masonry division of the estimate includes all the work done by bricklayers and bricklayers' helpers. The heading for this division is Masonwork, Masonry, or Brickwork. A more clearly descriptive term, although it is one that is never employed, would be Bricklayers' Work.

## *Materials*

Materials that are used in masonwork include common brick, face brick, firebrick, concrete blocks, terra-cotta blocks, gypsum blocks, terra-cotta flue lining, and terra-cotta coping. Incidental materials are sand, cement, lime, wall ties, and muriatic acid (for cleaning). The equipment needed includes mortar boxes, hoes, shovels, wheelbarrows, mortarboards, scaffolding, mortar mixers, etc.

The street-front walls and the returns from these walls for a specified distance are generally faced with one thickness, 4″, of face brick. Sometimes other portions of walls also are faced with face brick. Common brick is used for the backing up of the walls that have face brick, for rear-wall construction, and for brick chimneys and partitions.

**Terra Cotta.** Terra-cotta blocks, as considered under the masonry heading, commonly are the structural blocks and not the type that have specially treated surfaces. The latter are referred to as architectural terra cotta and often are put under a special heading. Structural terra-cotta blocks of the regular kinds are made in thicknesses of 2″, 3″, 4″, 6″, 8″, and 10″. They are all 12″ × 12″ in face size. These blocks, especially the 3″, 4″, and 6″ sizes, are used mainly to build partitions, and the 2″ and 3″ sizes are used for column covering and wall furring. Terra-cotta blocks are also used to form exterior walls and for backing up in exterior walls.

**Gypsum.** Gypsum blocks, sometimes referred to as plaster blocks, are made in thicknesses of 2″, 3″, 4″, 5″, and 6″. The 3″ and 4″ thicknesses are by far the most common. They are all 12″ × 30″ in size; therefore, one block equals 2½ square feet, whereas a regular terra-cotta block is 1 square foot. Gypsum blocks are used almost exclusively to form partitions. Being made of a porous plaster composition, they are much lighter in weight and of less strength than the terra-cotta blocks.

**Concrete Blocks.** Concrete blocks are described by nominal dimensions, especially their thickness. Thus one measuring 7⅝″ thick, 7⅝″ high, and 15⅝″ long is called an 8″ block or an 8″ × 8″ block. When laid with ⅜″ joints, this block will occupy a space 8″ high and 16″ long.

Solid concrete blocks are obtainable, but generally a solid block means one in which the open core area is not more than 25 per cent of the gross cross-section area of the block. A hollow concrete block means one with a core area greater than 25 per cent.

Concrete blocks are made with either heavy or lightweight aggregates, a heavy 8″ block weighing 40 to 50 lb and a light block 25 to 35 lb. Heavy blocks are made with sand, gravel, or crushed stone. Light blocks are made with cinders, slag, or other lightweight materials. The lightweight blocks provide insulation against heat and cold.

**Brick.** Brickwork is measured per thousand bricks. This is obtained by first computing the number of cubic feet of solid brickwork. All brickwork, whether common brick, face brick, or firebrick, is best handled by consistently finding the number of cubic feet of actual solid brickwork. All the brickwork of each kind is measured separately in cubic feet. This volume is multiplied by 20 in order to get the number of bricks involved. For all practical purposes, an allowance of 20 bricks per cubic foot is accurate enough, although the number will vary slightly according to the size of brick used and the thickness of the mortar joints. The number of bricks is separately extended for each item or group of similar items. This is priced at the cost per thousand bricks, laid in place complete, with all expense of material, labor, and incidentals considered in the unit cost. With this uniform measurement, it does not matter how thick a wall is to be or even whether the brickwork is to be in a wall or not; the same treatment may be consistently applied. Often brickwork is filled in around steel, as in spandrels and in piers, or is built up in odd shapes and thicknesses, as at chimneys and tank or machine foundations. Walks and drives and fire linings may be only 2" or 4" thick. The thickness of solid brick walls and partitions is regularly a multiple of 4"—the nominal thickness of a brick as laid. Thus, walls or partitions are usually 4", 8", 12", 16", etc., in thickness.

Of course, the entries for the estimate are from measurements taken accurately from the plans, and they are grouped in orderly fashion under appropriate minor headings. Even though many of these minor headings may refer to one kind of brick, the total is extended separately for each group, so that it may be given a separate unit price. This is necessary because, in the analysis of the cost, there may very well be different labor amounts or other considerations besides the cost of the material that will make the unit cost of one group greater than that of another group where the same material is concerned.

## Method of estimating

The method of measuring and entering the items for masonwork is illustrated by sample entries of several portions in Fig. 14-1. Note that the net amount of work is extended; that is, the gross volume of the brickwork or the gross area of the terra-cotta or gypsum partitions is first entered. Then the openings are deducted by entries that are technically called *outs*. These deductions are made before the final extensions are arrived at.

Terra-cotta and gypsum-block work of partitions, column covering, furring, etc., are measured and priced by the square foot, complete in place. Flue lining and coping are handled by the linear foot. The setting of stones and iron fittings is priced per piece.

All this estimating is subdivided into small parts and arranged neatly so that it may be properly priced and checked. Skip lines after each group in order that adjustments can be made and the estimate may be read more easily.

Figure 14-1 is similar to Fig. 12-1, but it has separate pricing col-

## BUILDING CONSTRUCTION ESTIMATING

| | | | | | QUANTITY | LABOR | MATERIAL |
|---|---|---|---|---|---|---|---|
| EST. 502 | | Mason Work | | | | | SHEET 7 |
| | Common Brick | | | | | | |
| | Basement | | | | | | |
| | Boiler room | | | | | | |
| | 69-0 | 11-10 | 1-0 | | 817 | | |
| | Outs | | | | | | |
| | 3-4 | 7-6 | 1-0 | 25 | | | |
| | 5-0 | 7-4 | 1-0 | 36 | −61 | | |
| | | | | | 756 | | |
| | | | | | ×20 | | |
| | | | | | 15120 Bk. | .06½  983 | .07  1058 |
| | First Floor | | | | | | |
| | North wall | | | | | | |
| | 102-8 | 17-3 | 0-4 | | 590 | | |
| | | | | | ×20 | | |
| | | | | | 11800 Bk. | .07½  885 | .07½  885 |
| | North parapet | | | | | | |
| | 20-0 | 3-6 | 1-0 | | 70 | | |
| | East wall | | | | | | |
| | 83-2 | 20-9 | 1-0 | | 1726 | | |
| | | | | | 1796 | | |
| | | | | | ×20 | | |
| | | | | | 35920 Bk. | .06  2155 | .07  2514 |
| | Garage front & rear | | | | | | |
| | 20-0 | 20-6 | 0-8 | | 273 | | |
| | 19-0 | 17-0 | 1-0 | | 323 | | |
| | | | | | 596 | | |
| | Outs | | | | | | |
| | 12-0 | 12-2 | 0-8 | 97 | | | |
| | 6-9 | 8-6 | 1-0 | 57 | −154 | | |
| | | | | | 442 | | |
| | | | | | ×20 | | |
| | | | | | 8840 Bk. | .07  619 | .07  619 |
| | | | | | | 4642 | 5076 |

Fig. 14-1  *A masonwork estimate.*

umns for the labor and material. This is preferred by some estimators, although the two unit costs could be added together and entered as one total unit cost just as well. Figure 14-1 shows the first of several estimating sheets used for the masonry of a business building. On it, the labor column provides for the cost of all labor and direct supervision and the insurances, taxes, welfare funds, and pension funds based upon all this labor. The material column provides for the cost of all the materials used in each item, delivered to the job, including any sales or other taxes involved. The scaffolding required is not included on this sheet because this estimator set up a separate item for scaffolding at the end of his masonry estimate, in which he showed all the wall areas with prices, per square foot, for the scaffolding. The brickwork is priced per brick instead of per thousand bricks. Either way is satisfactory.

## Masonry costs

Common brickwork costs the contractor between $125 and $150 per thousand complete, in place. The bricklayers lay between 1,000 and 1,200 bricks in a day. Face brickwork costs the contractor between $200 and $250 per thousand complete, in place. Bricklayers can lay only about 600 of these in a day. The amount of work produced in any given time will depend on the type and thickness of the wall and on conditions of working and other factors. From one-half to two-thirds of a cubic yard of mortar is required for each thousand brick. Terra-cotta partitions cost the contractor 50 to 90 cents per sq ft; and gypsum-block partitions, 40 to 60 cents per sq ft, in place complete, for regular work. Terra-cotta coping runs $1 to $2 and flue lining $1.25 to $5 per lin ft. A range for handling and setting isolated pieces of stone and iron would be $1.50 to $3 per single piece. All these prices are rough approximations, simply to furnish an illustration of the relation between them and to give the student some feeling for the cost of the work. Prices of labor and material and incidental items change from time to time, and each job has its own peculiarities that will affect the unit costs (Fig. 14-2).

The allowance of 20 bricks per cubic foot already provides about 5 per cent for waste; therefore, no further allowance is necessary. However, as has been stated before, on very large jobs the size of the actual bricks that are to be used, as well as the thickness of the joints, should be known, and then a study should be made of the actual number to be allowed per cubic foot. If such a study is considered necessary, a study of the mortar requirements also should be made. In fact, all the requirements for a large job should be analyzed more extensively than those for a small job. A large job is one involving more than 500,000 common brick. If a large number of face brick—say, 50,000 or more—

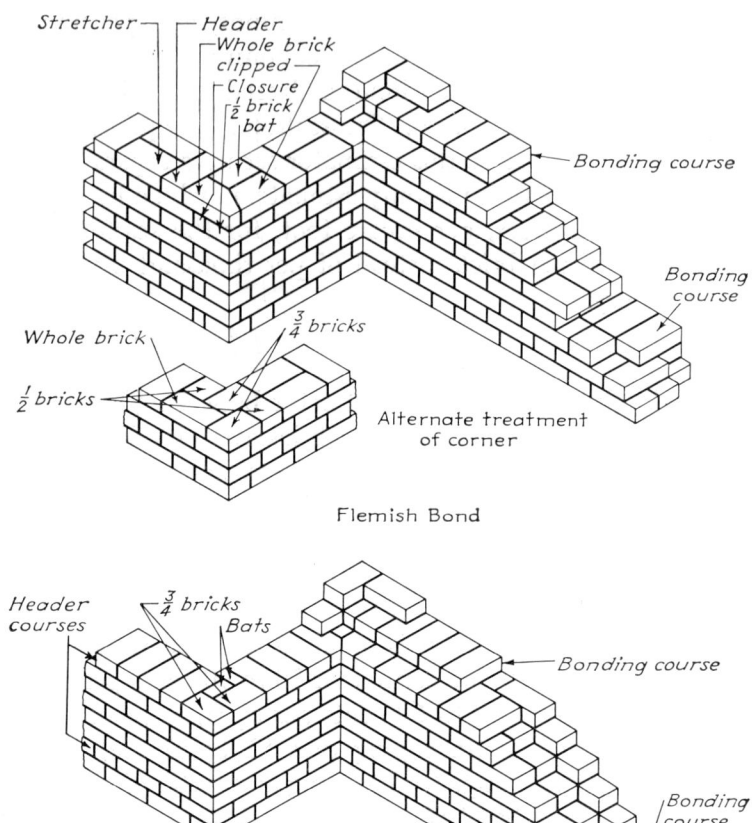
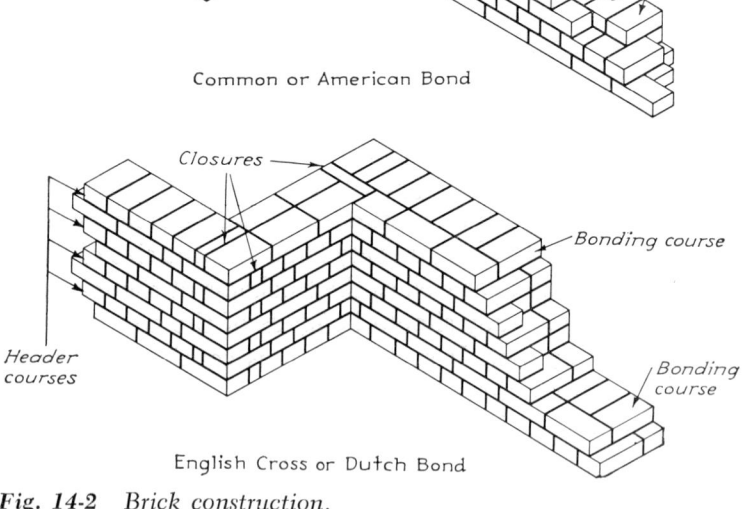

Fig. 14-2  Brick construction.

are required, these should be analyzed as to the number of full headers required, and an entry should be made for the adjustment of the figures.

The number of bricks to be added for headers is, of course, for only the number of full headers. The pattern of the brick appearing on

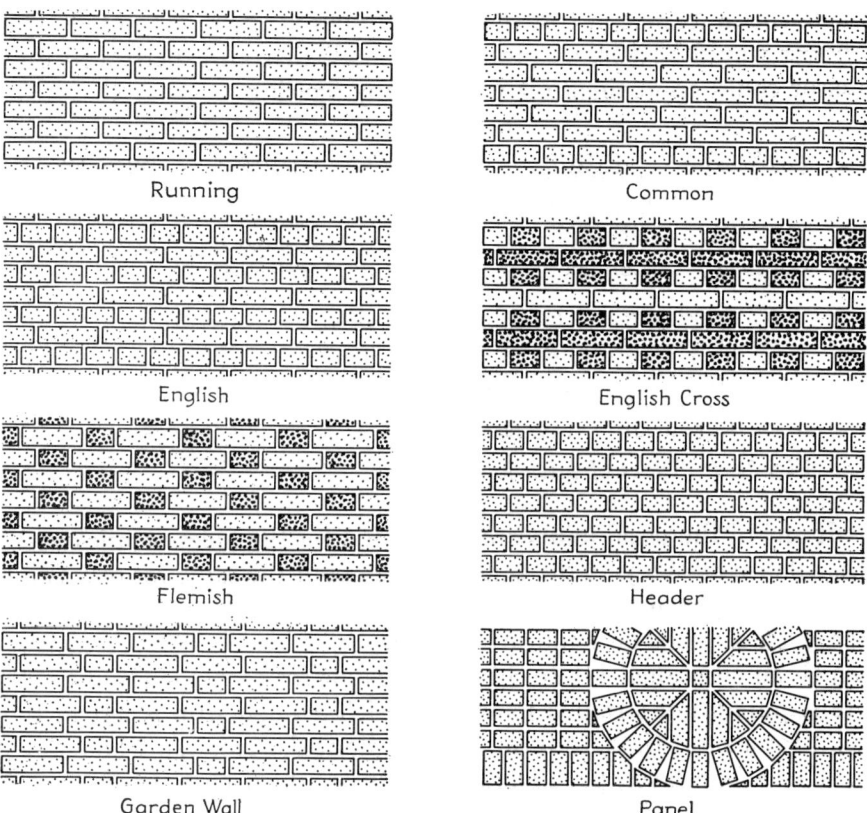

*Fig. 14-3   Brick bonds.*

the face of the wall may be in any bond, but most of the headers may be only half bricks, especially if the face brick are expensive, and therefore both halves would be used and no additional bricks required for such headers (Fig. 14-3). If Flemish bond is formed, for example, with full headers used only every sixth course, the additional brick for these headers would amount to only one-eighteenth of the total number of face brick. To make the adjustment for this additional cost it is necessary to take the difference between the cost of the common brickwork and the cost of the face brickwork and multiply it by one-eighteenth of the number of face brick. Thus, if 10,000 face bricks were counted for the job without considering headers, and the cost of the common and face brickwork were $140 and $220 per M, respectively, the adjustment would be an addition of one-eighteenth of $800, the difference in cost for 10,000 bricks at the prices noted. The answer would be $44, an amount so small that an experienced estimator could approximate it without any figuring.

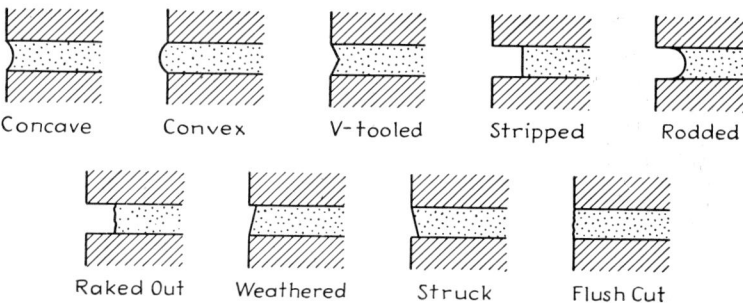

Fig. 14-4  *Types of brick joints.*

If any of the usual joints are to be used for the faces of brickwork, the labor cost of laying the brick will be about the same for one as for another. Sometimes, however, joints out of the ordinary are called for, such as raked-out, concave, convex, or rodded joints, for each of which an additional 1½ to 3 hours of labor for each thousand bricks are required. For stripped joints, in which a wood strip is left in the joint for a short time, to form a true rectangular recess, a whole day's time of a bricklayer, per M bricks, must be added (Fig. 14-4).

A bricklayer will clean down about 800 sq ft of wall per day, if the bricks are fairly smooth and not much repointing of the joints is necessary. The cost of this work, including the acid, pails, and brushes, runs about 8 to 10 cents per sq ft. To this must be added any scaffolding that may be required. In figuring cleaning areas, only very large openings are deducted.

The cost of scaffolding the exterior of ordinary buildings runs between 10 and 15 cents per sq ft of wall, measuring the entire length and height of wall to be scaffolded. This will depend upon the amount of scaffolding, the type used, the amount of changing, etc.

Scaffolding for masonwork is an important item of cost and must always be provided for (Fig. 14-5). Some contractors raise the unit price between $4 and $6 per M, to take care of all scaffolding for brickwork. This method is satisfactory, if the proper amount is used, for

*Fig. 14-5  Scaffolds.*

regular work. Construction out of the ordinary, however, should be separately studied as to the scaffold and hoisting requirements—with the amount of material, labor, and expense entered in detail, if necessary, in order to provide the money value of this work.

The quality of the work will affect the unit cost of labor. A bricklayer will lay 1,200 brick per day in ordinary work; but if high-quality workmanship is called for, he will not lay over 900 per day. This difference, together with the resultant extra cost of insurance, etc., will have a great effect upon the total cost of the job. Careless estimators think of the cost of common brickwork in terms of a certain number of brick per day for all jobs and, thereby, produce an estimate that is not worth the name of estimate; it is merely a rough approximation.

Face brick laid up in regular style is one thing; the same brick laid up in pattern or with special effects may cost two or three times as much for the labor.

Brick walls are cleaned down with a solution of about 15 parts of water to 1 part of muriatic acid. This is applied with stiff brushes by the bricklayers.

Dampproofing, in the form of a tar or asphalt composition, is sometimes applied to the unexposed faces of walls. Fabric dampproofing or waterproofing is applied to particularly vulnerable locations, such as spandrels, window heads, and parapets (Fig. 14-6). Window frames are generally caulked with an elastic caulking compound (Fig. 14-7).

Terra-cotta flue lining is clay pipe and is usually made in lengths of 2 ft. It is used to line the inside of chimneys and can be had in round, square, and rectangular shapes. Stock sizes run from 6" to 24" in cross section. Terra-cotta coping, sometimes called tile coping or crock or saddleback coping, also comes in 2-ft lengths. It is used to cap 8", 12", and 16" walls.

The handling and setting of stone window sills and lintels in brick walls come under the masonry heading, as do also the handling and setting of isolated pieces of stone trimmings and small iron fittings, angle-iron lintels, etc.

A bricklayer will lay about 200 brick per day in fireplaces and about 400 per day in chimneys. He should set about 80 lin ft of TC flue lining of the ordinary sizes. Thus, with the incidental laborer's time, the mortar, etc., fireplace brickwork should run between $375 and $450 per M; chimney common brickwork, $200 to $250, plus the cost of scaffolding; and ordinary flue lining $2 to $3 per lin ft. Larger flue lining is so heavy that the cost per foot rises rapidly for both material and labor, 24" costing about $12 per lin ft installed.

Fireplaces, chimneys, and other odd shapes can sometimes be best figured as though they were solid, and then the voids in them figured and subtracted from the gross volume, in order to arrive at the net volume of brickwork. Hollow walls may be calculated either in this way or by taking the actual thickness of the parts.

MASONWORK 185

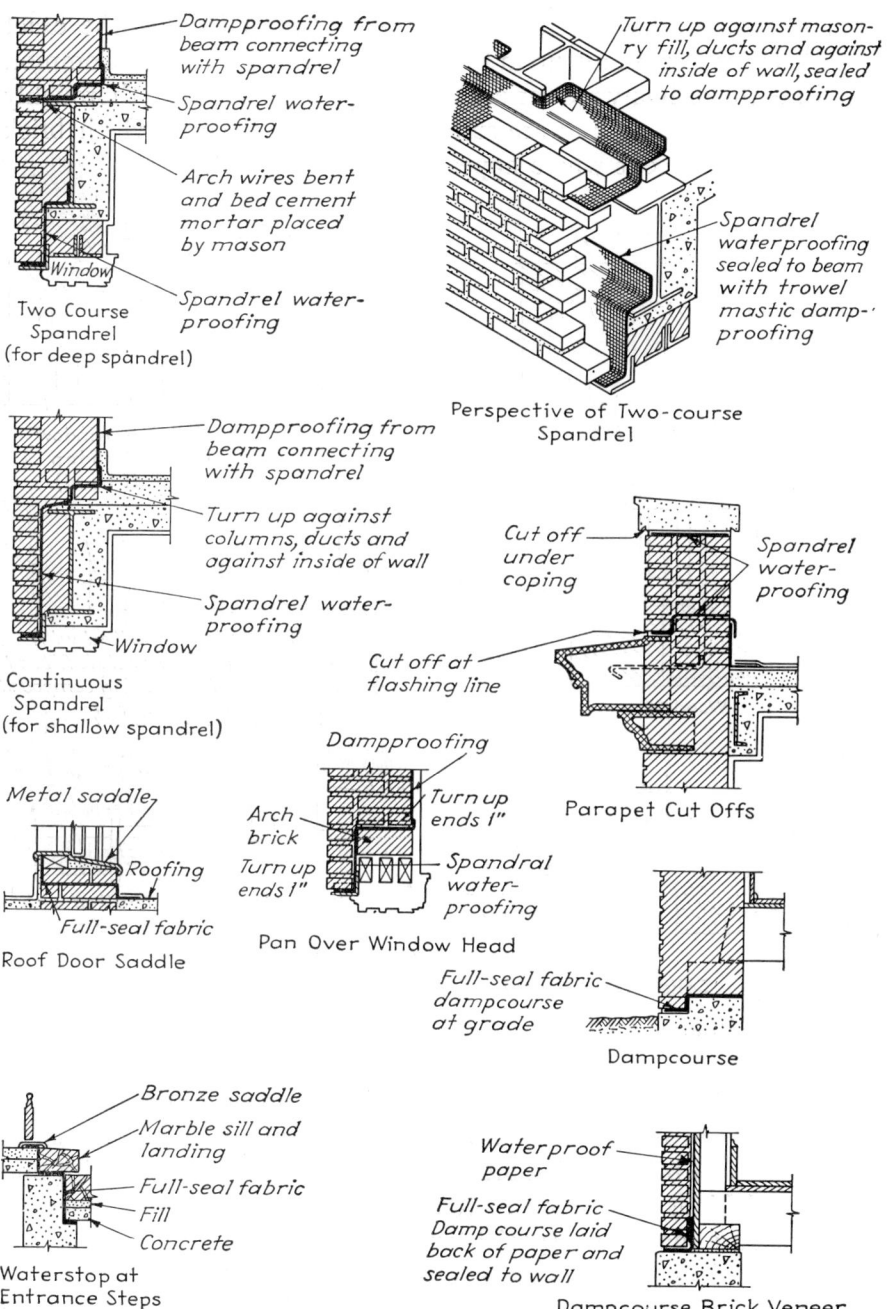

Fig. 14-6  *Waterproofing and dampproofing.*

The standard-size brick is 2¼" × 3¾" × 8". However, the exact size will vary, depending upon the manufacturer's standards. Common brick are the ordinary red variety seen everywhere. They are thoroughly fireproof and strong and are suitable for every place in which brick can be used, except where appearance is important.

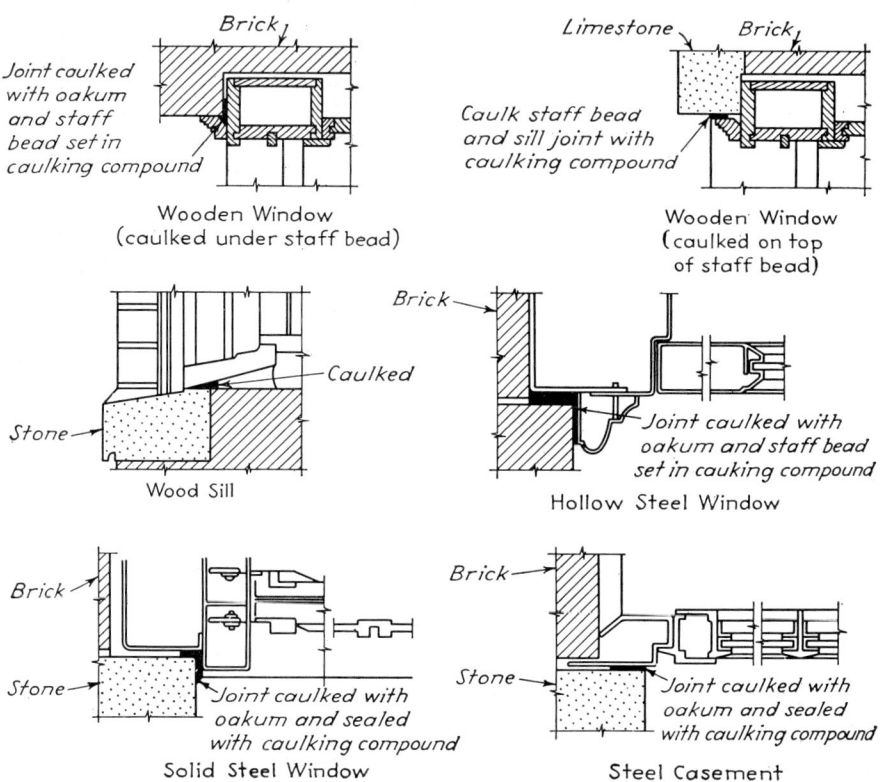

*Fig. 14-7* Window caulking.

Face brick, which are more uniform in size and shape than common brick, come in many colors and shades and surface finishes. Glazed and enameled brick, which are very true to size, are coated with one of several kinds of glazing material in the process of manufacture. Firebrick are made of a special type of clay, highly resistant to heat, and are used in large chimneys, fireplaces, fireboxes, etc.

Bricks appearing in the wall in common fashion are called *stretchers*; those laid with an end exposed are termed *headers*. Stretchers turned up on end are called *soldiers* and headers on end are called *rowlocks*. *Common bond* consists of five courses of stretchers and one of headers, alternately. *English bond* consists of alternate courses of stretchers and headers. *Flemish bond* consists of alternate stretchers and headers in every course.

Bricks used in arches are laid with tapering joints or, for exposed arches where the appearance is important, bricks ground to shape are used (Fig. 14-8).

Portland cement is composed of finely ground limestone and a clay containing silica, alumina, and iron. This mixture is burned at a high temperature and then pulverized. Standards of the American Society for Testing Materials are generally followed in the manufacture of this popular material. It is made in many parts of the country and comes in 94-lb lots, in cloth and paper bags. Special types of cement also are made, such as masonry cement, high-early cement, waterproofed cement, etc.

Mortar is a mixture of cement, sand, and water, usually about 1 part cement to 3 parts of sand. In order to make the mixture more workable, lime is generally added, especially when the mortar is to be used by bricklayers and spread with a trowel. A good mortar for brickwork is also made of 1 part cement, 1 part lime, and 6 parts sand. Mortar forms the material used for cement finish work, as well as for binding masonry.

When the specifications call for a certain grade of brick, hollow tile, or other material, the estimator must watch out to see that quotations are based on the quality called for. A quotation for just "brick," "hollow tile," etc., does not bind the dealer to deliver any but the poorest quality. If a better grade is required, the wide-awake estimator will check, to be certain that the prices quoted will cover the materials actually desired. Also, as there is usually more breakage in using poor-grade materials, he will consider this in his unit prices or make an entry for breakage or other forms of waste that the particular materials, of whatever quality, will probably entail. If face brick are specified to cost a certain price per thousand, it should be ascertained whether the specifications allow for taking the trade discount, if there is any, or whether this is to be passed along to the owner in the accounts of the job.

On important jobs it is customary to send samples of all the main materials to the architect or to testing laboratories before they may be

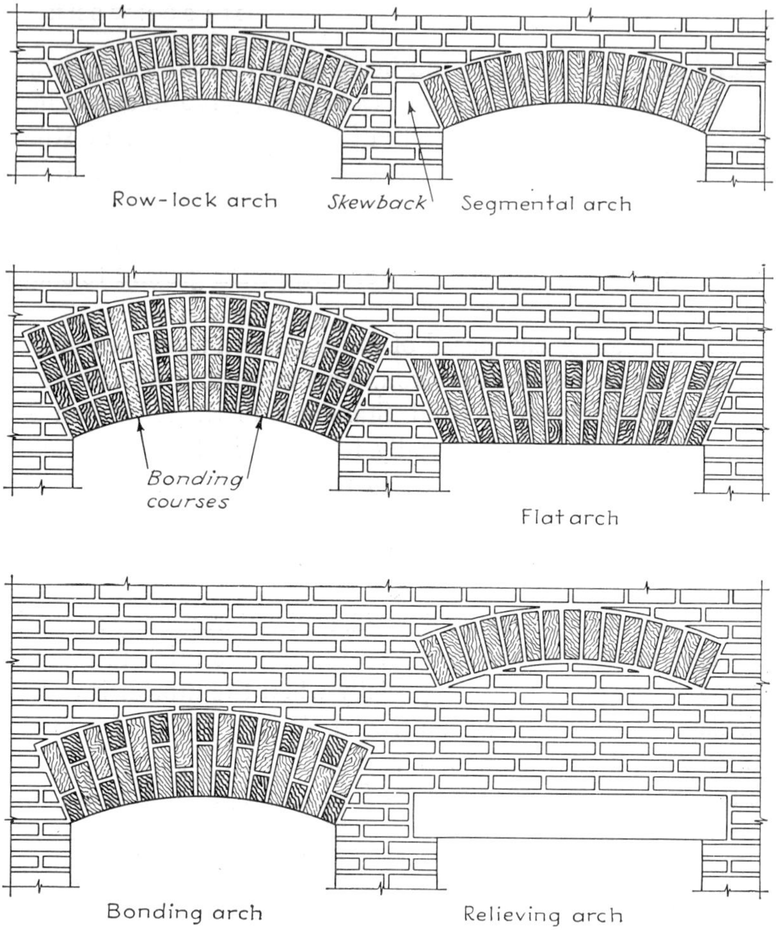

*Fig. 14-8* Brick arches.

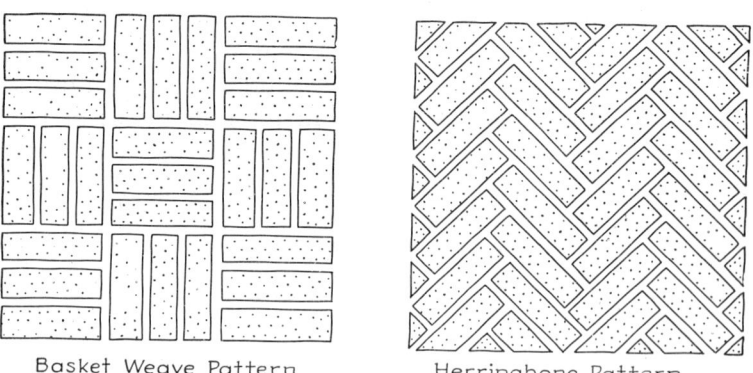

*Fig. 14-9* Brick paving.

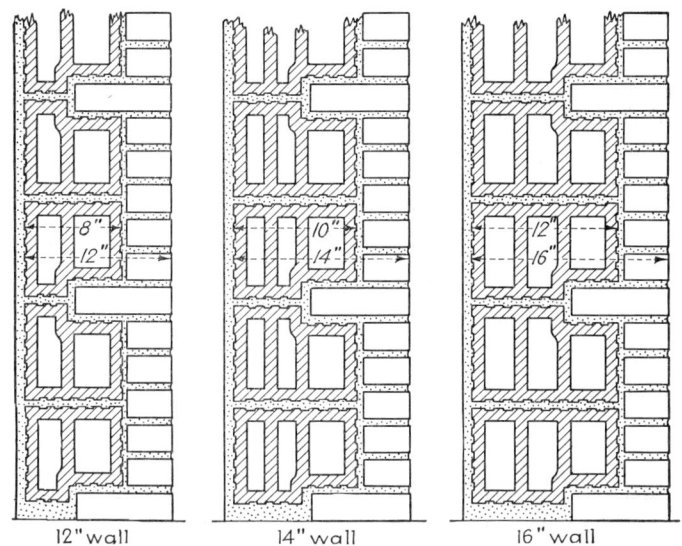

*Fig. 14-10* Combination walls.

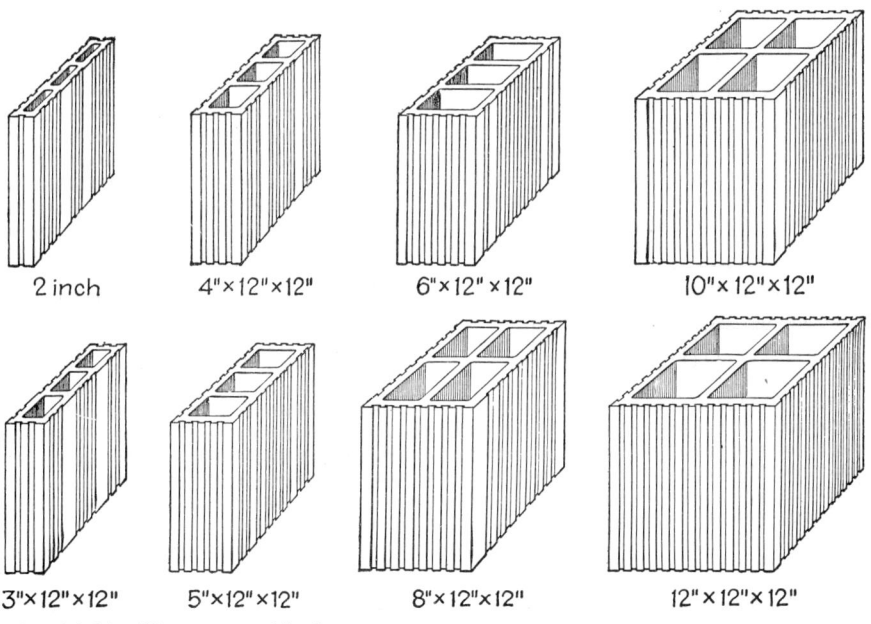

*Fig. 14-11* Terra-cotta blocks.

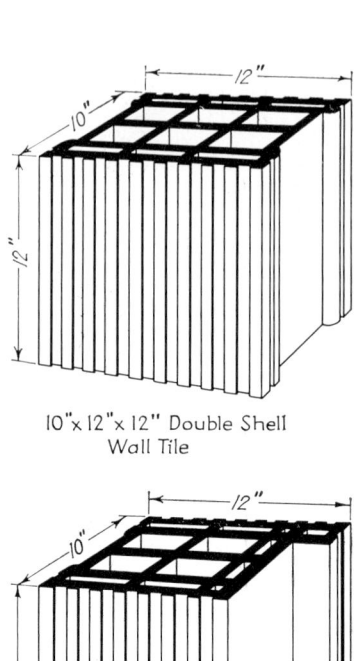

10"x 12"x 12" Double Shell
Wall Tile

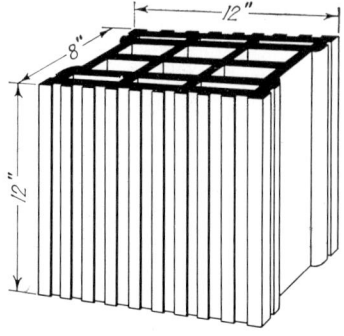

8"x 12"x 12" Double Shell
Wall Tile

10"x 12"x 12" Double Shell Full Jamb
for Box Frame Windows

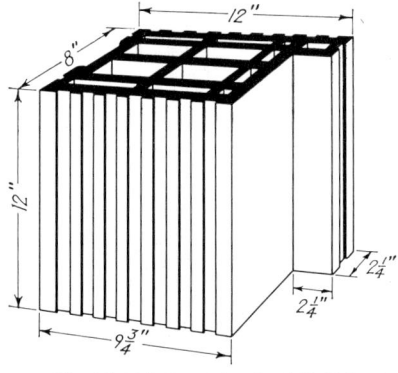

8"x 12"x 12" Double Shell Full Jamb
For Box Frame Windows

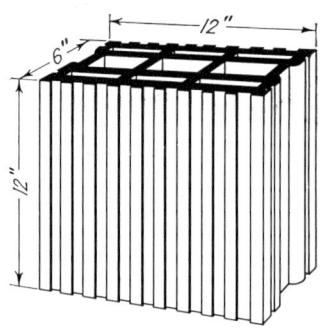

6"x 12"x 12" Double Shell
Wall Tile .

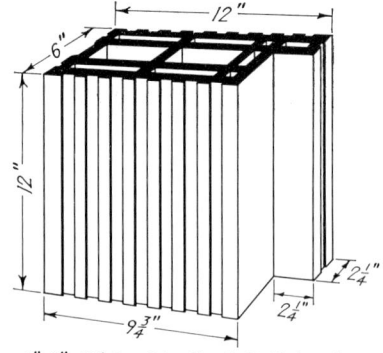

6"x 6"x 12" Double Shell Full Jamb
For Box Frame Windows

**Fig. 14-12**  *Load-bearing terra cotta.*

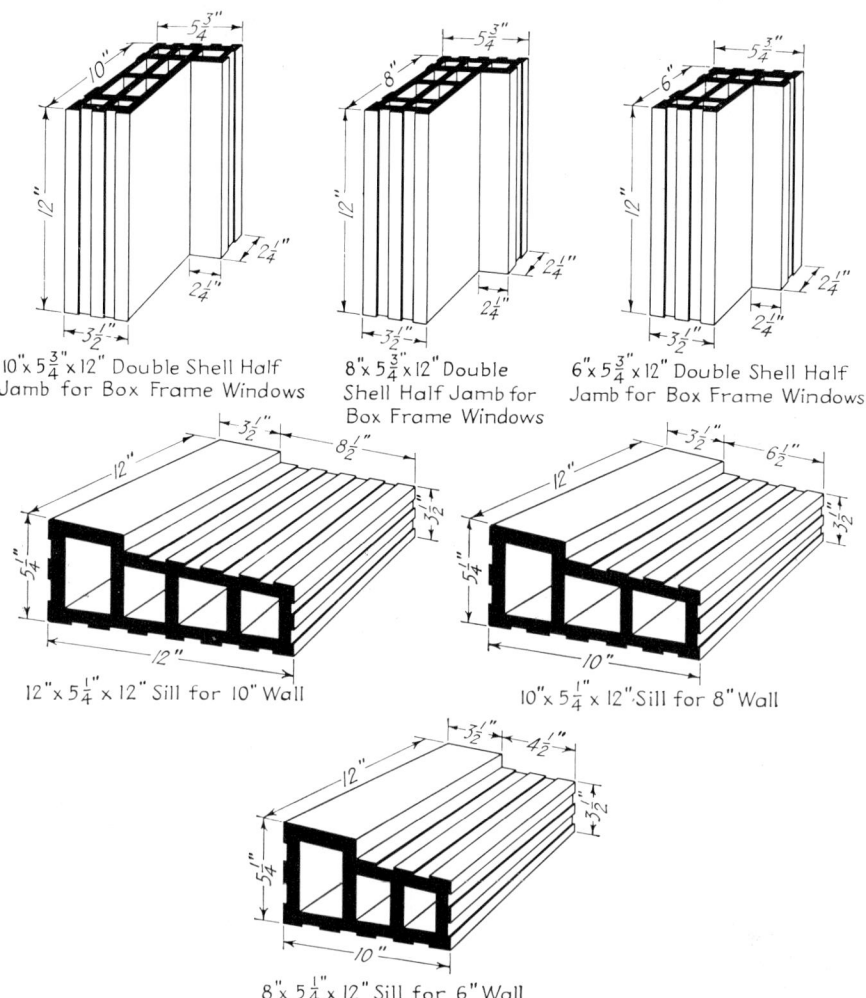

Fig. 14-13 Special shapes.

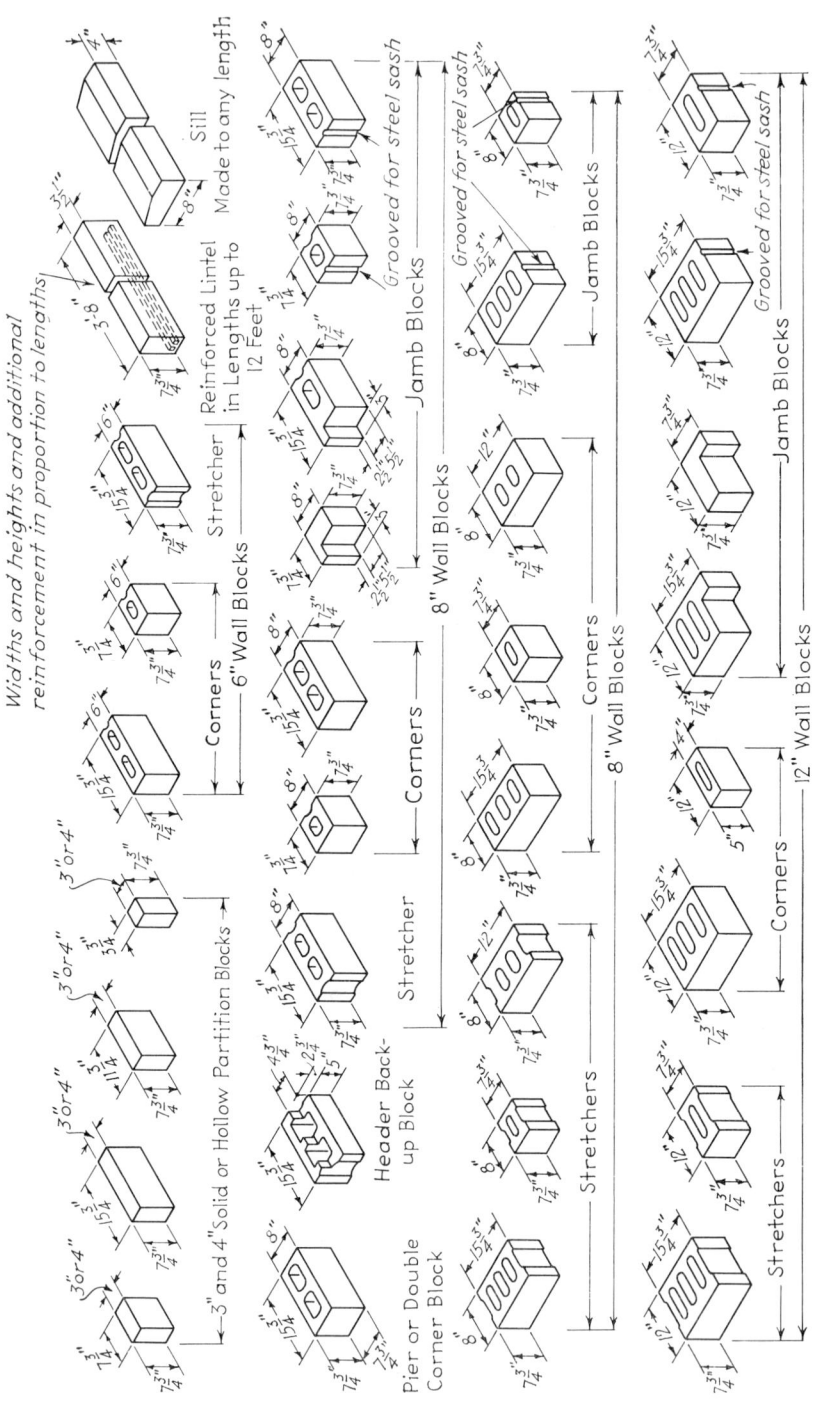

Fig. 14-14 Concrete blocks.

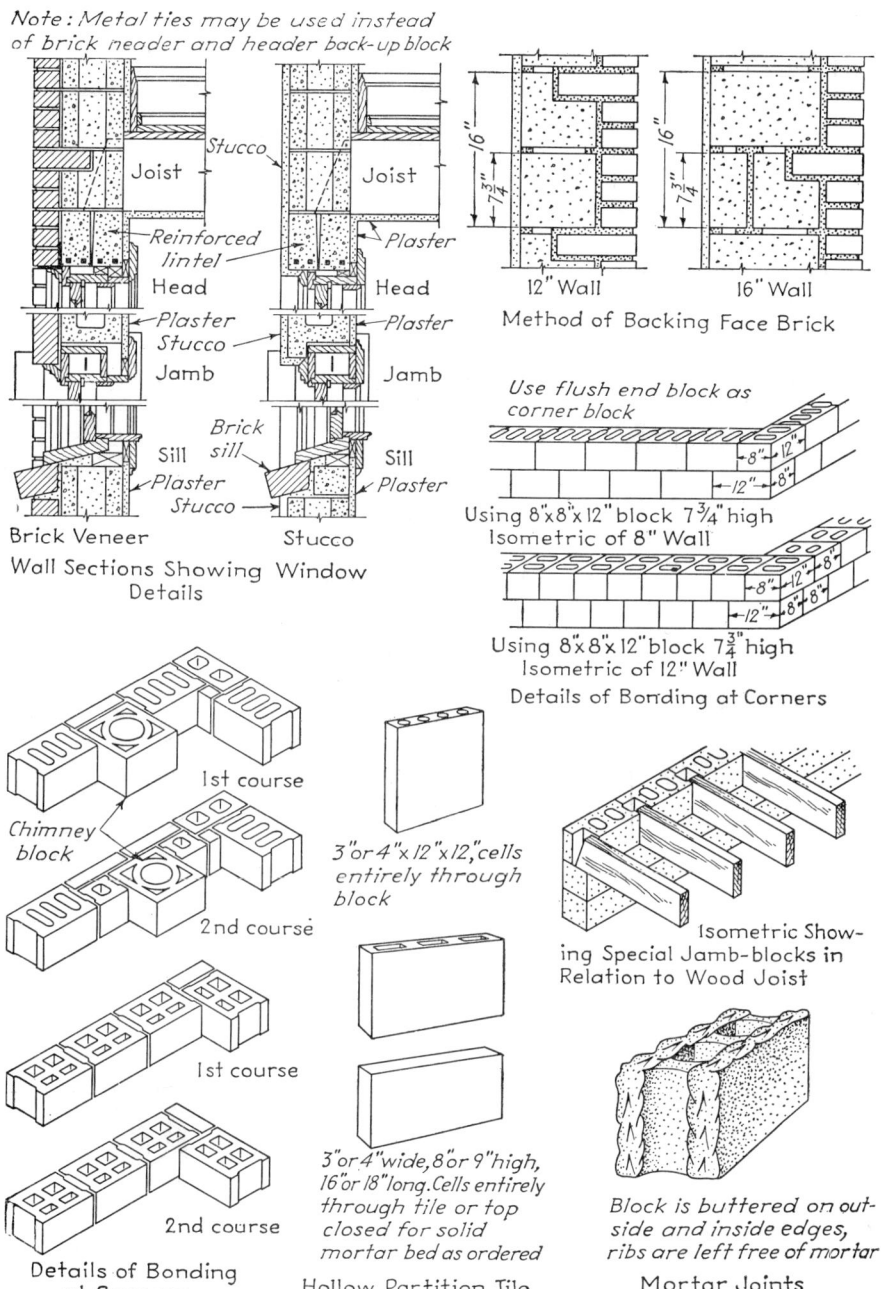

Fig. 14-15 Details of block construction.

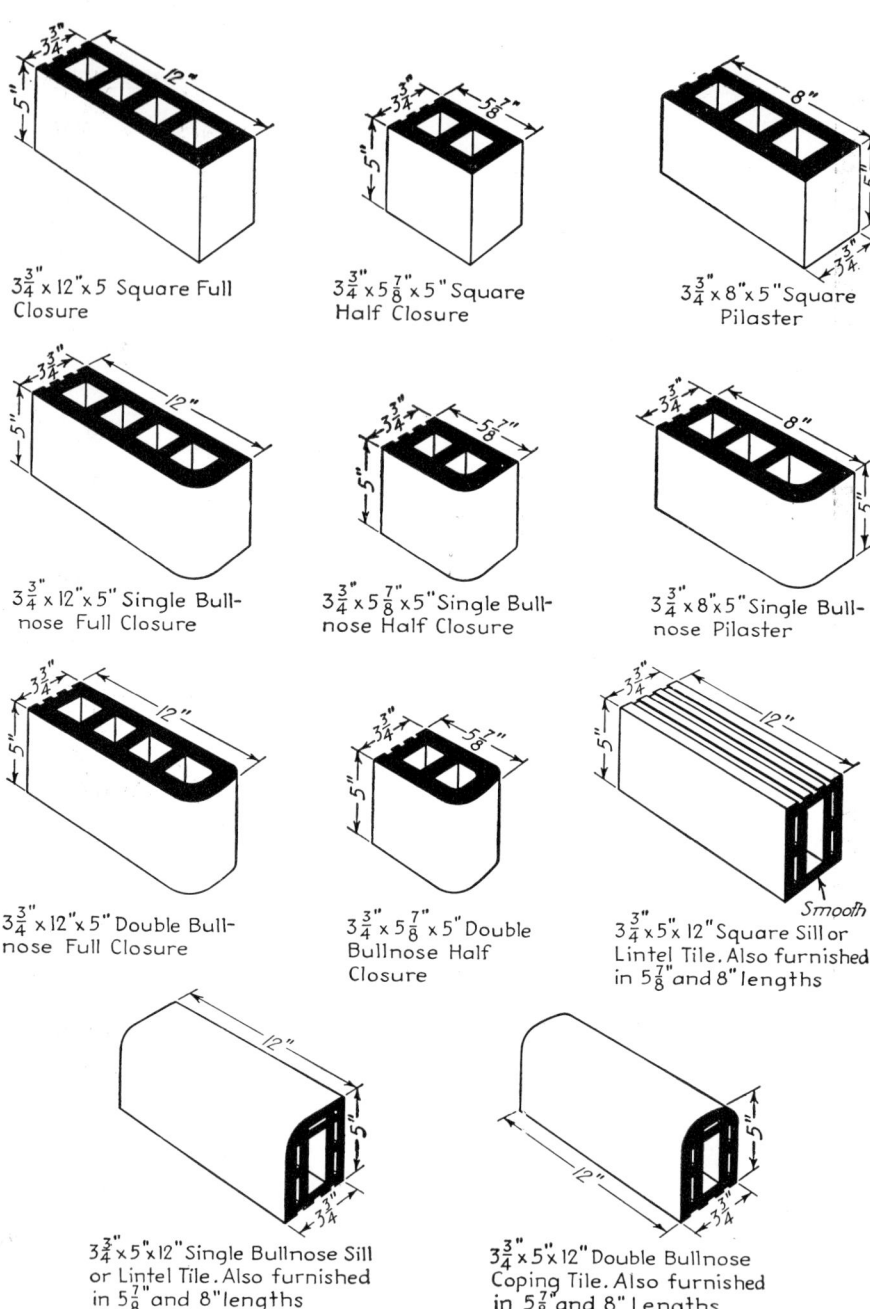

*Fig. 14-16* Glazed terra cotta.

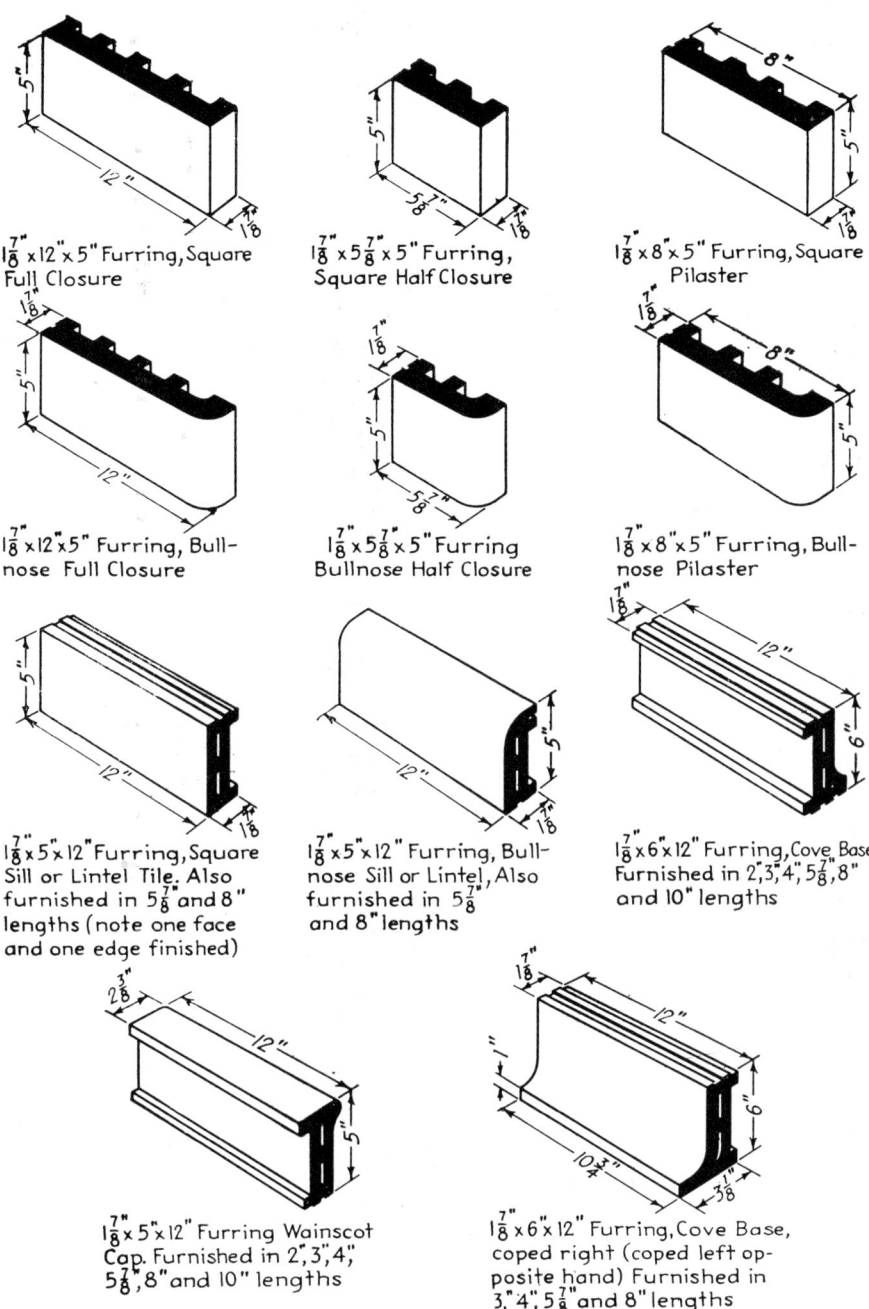

Fig. 14-17  Glazed terra cotta.

ordered. Often the cost of the tests is to be borne by the contractor. These call for entries in the estimate for packing and shipping, test fees, etc.

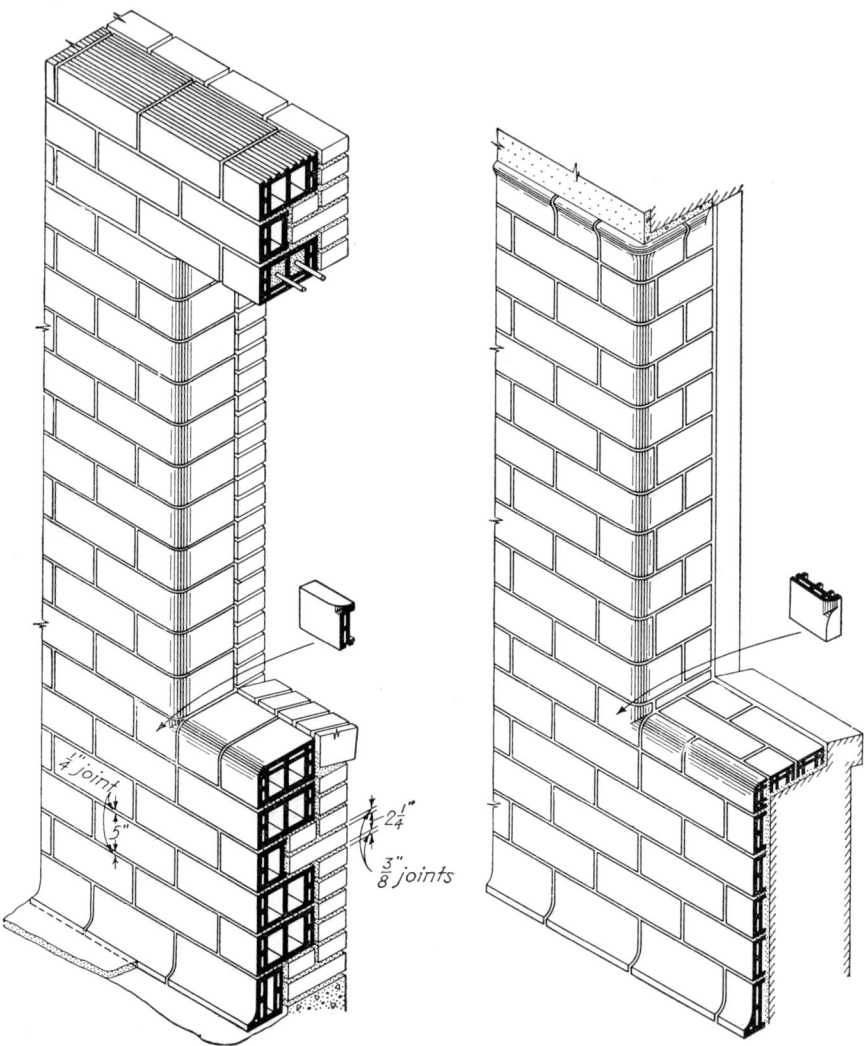

*Fig. 14-18  Details of glazed terra-cotta work.*

Under the heading of masonwork, rough bucks are sometimes estimated, or at least the setting of them and, perhaps, the side anchors. In any event, it is the estimator's responsibility to provide, in one or another division of his estimate, for bucks, anchors, and the setting of them.

The handling and setting of terra cotta with finished surfaces are, of course, not the same as for ordinary terra-cotta blocks. Particular care must be exercised at every stage. The finished material is lifted

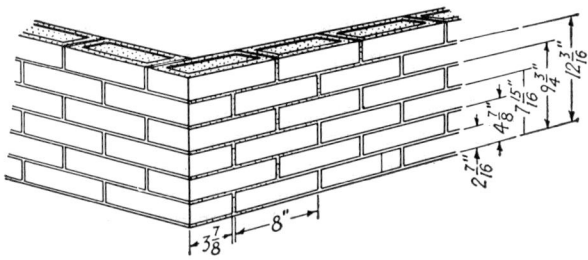

Standard Size
Approximately 7$\frac{1}{5}$ standard brick with $\frac{3}{16}$" Joint = 1 sq. ft.
5 brick and 5 joints lay up approximately 12$\frac{3}{16}$" high

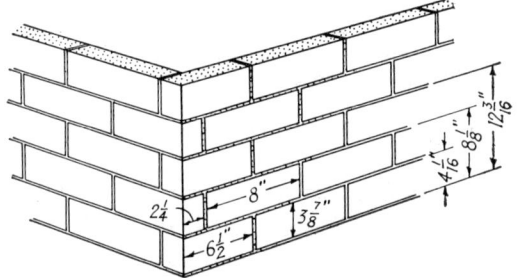

Flatter Size
Approximately 4$\frac{1}{3}$ flatter brick with $\frac{3}{16}$" joint = 1 sq.ft.
3 flatters and 3 joints lay up approximately 12$\frac{3}{16}$"high

Fig. 14-19 *Sizes of enameled brick.*

carefully from the trucks and stacked in accordance with the various sizes and shapes that are to be used. It is then picked out as needed and carefully carried or wheeled into the job to the point of setting. Here the bricklayers set it not nearly as quickly as they would the ordinary blocks. It must be treated more nearly like the ceramic tile employed in many shapes and colors in bathrooms, swimming-pool rooms, etc., each piece being carefully placed to suit the design and profile called for on the plans. In fact, it takes the bricklayers about three times as long to set this sort of terra-cotta block as it does to set the ordinary kind. Standard shapes and colors of many types are obtainable. Special blocks are made to order and, of course, for these, the manufacturers usually quote for each job separately. Since the blocks are special, the estimator knows that he must take special care in establishing the labor cost also, and not use prices that will fit only standard shapes.

Models are sometimes called for in connection with terra cotta and other materials. Even sample panels of brickwork may have to be built for the architect's selection. These cost the contractor money, and the amount involved must appear in the estimate.

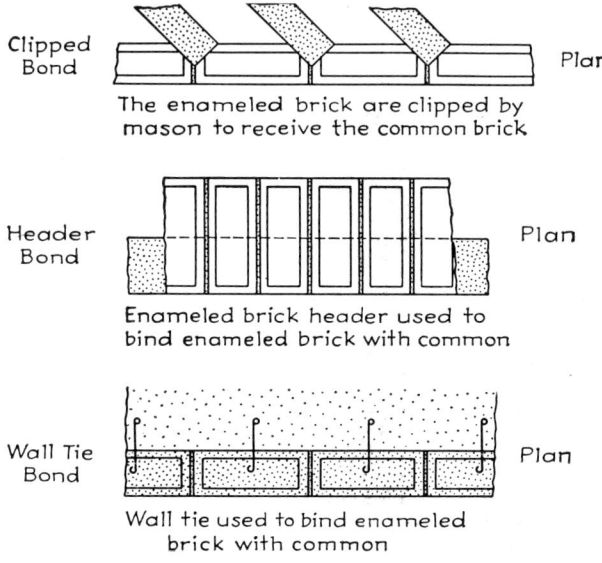

Fig. 14-20  Bonding enameled brick.

Figures 14-9 to 14-21 show various types of masonry construction and materials.

**Construction.** The following excerpts from the New York City Building Code describe the minimum requirements for good masonry construction:

## General

**Protection.** Masonry shall be protected against freezing until such time as the setting of the cementing material has advanced far enough to prevent any displacement of such masonry. It shall be unlawful to use any frozen material or to build upon any frozen masonry or frozen soil.

**Piers.** Masonry piers shall be built of solid masonry and shall be laid in cement mortar or cement-lime mortar, and the maximum unsupported height shall be ten times the least dimension. Sections of panel walls in skeleton construction shall not be considered as piers. It shall be unlawful to have openings or chases within the required area of any pier. Masonry piers shall be bonded.

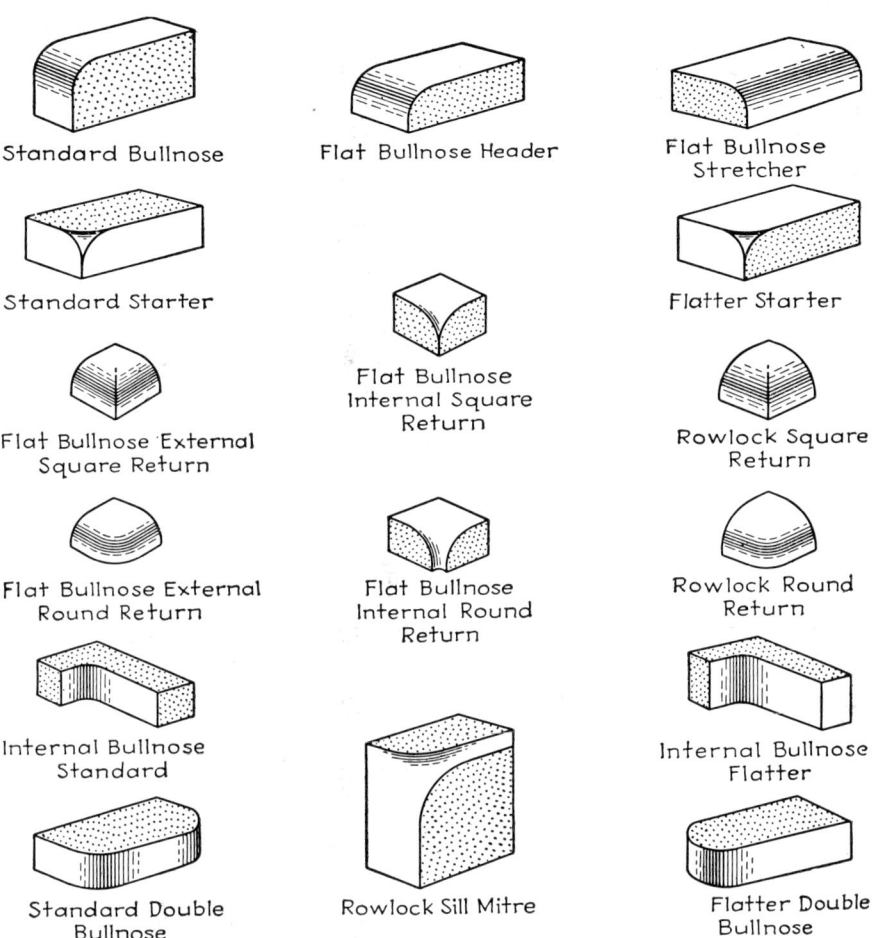

*Fig. 14-21  Bullnose enameled brick.*

**Anchorage.** Masonry walls shall be anchored, at maximum intervals of four feet, to each tier of joists or beams bearing on such walls, by metal anchors having a minimum cross-section of one-quarter inch by one and one-quarter inch, and a minimum length of sixteen inches, which anchors shall be securely fastened to the joists or beams and shall be provided with split and upset ends or other approved means for building into masonry. Masonry walls parallel to joists or beams shall be provided, at maximum intervals of six feet, with similar anchors engaging three joists or beams.

Girders shall be similarly anchored at their bearings. Upset and "T" ends on anchors shall develop the full strength of the anchor strap.

**Bracing.** Masonry walls shall be braced either horizontally or vertically at right angles to the wall face, at maximum intervals of twenty times the wall thickness. Horizontal bracing may be obtained by floors or roofs. Vertical bracing may be obtained by cross walls, wall columns or buttresses, or by increasing the wall thickness.

**Openings.** The area of openings in any horizontal section of bearing wall shall be fifty percent or less of the gross sectional area except that the thickness of the wall shall be increased four inches for each fifteen percent or fraction thereof of increased opening area in excess of fifty per cent and in all cases the total percentage of openings shall be less than seventy-five per cent of the horizontal sectional area of the wall.

**Buttresses.** Buttresses shall be bonded into the wall by masonry in the same manner employed in the construction of such wall.

**Lintels and Arches.** Openings shall be spanned by a lintel or arch of incombustible material. Where steel or reinforced masonry lintels are used, such lintels shall be of such strength that the maximum deflection is one-three hundred sixtieth of the clear span and such lintels shall have at least five inches of bearing on each end and shall rest on solid bearing. Masonry arches shall be properly designed to carry the superimposed load and proper provision shall be made for resisting lateral thrust.

**Walls.** Structures shall be enclosed by materials conforming to the requirements of the type of construction under which such structures are classified. Such structures shall be entirely within the property lines, except for such projections beyond the building line as are authorized by the code. Party walls may be considered to be enclosing walls.

**Parapet Walls.** Exterior or division walls of masonry, other than fire walls over fifteen feet high in non-fireproof structures, shall have parapet walls carried two feet above the roof, except in residential structures and detached structures with overhanging roofs or in places where such walls are finished with cornices or gutters. Parapet walls shall be of the same thickness as the wall below, except that in all cases the thickness shall be at most twelve inches, All parapet walls shall be coped with incombustible and durable material. Fire partition walls for the purpose of subdividing non-fireproof residence structures shall be carried to the top of the roof boards, be thoroughly grouted with cement mortar and fire-stopped, or carried through the roof.

**Furring.** Masonry materials used as furring shall be excluded in calculating the required wall thickness, and such furring shall be considered to lack any structural value.

**Chases.** It shall be unlawful to have chases in eight-inch walls or within the required area of any pier. The maximum depth of any permitted chase in any wall shall be one-third of the wall thickness. The maximum length of any horizontal chase without suitable structural support shall be four feet, and the maximum horizontal projection of any diagonal chase shall be four feet. Recesses shall have at least eight inches of material at the back. The maximum aggregate area of recesses and chases in any wall shall be one-quarter of the whole area of the face of the wall in any story, except that for stairs, elevators and dumbwaiters, the walls, including foundation walls behind such facilities, may be reduced to twelve inches. Chases and recesses may be built into hollow walls and walls constructed of hollow blocks, but it shall be unlawful to cut chases or recesses in walls of these types of construction.

**Corbelling.** It shall be unlawful to corbel walls less than twelve inches, except for fire-stopping. Corbelling of hollow masonry shall be supported by at least the equivalent of one full course of the hollow masonry in solid masonry. All corbelling shall be done with solid masonry. The maximum horizontal projection in any corbel shall be one inch for each two inches of vertical projection, and in all cases the total projection of the corbelling shall be one-third or less of the minimum thickness of the wall to be corbelled.

## Solid walls

**Bonding.** In solid brick walls there shall be the equivalent of at least one full header course for each six courses of each wall surface. Where facing brick of a different thickness from the brick used for backing is used, the courses of the facing brick and the backing shall be brought to a level at least once in each six courses in the height of the backing, and the facing brick shall be properly tied to the backing by a full header course of the facing brick or by some other approved method. Facing brick shall be laid at the same time as the backing. In walls more than twelve inches thick, the inner joints of header courses shall be covered with another header course which shall break courses with the courses below.

**Wetting.** All brick having appreciable absorption shall be thoroughly wet before laying.

**Above Roof.** Solid masonry walls above roof levels, twelve feet or less in height, enclosing stairways, elevator shafts, penthouses or bulkheads shall be at least eight inches thick and may be considered as neither increasing the height nor requiring any increase in the thickness of the wall below, provided the allowable working stress requirements are met.

## Hollow walls

**Height.** The maximum height of hollow bearing walls of masonry units or portions of such walls, in any class of structure, shall be forty feet above the support of such walls or portions of walls. Hollow bearing walls may be constructed to the maximum permissible height on top of a solid masonry wall whose maximum height is thirty feet above the first tier of beams. At points where wall thicknesses decrease in hollow walls, a course of solid masonry shall be interposed between the wall section below such point and the wall section next above.

**Bonding.** When hollow walls are built in two or more vertically separated withes such withes shall be bonded together with similar units as are used in construction of the wall, so that the parts of the wall may exert common action under the load, or with approved non-corroding metal ties one to every four square feet. In all hollow walls built only one unit in thickness, each unit shall break joints with those next above. When anchors are used in walls of hollow masonry, such anchors shall be galvanized or shall be of non-corroding metal of adequate size and substantial construction.

**Joints.** Where hollow units are set with cells horizontal, such units shall be set in a full bed of mortar one-half inch or less in thickness, with vertical joints buttered full on shells and webs. Where such units are set with cells vertical, the bearing members shall be buttered and vertical joints slushed full of mortar.

**Supports.** Wherever girders, beams, joists or other structural members frame into hollow masonry, such members shall rest upon such solid incombustible material as will properly distribute the load.

**Partitions.** Partition blocks shall be at least three inches thick for heights up to twelve feet, four inches up to sixteen feet, and six inches up to twenty feet. They shall rest on an incombustible support and shall be wedged or anchored to the ceiling construction.

## Veneered walls

**Anchorage.** For anchorage of brick veneering on masonry, one substantial non-corroding metal tie shall be used for each 300 square inches of wall surface. On frame structures one to every 160 square inches.

**Frame Structures.** Wood frame structures may be veneered with masonry laid up in cement or cement-lime mortar. The veneer shall be directly supported on the foundation. It shall be unlawful to use such veneer above a maximum height of thirty-five feet above the foundation. It shall be unlawful to use such veneer on frame structures having more than two stories and a gable.

# EXERCISES

1. Write a list of 10 materials used in masonwork.
2. State the general method of estimating brickwork.
3. Compute the amount of brickwork in the following solid brick walls:
    a. 119'0" × 11'6" × 1'0"
    b. 63'0" × 8'9" × 1'4"
    c. 82'9" × 10'10" × 0'8"
    d. 121'0" × 11'6" × 1'0"
    e. 68'0" × 14'3" × 1'8"
4. Make a schedule of crosshatching indications suitable for masonwork.
5. Make sketches showing brick stretchers, headers, soldiers, and rowlocks. Mark each by name.
6. Make sketches showing Common, Flemish, and English bonds. Mark each by name.
7. Write at least 300 words of notes on masonwork, taken from one or two of the books listed in the Bibliography. Name the books used.
8. Make a complete estimate of all the mason work shown on the house plans in Chap. 6. The specifications are as follows:

## MASONWORK

Furnish all labor, materials, and appliances, and perform all operations in connection with the masonry work, complete, in strict accordance with the drawings and as specified herein.

The following items of work are included in this division of the specifications:

    Hollow-concrete-block foundations
    Concrete fill in certain blocks
    Brick piers in garage
    Brick on top of block wall at crawl space
    Cinder-concrete-block partitions in basement
    Brick chimney with terra-cotta flue lining
    Firebrick fireplace lining
    Clean-out doors, damper, and chimney cap
    Brick hearth and concrete slab under it
    Brick veneer on front and left side walls
    Brick pedestal for flower box
    Brick walls under front porch
    Cement window sills in basement
    Setting of bearing plates and anchor bolts
    Cutting, patching, and chases
    Cleaning and pointing
    Mortar and masonry scaffolds

The following work is specified in other sections of the specifications:

Concrete footings
Cement floors, steps, and copings
Stonework

The concrete blocks shall be high-pressure, steam-cured, and shall conform to the current edition of the ASTM Standard Specifications for Concrete Units, with gravel or crushed-stone aggregate. Provide special blocks for corners and elsewhere as required. Fill with 1:2:4 concrete all the blocks under the chimney and the top course of all the foundation walls. All exposed joints shall be uniform, smooth flush joints.

All exposed brick, both interior and exterior, shall be textured brick. The contractor shall allow the sum of $80 per M for these brick, delivered to the job. All other brick shall be sound, well-burned, red common brick. All brick shall be shoved into place in a full bed of unfurrowed mortar and all joints shall be completely filled with mortar.

All mortar shall be 1 part portland cement, 1 part hydrated lime, and 6 parts clean, sharp, well-screened sand. Lehigh, Lone Star, or Brixment masonry cement may be used in lieu of this mixture.

The brick piers shall be well bonded within themselves and to the adjoining block walls.

The 4″ block partitions in the basement shall be composed of approved cinder-concrete blocks neatly laid up with smooth flush joints.

Terra-cotta flue lining shall be sound and well burned, properly installed, and with a thimble unit for the boiler smoke pipe. The fireplace and the back hearth shall be lined with approved firebrick laid flat. The flue lining and firebrick shall be laid in approved fire-clay mortar.

Provide an approved cast-iron fireplace damper and cast-iron, hinged clean-out doors with frames.

The front hearth shall extend the full width as shown on the drawings, on a slab of 1:2:4 concrete over solidly tamped fill. The front hearth and the fireplace chimney breast shall be face brick laid up in a simple pattern as will be directed by the architect.

The brick veneer and the flower-box pedestal and front-porch walls shall be neatly laid up in Flemish bond with uniform "weathered" joints. Provide substantial galvanized iron ties as required. Construct rowlock sill course sloped to shed water.

Construct neat, smooth-finished cement sills under the basement windows and a sloping cement cap on the chimney.

Install and build in all anchor bolts, bearing plates, door frames, etc., in masonry. These items will be furnished by others. Build in all girder and beam ends with solid masonry.

Cut and patch masonry and form chases, etc., as required by the work of all trades.

Clean all exposed masonry, both interior and exterior, with muriatic acid and water and repoint joints where required.

CHAPTER 15

# rough carpentry

Most estimators prefer to have at least two headings for carpentry work—one for rough carpentry and one for finish carpentry. On large jobs, special headings are set up, besides, for millwork, wood stairs, wood flooring, kalamein work, and other divisions of carpentry, in order to provide for separate total prices for these items or to make comparison with subcontract estimates that may be received for them. Even on small jobs, it is good practice to split up the work in this way.

Rough carpentry takes in all the framing and plain boarding. Finish carpentry includes the trim, mouldings, baseboard, windows, doors, cabinets, shelves, etc. Wood flooring and wood stairs are included in finish carpentry unless special headings are set up for them. Finish hardware also is placed under Finish Carpentry, but on larger buildings this is generally given a separate heading. Frequently, a lump-sum allowance of money is specified for the purchase of the hardware. Of course, any item that is provided for elsewhere in the estimate does not require handling under any of the carpentry headings.

## Lumber

Lumber is a general term. Pieces of large cross section are called *timbers*. Small stuff is referred to as boards, studs, planks, flooring, mouldings, etc. Lumber comes in many grades, which are established mainly by trade associations. Spruce is used for beams, studs, rafters, sheathing, etc. It is nearly white in color and withstands exposure to weather. Shortleaf pine, often called N.C. *pine*, is used for the same purposes as spruce. It is yellowish white with noticeable annual rings and is strong and durable. Fir, a wood somewhat like pine, is used for the same purposes as pine. Other kinds of wood used in building work are longleaf pine, white pine, maple, oak, and cypress. Oak, maple, and pine flooring are the most commonly used of the wood floorings.

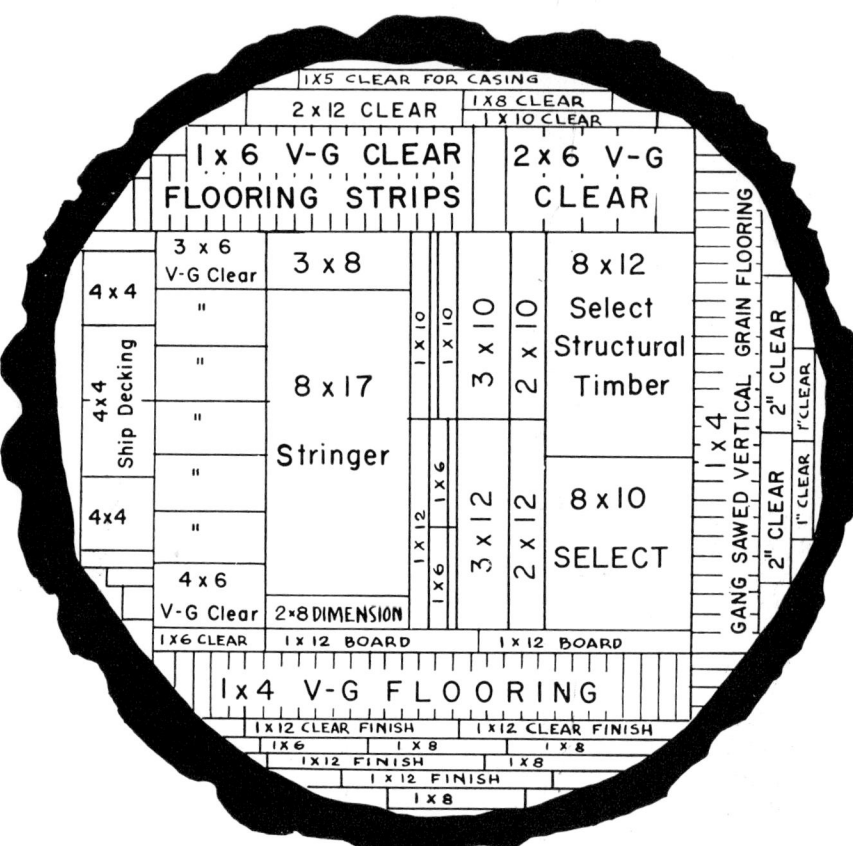

*Fig. 15-1* The lumber content of a Douglas fir log. This log might be cut in a number of ways. The purpose of the illustration is to show the portion of a log from which various items are cut. This tree reached a diameter of 14 in. at an age of thirty years, at an age of one hundred years it was 27 in. in diameter, and at time of cutting was about three hundred years old and 42 in. in diameter. Its rate of growth decreased from an inch of wood in six years at an age of thirty years, to an inch of wood in twenty-five to thirty years at time of cutting.

Standard stock sizes and lengths are well established in carpentry. Beams and rafters for general building construction are $2'' \times 8''$, $2'' \times 10''$, $3'' \times 8''$, or $3'' \times 10''$ in cross section, and regular stock ranges from 10 to 20 ft in lengths of 10, 12, 14, 16, 18, and 20 ft. Longer lengths are not readily obtainable in all lumberyards. Studs measure $2'' \times 3''$ or $2'' \times 4''$ and are stocked in the same lengths and also in 9- and 13-ft lengths.

Boards are generally of stock 1" in thickness and from 2" to 10" in width. If they are surfaced (dressed) on one side, they finish about ⅞" thick; and when they are surfaced on two sides, they become about $1\frac{3}{16}''$ thick. Boards less than 1" thick are counted as 1" thick.

Tongued-and-grooved (T&G) sheathing boards are termed $1'' \times 4''$, $1'' \times 6''$, and $1'' \times 8''$, but they actually measure less than these dimensions and, when laid, they cover about ¾" less width of space, per board. This is because of the milling to form the tongues and grooves and dressing to form straight and true surfaces. The narrower the board, the higher will be this percentage of loss. With T&G sheathing, underflooring, and finish flooring, and with shiplap boards, shingles, etc., the milling, overlaps, and other elements of waste must be taken into account, as they run up to a large percentage in some cases.

The Federal Government and several of the large national lumber associations have developed official grading rules for lumber. Thus one speaks of select structural material in certain types, and No. 1 Common, 1200# Douglas fir, B-Grade lumber, etc. If it is so ordered, the material is officially grade-marked.

The ordinary yard lumber obtainable at all lumberyards is graded as No. 1 Common and No. 2 Common shortleaf pine or fir and is used for ordinary framing purposes. Structural material, such as is graded structural Douglas fir or structural Southern pine, is used where a stronger material is desired, as for girders and beams. Shop lumber, used for doors, windows, trim, etc., is graded A, B, C, or D.

**Framing.** Framing lumber is 2", 3", and 4" thick for ordinary studs, beams, etc., although, as a result of the sawing, the actual thickness is about ½" less. Likewise, 1" boards measure nearer to ¾" thick. The depth of beams, etc., also is short of the figured dimensions, because of the sawing in the mill, and a 10" beam usually measures only about 9½". However, for figuring the amount of board feet, the full dimensions of the lumber stock are always taken.

Southern yellow pine and Douglas fir (Fig. 15-1) are the two most commonly used kinds of lumber for general building purposes. White pine, whitewood (poplar), and birch are in considerable demand for trim and other exposed work, white pine being preferred for outside items. Oak, pine, and maple, as has been noted before, are the most popular woods for flooring.

The highest grade of yellow pine or Douglas fir is called *select structural*, the next is *dense heart*, and the next lower grade is *structural square edge and sound*. Unfortunately, the lumber associations have made a poor job of names and they are very misleading and tricky. The estimator must beware and note the full wording of specifications, quotations, etc., referring to the grades of the material.

The lowest grades of framing lumber, even if specified, are seldom worth using, because of the great waste in them. This waste involves handling in order to unload and stack the material that will later be discarded, the culling out that must be done by the carpenters, the rehandling, and the final disposal. Such poor material is often the cause of friction between the contractor and the architect, as the architect desires to see good materials used on the job. This causes extra expense for taking out the poor work and replacing it with new material of better quality.

Lumber that is 5″ and larger in its least dimension is classed as timber; 2″ and under 5″, as dimension lumber. It comes to the job usually planed either on one side and one edge or on all sides. Terms such as D1S2E, S4S, meaning that the material is dressed, or surfaced, as noted, are used to designate the faces that are finished faces. Because of the saving in shipping space, very little additional is charged for dressed lumber over rough lumber. For this reason, and also because it is easier for the men to handle smooth material, lumber dressed on all four sides is preferred.

Floor beams 2″ × 10″ × 14′0″, wall and partition studs 2″ × 4″ × 8′0″, boards 1″ × 6″ × 16′0″, and other such expressions are used regarding the size of lumber. The dimensions in inches of cross section by the length in feet are given. Thus 2″ × 10″ is the cross section of the floor beams mentioned above, and 14′0″ is the length. This is the standard way of expressing the size of lumber.

By writing the standard dimensions of lumber over 12, in the form of a fraction, the number of board feet of lumber in the piece is readily found. Thus the board feet in a floor beam of the size noted would equal

$$\frac{2 \times 10 \times 14}{12} = \frac{280}{12} = 23\tfrac{1}{3} \text{ BF}$$

In cases of more than one piece, simply multiply the amount found in one piece by the number of pieces concerned.

The rule is: place the standard dimensions over 12, and multiply the answer by the number of pieces concerned. If there are 50 beams $2'' \times 8'' \times 16'0''$, the expression would be as follows:

$$50 \times \frac{2 \times 8}{12} \times 16 = 1{,}067$$

For every linear foot of $2'' \times 8''$ there are $1\tfrac{1}{3}$ BF, for every linear foot of $2 \times 6$ there is 1 BF, for every linear foot of $2 \times 4$ there is $\tfrac{2}{3}$ BF. This can readily be seen, because

$$\frac{2 \times 8}{12} = 1\tfrac{1}{3} \qquad \frac{2 \times 6}{12} = 1 \qquad \frac{2 \times 4}{12} = \tfrac{2}{3}$$

The same holds true for all lumber.

**Carpentry.** Rough carpentry is the name given to framing, or structural, carpentry. However, there is no definite line of demarcation between rough and finish carpentry. An estimator arbitrarily puts the carpentry items under either heading or under other headings for special portions of the carpentry work, to suit his own convenience, although, if a specification is well arranged, he is likely to follow it in arranging his estimate.

Architects frequently include many miscellaneous items under the heading of carpentry in their specifications, for want of a better place for them or for the sake of avoiding the necessity of setting up special headings. Metal doors, insulation, and fixtures of various kinds are often found in the carpentry specifications. This is one division of the specifications that estimators practically always expect to break up into several estimating divisions—some of the items even finding their place in the general job-expense division.

## House framing

A common method of framing a house is called *balloon* framing (Fig. 15-2). In this type, the wall studs extend all the way from the foundation sill to the plate that supports the rafters and the attic floor beams (country carpenters use the term *joists* for beams and rafters, and sometimes also for studs). In the braced-frame type (Fig.

15-3), the wall studs are only one story in height and the second-floor beams (or joists) are carried on a regular plate, formed usually of two 2 × 4's, instead of being carried on a ledger board cut into the studs, as is the case in balloon framing. There are other differences between

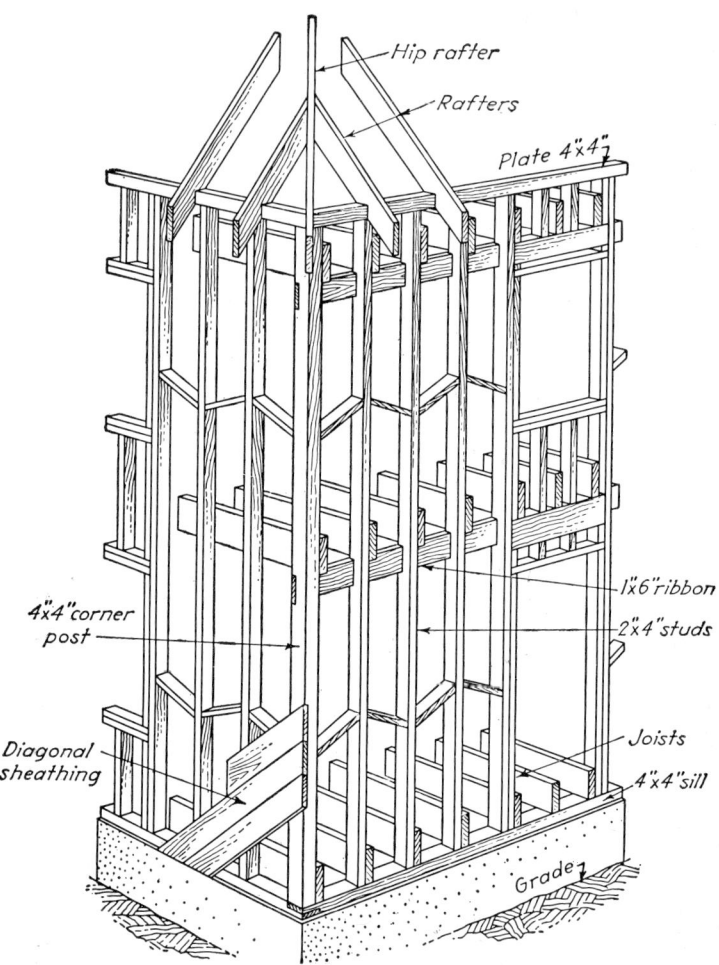

**Fig. 15-2** *Balloon frame.*

these two types of framing and, of course, a combination of methods can be used too. There is usually no difference of opinion about the balloon type's being satisfactory, provided that it is well put together and that the sheathing is applied diagonally, so as to brace the structure better than would be the case if the sheathing were applied horizontally.

Another type of framing is called *mill construction* (Fig. 15-4). As the name implies, this type is used principally for mills and similar

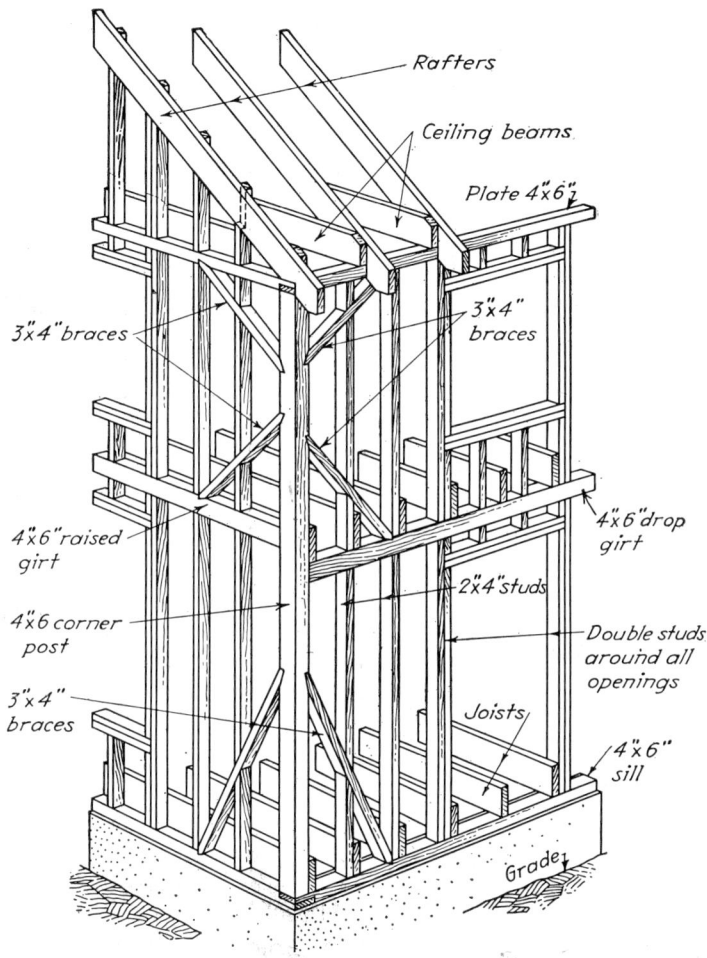

**Fig. 15-3** *Braced frame.*

structures. Genuine mill construction consists entirely of heavy timbers of large cross section, not only for the framing but for every member used in the building. The flooring is 3″ or 4″ thick T&G, or the same thickness of square-edged material laid on edge.

The construction of the framework for all houses is practically the same, regardless of whether the exterior is wood, stucco, or brick veneer. Upon good lumber properly used, together with accurate workmanship, depends the ultimate success of the completed building.

In localities subject to termites and at points of construction exposed to fungus growth, chemical treatment may be necessary for the

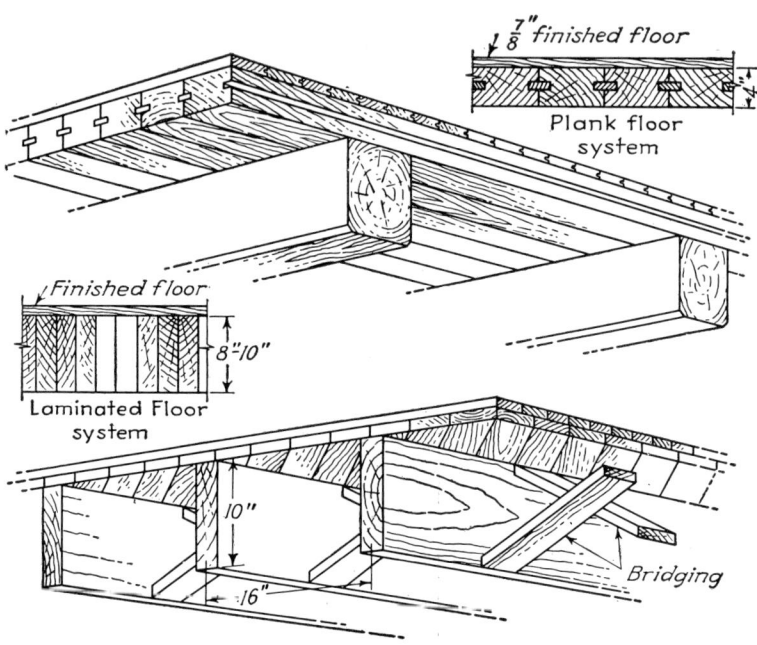

*Fig. 15-4* Mill construction.

lumber, particularly any that comes into contact with masonry foundations.

Wood girders may be built-up or solid. While either type may be used, built-up girders are preferable, because of greater strength, size for size; less shrinkage, owing to the complete seasoning possible with thinner stock; and greater ease of installation. Members of built-up girders should be spiked together securely.

Joists are normally spaced 16″ O.C. (on centers), except where additional strength is required for some particular area, in which case the joists (beams) are doubled or extra joists are set between the regular ones so as not to interfere with lathing. Spacing of 12″ will also conform to lathing requirements. Floor and roof beams must always be doubled or tripled around openings, such as those for stairways, shafts, hearths, etc. This does not always show on the plans, nor is it mentioned in all specifications, but it is required by good construction and by building codes. Similarly, floor beams are doubled under partitions, or larger beams are used there, to carry the extra load imposed by the partitions, unless, of course, the partitions are low or of very light construction, like wood-and-glass office partitions.

In exterior walls of one- and two-story houses, the studs are customarily 2″ × 4″. In buildings over two stories high and in cases of exceptionally high ceilings, studs should be 2″ × 6″ or larger.

Rafters are normally spaced 16″ O.C., since this permits setting them directly over the wall studs and nailing each one to the side of the ceiling beams. Thus the frame is tied together and spreading of the walls is prevented. This spacing also facilitates lathing the underside of the rafters, if a plaster ceiling is required under the roof.

All partitions should be framed with 2″ × 4″ studs, spaced normally 16″ O.C., except where thicker walls are required on large work for structural purposes, or where plumbing and heating pipes and vent ducts are to be concealed, or where architectural effect must be considered.

All anchors for sills, plates, etc., built into the masonry before the framing is erected, should be included in the concrete or masonry divisions of the estimate. All the plates, ties, stirrups, etc., that are attached to the wood framing, however, should be listed under Rough Carpentry.

Often there is a call for studs to be doubled or tripled at door and window openings and at corners and intersections. If this is the case and there are many openings, this will involve a large increase in the number of studs required. The estimator who is careful will not merely allow one linear foot of stud for each square foot of wall, as some quick estimators do, but will actually mark off the studs required on the drawings, then count them and figure separately the sills, bracing, and plates required. Of course, for a single building of small size the method is not very important, as the error would not amount to very much.

Corner braces cut in between the corner posts and the studs should be used in all buildings, as they add about 50 per cent to the strength and stiffness of the wall construction.

Diagonally applied sheathing on walls adds greatly to the strength of the wall, over the use of horizontally applied sheathing, especially if at every bearing it is well nailed with three nails instead of the more customary two nails.

Figure 15-5 shows the beginning of a rough-carpentry estimate for a house. For frame construction the number and sizes of all the pieces

|  |  |  |  |  |  |  |  |  |  |
|---|---|---|---|---|---|---|---|---|---|
|  | 402 Park Ave |  |  | April 1, '59 |  |  |  |  |  |
|  | Rough Carpentry |  |  |  |  |  |  |  |  |
|  | Girders 3 - 3" x 10" |  |  |  |  |  |  |  |  |
|  | 3/16  3/10 | 78' | 195 BF | .27 | 53 |  |  |  |  |
|  | Sill 3 x 8 |  |  |  |  |  |  |  |  |
|  | 34+26+28+24+10 | 128 | 256 | .25 | 64 |  |  |  |  |
|  | Corner posts 3 - 2 x 4 |  |  |  |  |  |  |  |  |
|  | 12/18  6/9 | 270 | 180 | .23 | 41 |  |  |  |  |
|  | Wall studs 2 x 4 |  |  |  |  |  |  |  |  |
|  | 81/18  12/9 | 1566 | 1044 | .26 | 271 |  |  |  |  |
|  | Plate 2 - 2 x 4 |  |  |  |  |  |  |  |  |
|  | 2 x 128 | 256 | 170 | .26 | 44 |  |  |  |  |
|  | Short studs & bracing 2 x 4 |  |  |  |  |  |  |  |  |
|  | 22/5  15/10  10/12 | 380 | 254 | .32 | 81 |  |  |  |  |
|  | Floor beams 3 x 8 |  |  |  |  |  |  |  |  |
|  | 6/18  22/16  11/14 | 614 | 1228 | .23 | 282 |  |  |  |  |
|  | Flr. beams 2nd flr. 2 x 10 |  |  |  |  |  |  |  |  |
|  | 6/18  11/16  9/14  4/10 | 572 | 954 | .22 | 210 |  |  |  |  |
|  | Bridging 1 x 3 |  |  |  |  |  |  |  |  |
|  | 26 +18 + 18 | 62 | 16 | .42 | 7 |  |  |  |  |
|  | Partn. sills, cats, plates 2 x 4 |  |  |  |  |  |  |  |  |
|  | 40 + 30 + 70 | 140 | 94 | .35 | 33 |  |  |  |  |
|  | Partn. studs 2 x 4 |  |  |  |  |  |  |  |  |
|  | 28/9  40/8 | 872 | 582 | .30 | 175 |  |  |  |  |
|  |  |  |  | Fwd. | 1261 |  |  |  |  |

*Fig. 15-5  Rough carpentry estimate.*

are listed wherever possible. Thus, the sill that is placed on top of the foundation wall is listed as so many linear feet of the size of cross section that is called for, as 126′ of 4″ × 6″. The corner posts, which occur at every angle of the building, are listed separately from the ordi-

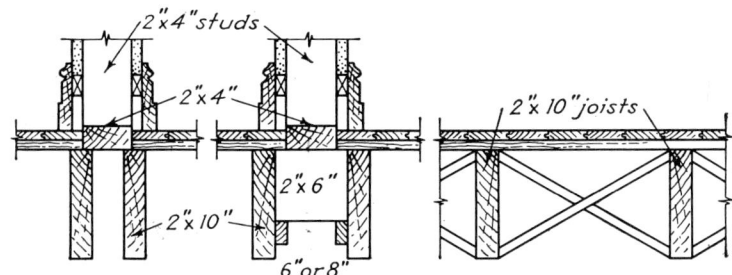

*Fig. 15-6* Floor bridging.

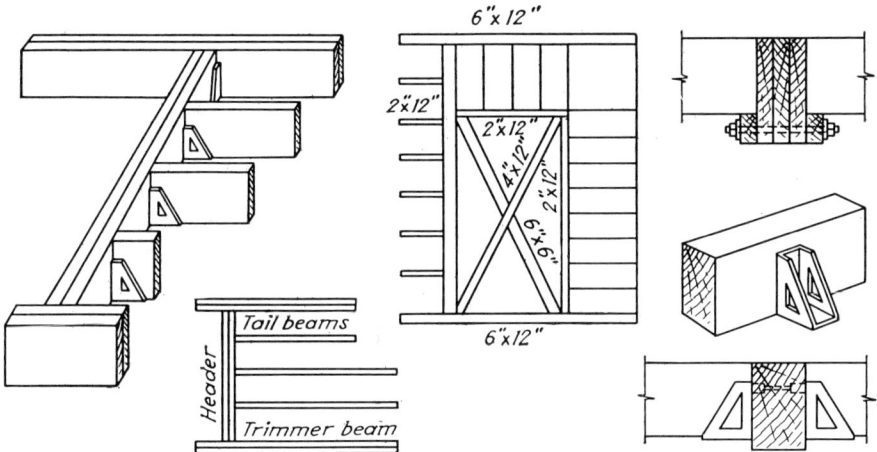

*Fig. 15-7* Framing at openings.

nary wall studs. Corner posts are sometimes specified to be solid pieces, as 4 × 6 or 6 × 6, or made up of several pieces spiked together. The studs are usually 2 × 4 or 2 × 6. Corner bracing, plates, short studs at windows, etc., must all be listed. Floor beams are counted and listed separately as to cross section and length, care being taken to note those that are doubled under partitions and around stair openings, etc. Sometimes tripled beams are called for.

Studs, both in walls and in partitions, are commonly spaced 16″

O.C.; that is, 16" from the center of one to the center of the next one. However, owing to the spacing of the doors and windows, exact spacing is not always possible and additional studs must therefore be counted, to give bearing for the lath, etc. Soles, bridging, and plates for partitions should be listed separately from the studs, and care be

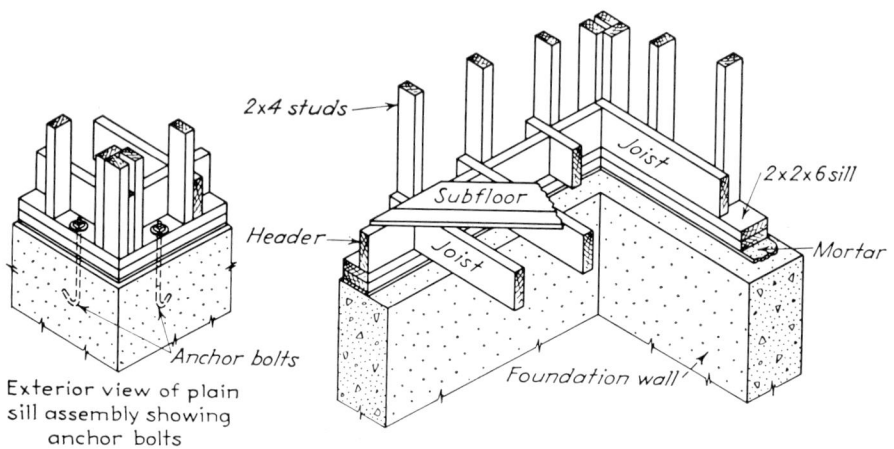

Fig. 15-8  Corner conditions.

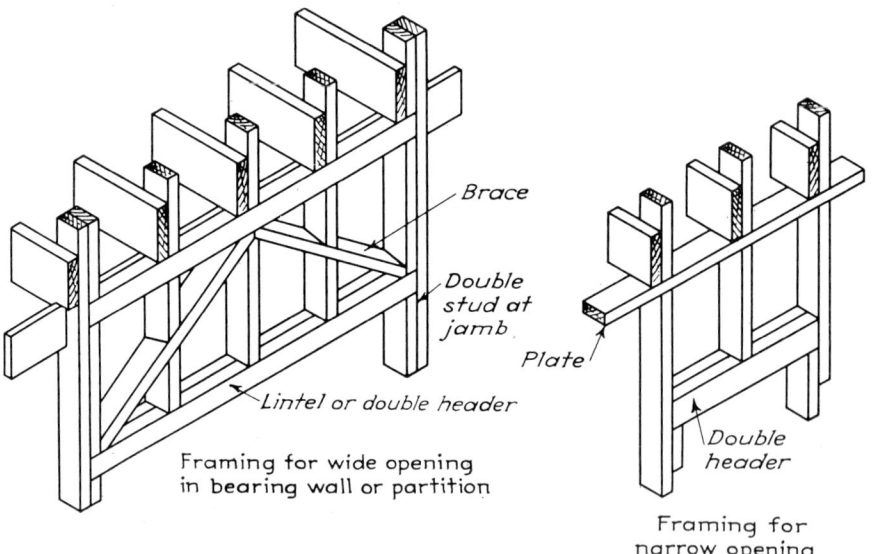

Fig. 15-9  Framing over openings.

taken to include them as single or double, in accordance with the particular job being estimated (Figs. 15-6 to 15-11).

Wood framing is usually measured by the board foot and priced per thousand board feet (MBM).

Bridging is measured by simply taking twice the length of each row, because floor bridging is always doubled, or crossed. This gives the total linear feet of material required, if it is for beams not more than 10″ deep. For beams 12″ deep, multiply by 2¼ instead of by 2, and for 14″ beams multiply by 2⅔. Bridging is often priced by the linear

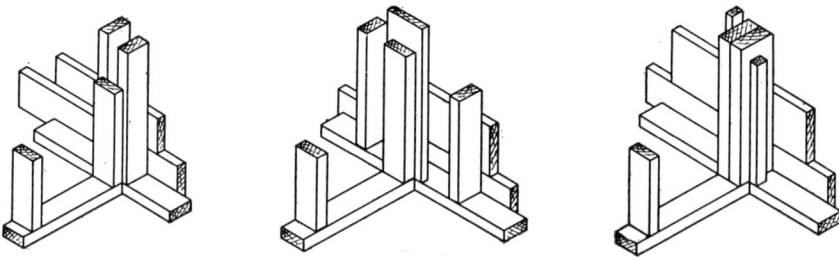

*Fig. 15-10* Partition corners.

foot, but if board feet are desired the calculation can be readily made by multiplying the total length thus found by the cross-section dimensions of the material specified and dividing by 12. Thus, 870 lin ft × 1″ × 3″, divided by 12 = 218 BF.

Building paper is measured by taking the net area to be covered and adding 5 per cent to allow for the laps.

If scaffolding is required for the outside work on walls, even on frame walls, take the gross area of the walls as the quantity, measured from the ground to the top of the walls. This is priced per square foot.

## Costs

Framing lumber costs the contractor between $110 and $140 per MBM for regular kinds, delivered to the job. The labor of handling, framing, and installing it costs $90 to $150 per MBM, including the cost of insurances, taxes, and union funds based on the labor. Heavy and intricate framing costs much more in both material and labor. Table 15-1 gives the time required, in hours per MBM, for ordinary framing work.

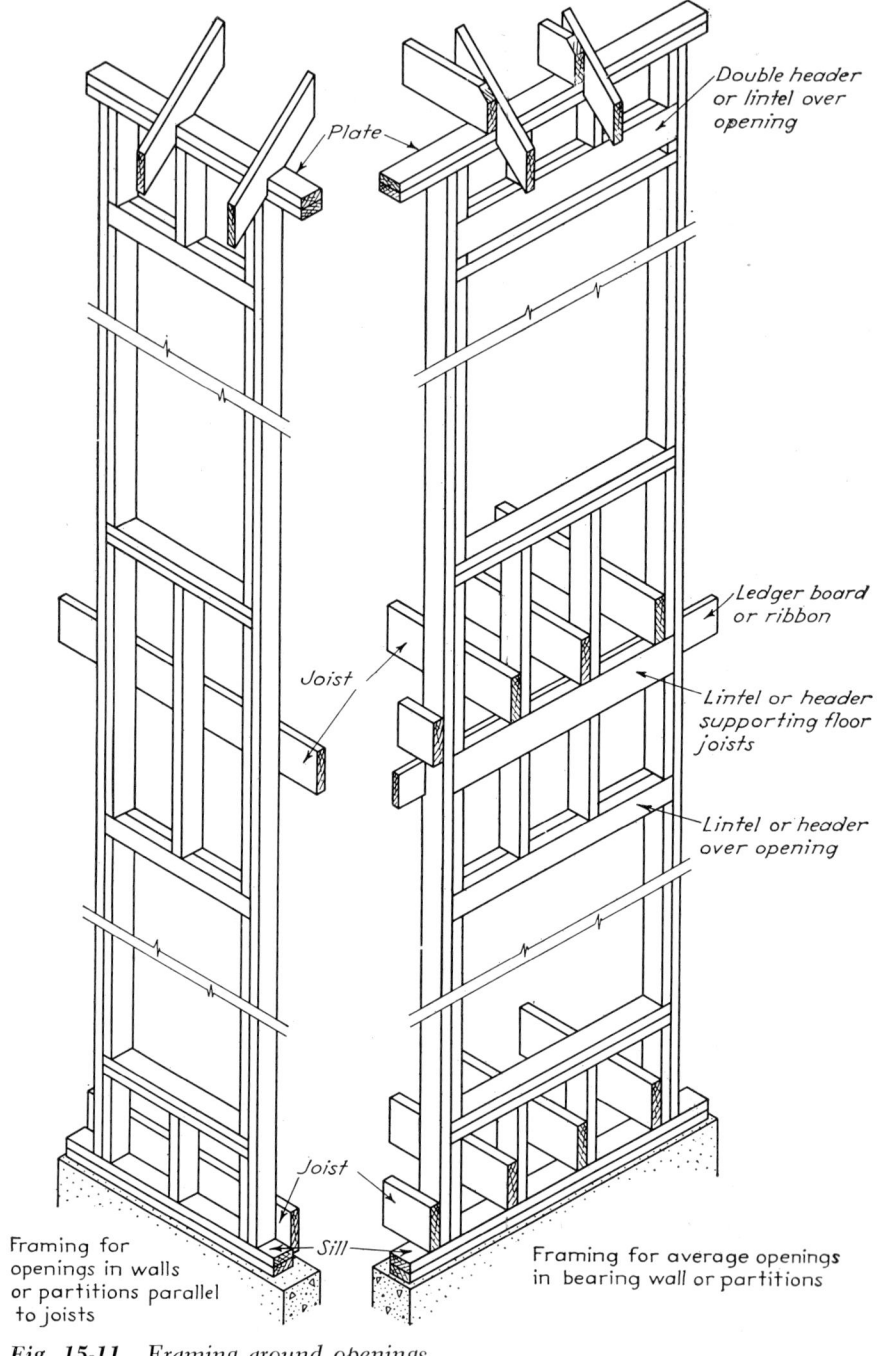

Fig. 15-11  Framing around openings.

## ROUGH CARPENTRY

TABLE 15-1

| | | | |
|---|---|---|---|
| Girders | 24 | Ribbon | 55 |
| Posts | 24 | Rafter plates | 30 |
| Wall plates | 28 | Rafters | 32 |
| Box sills | 25 | Wall studs | 30 |
| Beams | 20 | Partition studs | 40 |
| Bridging | 70 | Miscellaneous framing | 60 |

Wall and roof sheathing and rough flooring cost the contractor between $90 and $115 per MBM delivered to the job. Labor, insurances, taxes, and union funds cost $45 to $85 per MBM. Table 15-2 gives the approximate time required, in hours per MBM.

TABLE 15-2

| | | Walls | Roof | Floors |
|---|---|---|---|---|
| 1 × 6 SE | Straight | 13 | 12 | 10 |
| | Diagonal | 16 | | 13 |
| 1 × 6 T&G | Straight | 15 | 14 | 12 |
| | Diagonal | 18 | | 17 |
| 1 × 8 SE | Straight | 12 | 11 | 9 |
| | Diagonal | 14 | | 11 |
| 1 × 8 T&G | Straight | 14 | 13 | 11 |
| | Diagonal | 16 | | 13 |
| 1 × 4 SE | Straight | | | 11 |
| | Diagonal | | | 15 |

Table 15-3 gives the approximate percentage of waste allowance to be added to the net area before arriving at the number of board feet required.

TABLE 15-3

| | | Walls | Roof | Floors |
|---|---|---|---|---|
| 1 × 6 SE | Straight | 22 | 20 | 21 |
| | Diagonal | 25 | | 24 |
| 1 × 6 T&G | Straight | 27 | 25 | 26 |
| | Diagonal | 20 | | 29 |
| 1 × 8 SE | Straight | 18 | 16 | 17 |
| | Diagonal | 21 | | 20 |
| 1 × 8 T&G | Straight | 23 | 21 | 22 |
| | Diagonal | 26 | | 23 |
| 1 × 4 SE | Straight | | | 25 |
| | Diagonal | | | 28 |

To the cost of rough carpentry must be added about ⅓ keg per MBM for the nails required, if the unit cost does not include the cost of nails (Fig. 15-12).

The element of waste is more important in carpentry than in most other divisions of the estimate, especially on account of the grading and trade customs that are established among the manufacturers and

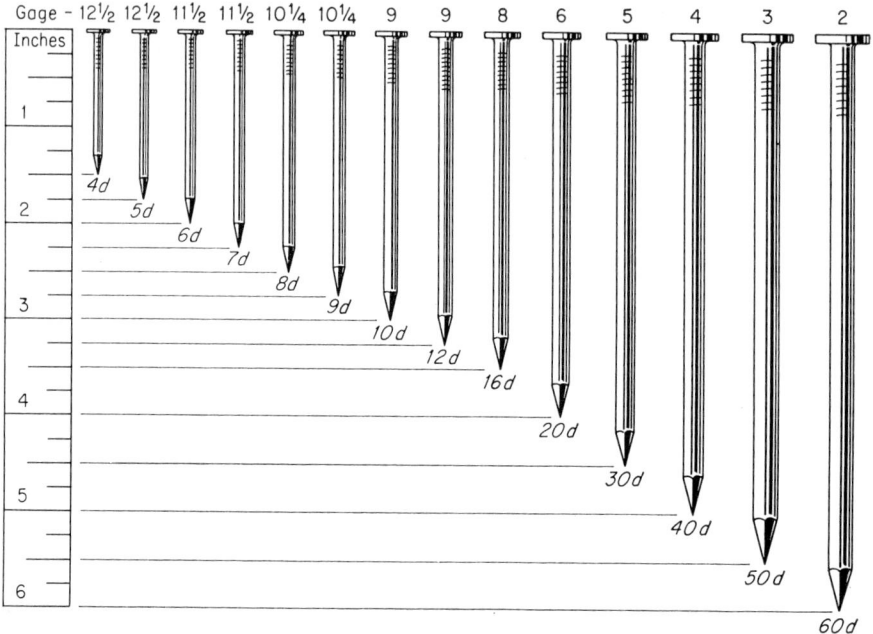

*Fig. 15-12  Common wire nails.*

dealers in the materials involved. Even lumber that is bought as good quality, free from large knots, not warped, straight, etc., may become damaged as a result of careless handling in transportation or at the job. Beams and other structural members that are intended to be straight get twisted if they are allowed to lie in a pile dumped from a truck. An estimator learns from experience that these things happen on the job, especially on a job that is not properly managed, and he adds to his estimating units because of them.

**Temporary Construction.** Temporary enclosures and protection, temporary arch centers, and other temporary construction items are sometimes specified under Carpentry. Sometimes they are not specified at all; but they will be found necessary nevertheless. The estimator must have these in mind on every job, and for this reason some offices use a printed general job-expense estimating sheet on which these and other usual expense items are printed, so that they will not

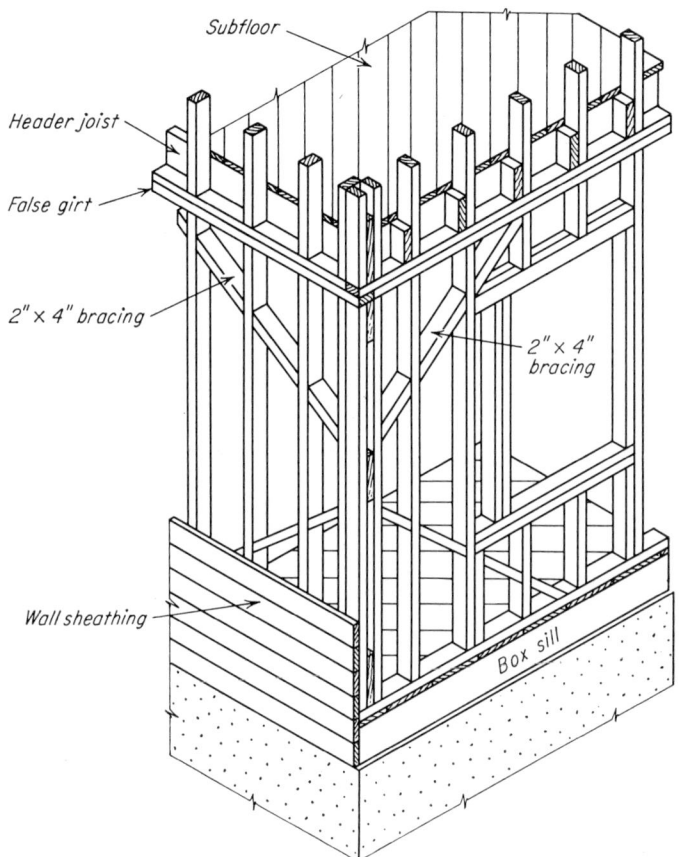

*Fig. 15-13   Corner bracing.*

be overlooked. In any event, the estimator should figure these items just as he would any items of permanent carpentry, and apply unit prices per board foot, per square foot, or in any other suitable manner.

**Special Framing.** Special framing or work out of the ordinary is to be watched carefully and analyzed closely in estimating. Roof trusses, tower framing, and odd-shaped structures belong in this class. New methods of timber framing developed in recent years call for special fastenings that require special tools. Every piece of lumber in special or heavy framing must be listed; and all the iron bolts, shear plates, connectors, etc., must be counted and listed separately. Special scaffolds, hoisting, and bracing are usually involved also (Figs. 5-13 to 5-21).

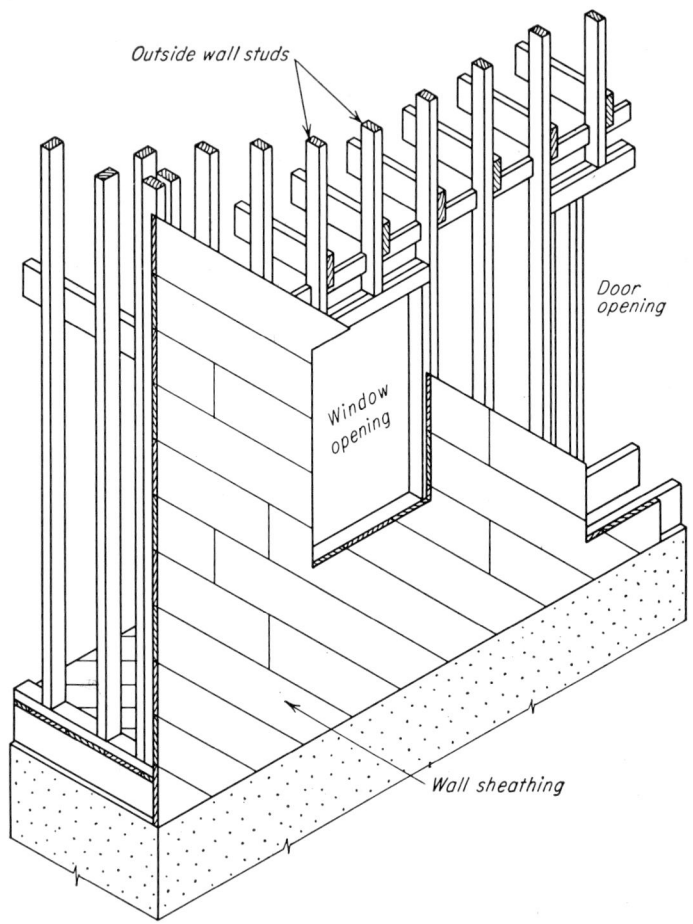

*Fig. 15-14* Diagonal sheathing.

Split-ring timber connectors are made of hot-rolled carbon steel and must fit snugly in prebored holes. Toothed-ring connectors are stamped from No. 16 gauge rolled sheet steel to form a circular, corrugated, sharp-toothed band, welded into a solid ring. Claw-plate connectors are malleable iron castings, each plate consisting of a perforated, circular, flanged member, with three-sided teeth on one side. Shear plates are of pressed steel or malleable iron, with a flange around the edge on one face. Bolts and washers of many sizes also are required in heavy framing (Figs. 15-22 to 15-24).

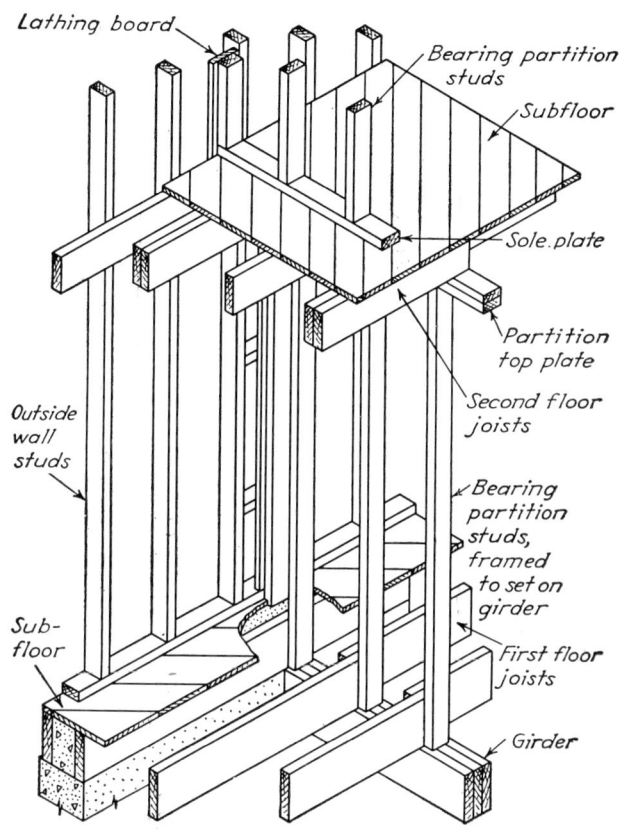

*Fig. 15-15   Bearing partition*

## Insulation

Insulation is fast becoming an essential part of building work. If a separate heading is not provided for it in the estimate, it is usually placed under Carpentry.

The structural parts of a house, designed chiefly for strength, weatherproofness, and appearance, have some insulating value, which may be ample for ordinary purposes in moderate climates. However,

224  BUILDING CONSTRUCTION ESTIMATING

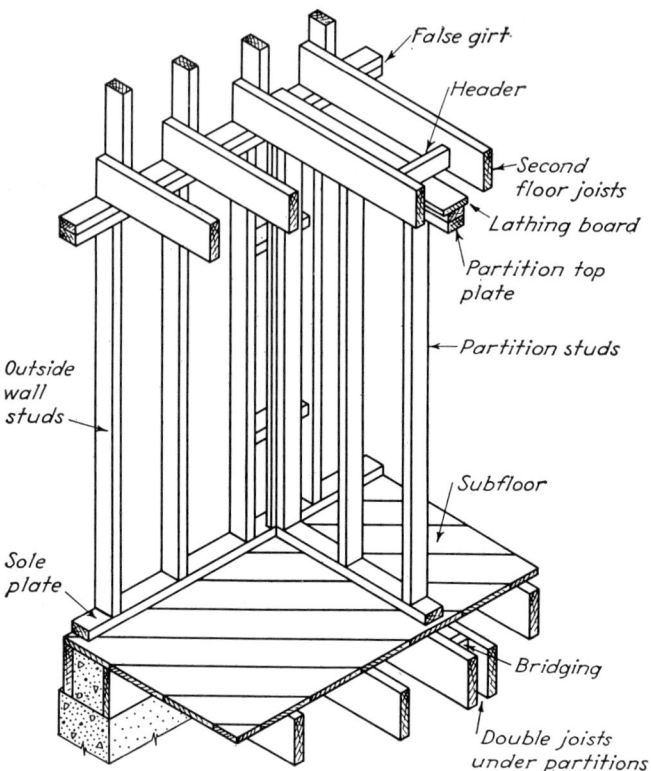

Fig. 15-16  Non bearing partition.

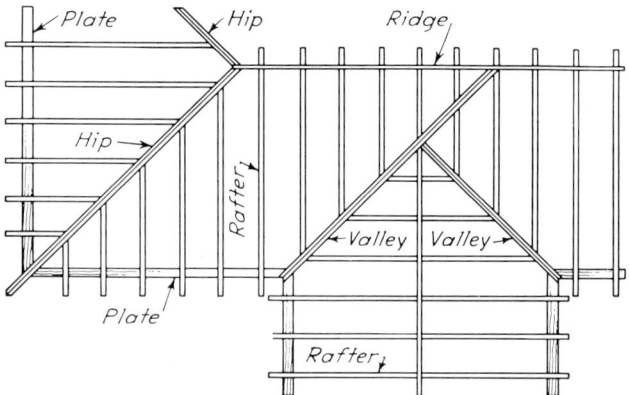

Fig. 15-17  Roof framing.

in cold climates and where heating seasons are long, it is desirable to incorporate or install in the structure of walls and other exposed parts of houses materials especially made to provide thermal insulation.

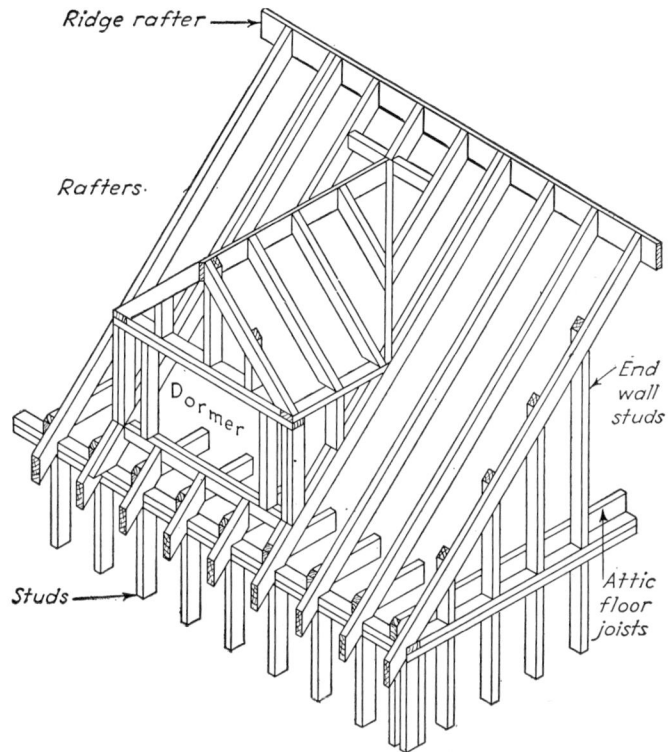

*Fig. 15-18* Dormer framing.

While no material will entirely prevent the passage of heat through a wall or roof, there are insulating materials which are efficient in reducing the flow. Some are made of soft, flexible fibrous materials such as mineral wool and vegetable or animal fibre, or blankets or batts which are designed to be stretched and tacked between the studs, floor beams, and rafters. Others are available in loose or shredded form to be blown or packed into hollow spaces, while still others are made in stiff board form having some structural strength to be used as plaster base, sheathing, or merely as insulation. The batt and blanket forms are usually furnished with a vapor-barrier facing made of asphalted paper and should always be installed with the

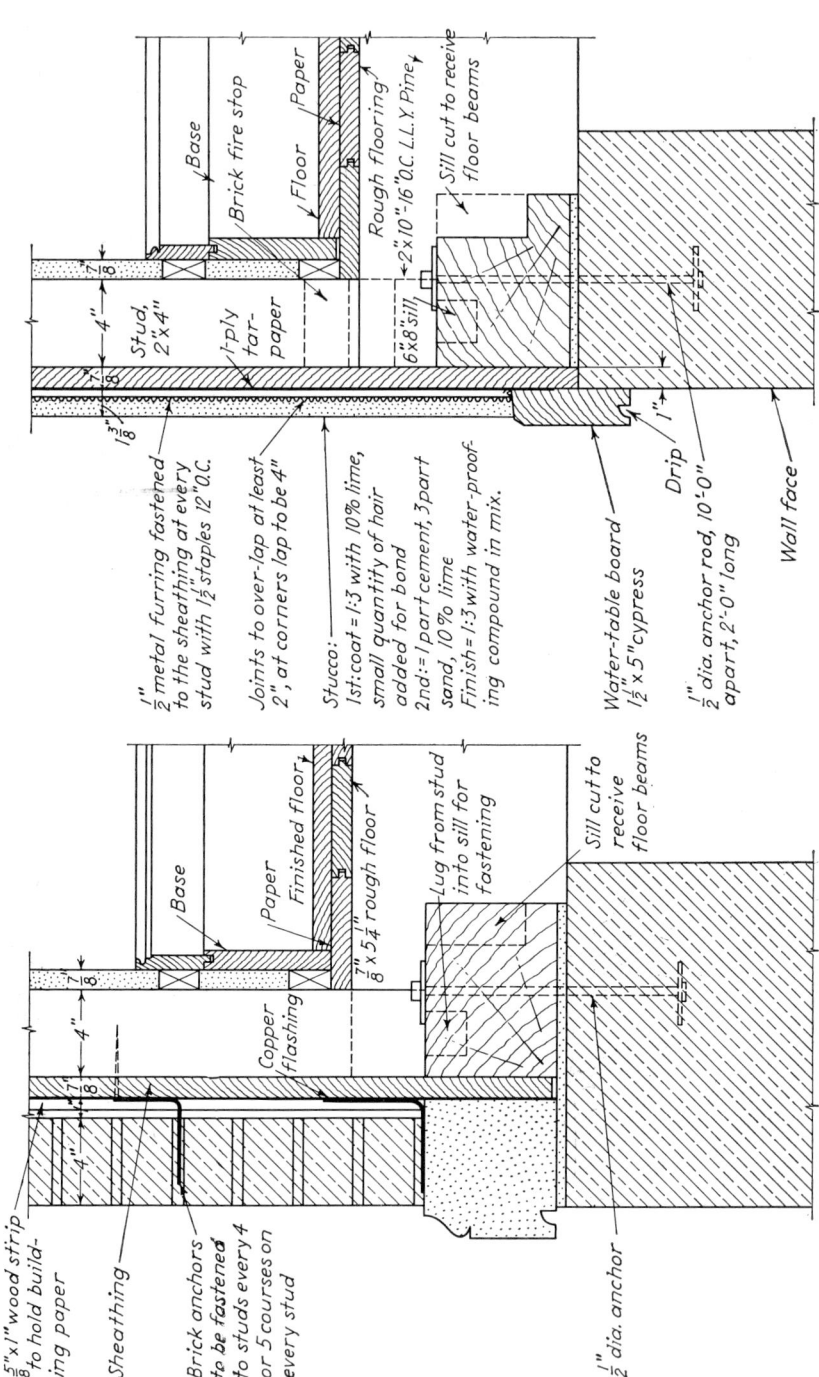

Fig. 15.19  Water tables, masonry.

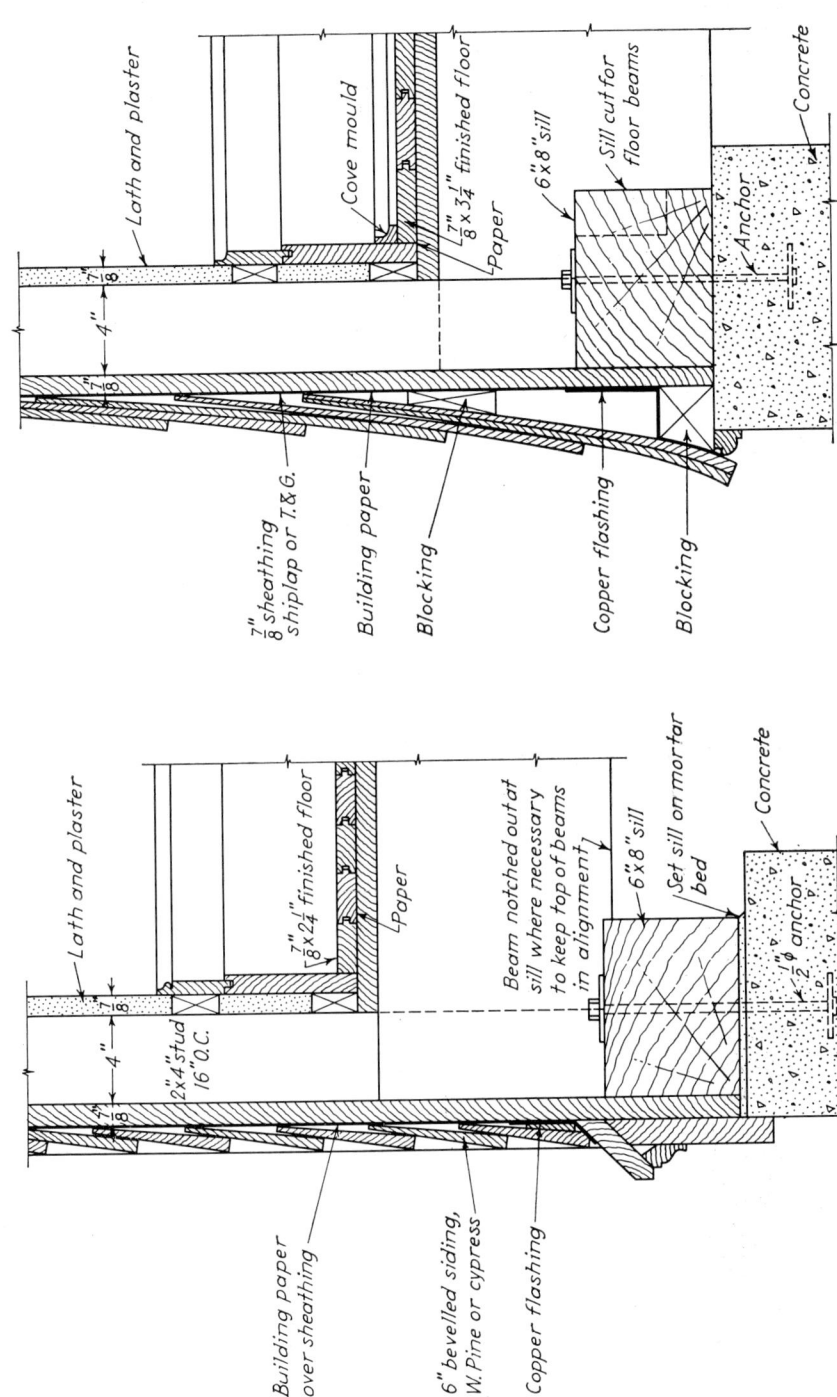

Fig. 15-20 Water tables, wood buildings.

vapor barrier on the warm side of the insulation. It is preferable to place reflective surfaces which are also vapor barriers, such as metal foils, on the warm side of the air space. To be effective, the barriers should form a continuous, unbroken membrane, which in most cases is behind the lath and plaster or other finish material.

The various kinds of fibrous insulation are substantially alike in insulating value, which varies directly with the thickness of insulation installed. Batt, blanket, and loose-fill fibrous insulations are ap-

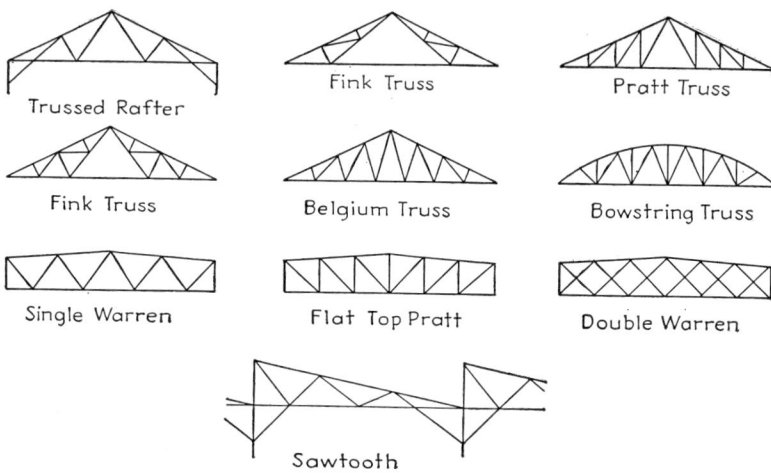

Fig. 15-21  *Types of trusses.*

proximately equal in insulating value when used in the same thickness. Insulating fibreboard has less insulating value per inch of thickness than fibrous insulations but is preferred in cases where structural properties are needed.

Brick walls which have been furred on the inside may be satisfactory without further insulation unless the climate is severe. If more insulation is required for fuel economy or comfort, blanket or other type of insulation may be placed between the furring strips. Lath, composed of either gypsum board or fibreboard with or without aluminum foil on the back, may be used as a plaster base. Various types of wallboard, with or without metal foil on the back, may be used instead of lath and plaster.

The roof, subject to strong winds in winter and to the direct rays of the sun in summer, is the most exposed part of the house. If an attic is insulated, less heat is lost through the roof during the winter months. If the attic is unoccupied and it is unnecessary to keep the temperature in that space at a comfortable level, insulation in the form of batts, blankets, or loose-fill material may be laid between the floor beams of the attic on top of the ceiling below. Attics should be provided with screened louvers or vents to permit good cross ventilation both in winter and in summer.

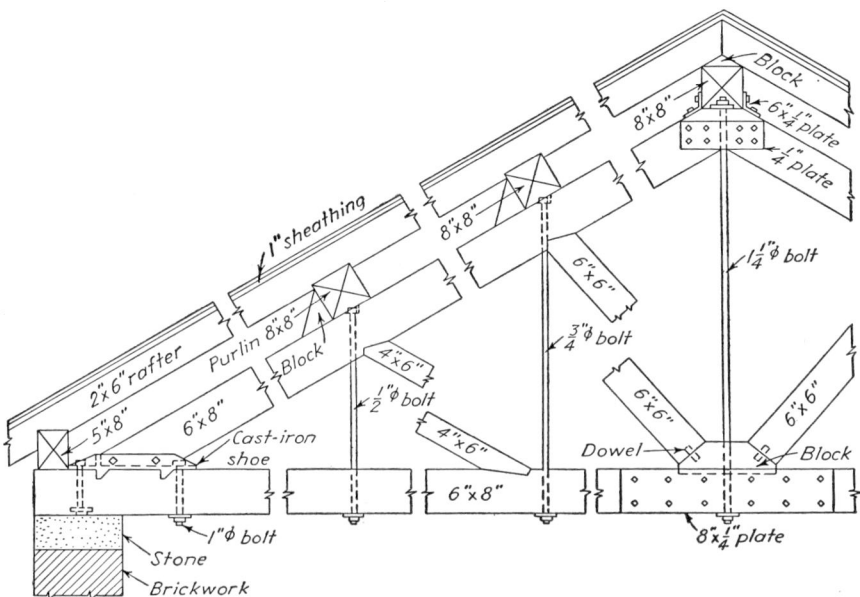

*Fig. 15-22   Truss with bolts.*

Insulating materials are of four types: flexible, rigid, fill, and reflective. The first three are composed of a large number of small cells, amounting to minute air spaces, which resist the passage of heat or cold. The fourth depends upon the reflection of radiated heat.

Insulation should be protected against the penetration of moisture. When there is a great difference in temperature between the inside and the outside of a building, moisture condenses. If the moisture is absorbed by the insulating material, serious damage to plaster and inside finish may result. Some insulating materials lose their efficiency because their porous nature is unprotected against air infiltrating and becoming moist in the insulation, owing to condensation within the wall. Furthermore, loose insulating materials settle and form open spaces or leaks at the top of the walls. Therefore, insulation must be carefully installed so as to remain permanently in place.

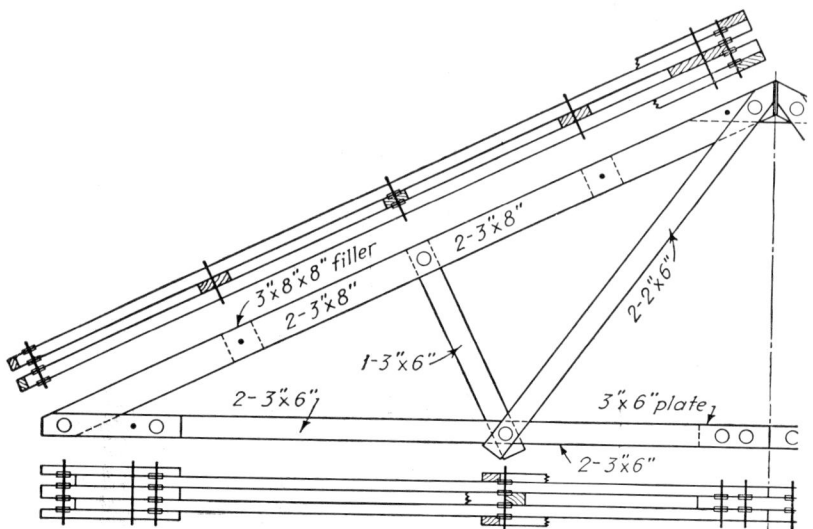

Fig. 15-23  Truss with connectors.

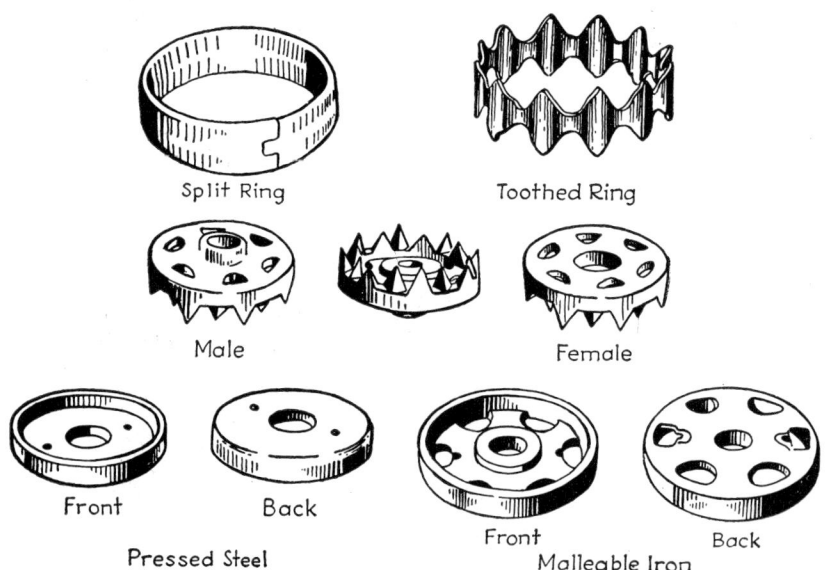

Fig. 15-24  Timber connectors

Rigid insulating sheets are sometimes used in place of wall or roof sheathing. Many varieties of rigid insulating boards are made, some of which are not suitable to serve as substitutes for sheathing.

Insulation is generally measured by finding the number of square feet of each different kind and thickness required, and building up a price per square foot to suit the material, labor, and conditions in each case.

## Building code

Building codes specify only the minimum requirements. The following notes are from the New York City Building Code:

**Support.** The ends of wood beams and rafters resting on masonry walls shall be cut to a bevel of 3″ in their depth and shall have a bearing of at least 4″ on the masonry.

It shall be unlawful, except in the case of one and two family dwellings, to support either end of a beam on stud partitions. Tail beams over eight feet long and trimmer beams and header beams shall be hung in approved metal stirrups or hangers and shall be spiked unless supported on a wall or girder. It shall be unlawful to notch or cut beams or rafters unless they are suitably reinforced.

Built-up girders shall be securely bolted together. Other built-up members shall be securely spiked or bolted together. Spiked trusses shall be of types which have been tested and approved.

**Bridging.** Wood floor beams and beams in flat roofs exceeding eight feet in clear span shall be braced with mitred cross bridging measuring at least 1″ × 2½″ and nailed twice at each bearing or, if metal bridging is used, it must have equivalent effective strength and durability. The maximum distance between bridging or between bridging and bearing shall be eight feet.

**Anchoring.** The ends of wood beams and girders shall be anchored, shall lap each other at least six inches and be well bolted or spiked together, or shall be butted end to end and fastened by approved metal straps, ties, or dogs in the same beams as the wall anchors. The ends of wood beams framing into girders shall be tied together with approved metal straps or dogs.

**Fire Prevention.** Wood beams shall be trimmed away from flues and chimneys. The header and trimmer beams shall be at least four inches from the face of chimneys and backs of fireplaces. In front of a fireplace an opening shall be trimmed to support a trimmer arch or approved masonry hearth at least sixteen inches from the face of the breast and at least twelve inches wider than the fireplace opening on each side. Combustible members entering a masonry wall shall be separated from each other and from the outside of the wall by at least four inches of solid masonry.

**Wood Posts.** Wood posts shall have level bearings and shall be supported on properly designed metal bases or base plates. Where timber posts are

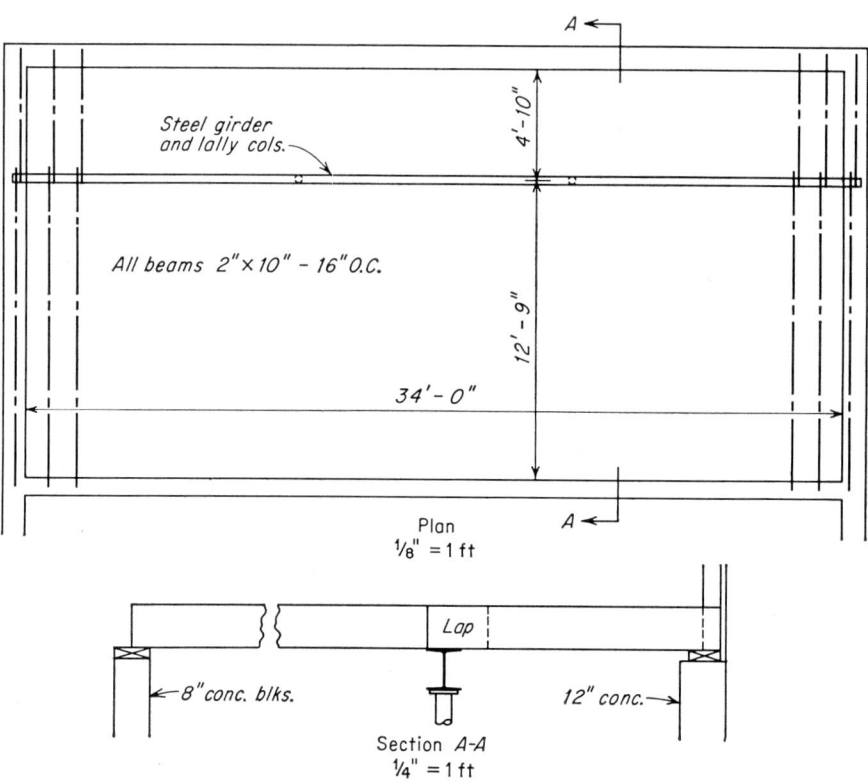

*Fig. 15-25  Floor beams.*

superimposed they shall be squared at the ends perpendicular to their axes and supported on metal caps with brackets or shall be connected by properly designed metal caps, pintles, and base plates.

**Bearing Partitions.** Stud-bearing partitions which rest directly over each other and are not parallel with the floor beams shall run down between the beams and rest on the top plate of the partition, girder or foundation below. Stud-bearing partitions parallel to the floor beams shall be supported on doubled beams.

**Fire-stops.** Exterior stud walls and stud-bearing partitions shall have the studding filled in solid between the uprights to the depth of all floor beams with suitable incombustible materials. Where walls are furred, the space between the inside of the furring and the wall shall be fire-stopped from the ceiling to the under side of the flooring or roof above with incombustible materials.

**Wood Frame Details.** The framework of wood-framed structures shall conform to the balloon frame, braced frame, or platform types and shall consist of sills, posts, girts, or ribbon strips and plates mutually braced at all angles or by wood sheathing laid diagonally and nailed twice at each bearing. The corner posts shall be at least the equivalent of three $2 \times 4$'s and sills shall be at least $4 \times 6$ or $3 \times 8$.

Approved fibreboard sheathing at least one-half inch in thickness and four feet in width may be used instead of wood sheathing when bearing on four studs and fastened to each bearing with nails spaced six inches or less apart. Gypsum sheathing board, at least one-half inch in thickness and two feet in width, may be used instead of wood sheathing when set horizontally and fastened to each bearing with $1\frac{3}{4}''$ galvanized flat head roofing nails spaced four inches apart.

■ **EXERCISES**

1. List the floor beams shown on the plan in Fig. 15-25 and compute the board feet, using regular stock lengths.
2. Compute the board feet of lumber in the following list:

    | | |
    |---|---|
    | Sills | 86 lin ft $4'' \times 8''$ |
    | Girders | 4 pcs $6'' \times 10'' \times 14'0''$ |
    | Beams | 42 pcs $3'' \times 12'' \times 18'0''$ |
    | Studs | 60 pcs $2'' \times 4'' \times 8'0''$ |

3. Describe briefly the balloon and braced-frame types of house framing.
4. Name 10 items of rough carpentry.
5. Make sketches showing square-edge, T&G, and shiplap boards.
6. Write at least 300 word of notes on rough carpentry, taken from two of the books listed in the Bibliography. Name the books used.
7. State the meaning and use of the following terms, giving sketches where possible: board foot, girder, beam, bridging, tail beam, header beam, trimmer.
8. Make a complete quantity survey for the rough carpentry required for the house shown in Chap. 6. The specifications are as follows:

    ROUGH CARPENTRY

    Materials shall be new, thoroughly seasoned, and protected from the weather until placed in the building. All lumber shall be No. 1 Common

Douglas fir dressed on four sides for framing. All wall and roof sheathing and underflooring, except the underflooring in living room and kitchen, shall be 1" × 6" T&G No. 2 Common shortleaf pine or fir. Underflooring in living room and kitchen shall be ¾" thick fir plywood. The 1" × 6" wall sheathing and underflooring shall be laid diagonally, the roof sheathing straight.

Wall and partition studs shall be 2" × 4" set 16" O.C., in one length where possible, and doubled around and trussed over openings.

Beams shall be doubled under all partitions running parallel thereto, and doubled or tripled elsewhere as required. Bridging shall be 1" × 3".

Provide all required special framing, blocking, and grounds, etc.

All work shall be done in accordance with the requirements of the building code.

CHAPTER 16

# cement work

Cement work, sometimes called *cement finish*, is frequently included under the Concrete heading. On small jobs it is likely that only one heading, such as Concrete and Cement Work, will serve for all the concrete and cement items on the job. On large jobs a number of headings may be necessary, because entirely different types of concrete construction are involved or because subcontractors are expected to figure certain portions. On very small jobs all the items may be thrown together under the head Masonwork.

Cement finish is applied as the top-finish wearing surface on floors, usually 1" in thickness. The specifications state the mix desired and the method of treating the top surface. The mix is generally about 1 part cement to between 2 and 3 parts of sand, but special mixtures and special ingredients may also be specified.

The cement base, run along the walls at the floor, comes under this heading. It is generally about 6" high and may be straight or have a cove at the floor. The top may finish flush with the wall surface above or project slightly and have a bullnose edge.

The floor areas are listed in square feet and the base in linear feet. Care must be taken to state the thickness and the mix specified, as well as any special treatment that may be called for. Each thickness and each different kind of work is kept separate, so that proper unit prices may be applied to the quantities.

When subcontract estimates are received for any part of the concrete or cement work, the estimator is required to check them carefully and see that nothing is figured more than once among the several subcontractors whose figures may be used. For example, the subcontractors for the floor and roof construction may include in their figures the floor fill that is placed on top of the floor arches and the fill that is placed on the roof construction for drainage. This fill may be found also in the estimates of the cement-finish subcontractors or on the estimator's own sheets, and an adjustment would therefore have to be made in all the figures for the purpose of making proper comparison among them.

Cement floor finish is applied on top of slabs either immediately following the pouring of the concrete slab, before the concrete has set, or else as a separate coating some time afterward. The first method, which is called *integral* or *monolithic*, is the one more commonly used. It consists of a layer of mortar, from a simple sprinkling to a coating 2" thick, which is carefully spread over the base slab. This is then floated smooth with a wood trowel and sometimes also burnished with a steel trowel for a fine finish. The second method, sometimes called *bonded finish*, consists of a coating of mortar 1" to 2" thick, applied at a later date to the concrete slab and then finished as above noted or as may be called for in the specifications.

Sometimes no finish at all is applied and the concrete slab is itself finished by floating or troweling. Walks, roads, and other outside pavements of concrete are usually given no special top coating. Interior floors are specified to be treated in one or more of many ways. The estimator must take care to provide for the mixture and the thickness required and also for any hardener or coloring material that is to be incorporated in the finish or applied to it.

Various types of hardeners, waterproofings, and colorings are in use. When these are called for, suitable entries must be made to cover the additional cost of them. The number of square feet of area to be given special treatment is sometimes separately listed and this additional cost is applied to the quantity thus expressed. A better method, however, is to keep each kind of complete floor finish by itself and then price each one with a complete price per square foot. In this way the total cost of each floor area may be readily seen and comparisons and adjustments easily made. This will include all the elements of materials, labor, and incidental expense.

Walks, cellar floors, driveways, and other pavements that are laid directly on earth are customarily included under the heading of cement-finish work. In this kind of work the entire construction of fill, slab, and top finish is figured complete, by the square foot.

The cement treads and platforms that are laid on iron stairs are included in the cement-finish division of the estimate, and these are also measured by the square foot. Light wire mesh is sometimes called for to be embedded in this tread and platform finish, and the unit price is then adjusted so as to include the mesh.

The unit cost of plain cement floor finish runs between 25 and 30 cents per sq ft. If cinder-concrete fill 3" thick is placed under the finish, then 20 to 25 cents per ft, additional, must be included, which would make a total of 45 to 55 cents. This is a typical floor, such as is used in small house cellars or on top of floor arch construction. If, as in a cellar floor, 6" of tamped dry cinders is included as a base under the concrete, a further cost of between 12 and 15 cents per sq ft is incurred. Thus a pavement composed of a dry cinder bed, a cinder-concrete slab, and a top cement finish will cost the contractor about 65 cents per sq ft, without any general-expense items or profit being considered. If broken stone or gravel were used in place of the cinders as the coarse aggregate in the concrete slab, this would increase the cost still further about 10 cents per sq ft.

Sills and copings and expansion joints are measured and priced per linear foot, each size and type separately.

**Cement.** Practically all mason-supply dealers handle cement. It is packed and shipped in cloth sacks or paper bags. One cubic foot of cement (94 lb) is contained in each sack or bag. A refund may be obtained for any cloth sacks that are returned in good condition. Paper bags are not returnable. All cement must be kept in a dry place until used.

**Construction.** The general discussion of concrete, here and in Chap. 12, applies to all concrete and cement work. The term *cement work* is a misnomer anyway, as much of the work included under this heading is really concrete construction, even including, at times, reinforcing mesh or rods.

Concrete is a mixture of cement, water, and inert materials, such as sand, gravel, broken stone, or cinders. After it is mixed and placed, it turns from a plastic mixture into a hard artificial stone, owing to the hydration of the cement. When concrete is properly cured, hardening continues for a long time after it has acquired sufficient strength for the purpose intended. The correct proportions of water, cement, and coarse aggregate to use in the mixture are governed by the character of the work for which the concrete is intended. In a concrete mixture, cement and water form a paste, which, upon hardening, acts as a binder cementing the particles of sand and coarse aggregate together into a permanent mass. The use of too much water thins or dilutes the paste, weakening its cementing qualities. Consequently, to get the best results, it is important that the proper proportions of cement and water shall be used, the exact proportions depending upon the work.

A workable mixture of concrete is one of such plasticity and such a degree of wetness that it can be placed readily and, with spading and tamping, will result in a dense concrete. In a workable mixture there is sufficient cement-sand mortar to give good smooth surfaces free from rough spots, called *honeycombing*, and to bind the pieces of coarse aggregate into the mass so that they will not separate out in handling. Mixtures with a deficiency of mortar will be harsh, hard to work with, and difficult to finish. On the other hand, oversanding, or the use of too much sand, increases porosity and reduces the amount of concrete that can be produced with a sack of cement. For given materials and conditions of handling, the strength of concrete is determined solely by the ratio of the volume of mixing water to the volume of cement, as long as the mixture is plastic and workable. The ratio of water to cement also governs the watertightness or impermeability of the concrete, about 6 gal of water per sack of cement usually producing watertight concrete.

Sand or rock screenings, the fine aggregate, include all particles that will pass through a screen having meshes ¼" square. Coarse aggregates range from ¼" up to 1½" or 2". In thin slabs or walls the largest pieces of aggregate should not exceed one-third the thickness of the section of concrete being placed.

Concrete should be placed as soon as it is ready, in no case more than 45 minutes after mixing. It should be deposited in layers of uniform depth, usually not exceeding 6" in depth. It should be continually tamped or spaded so that it will settle thoroughly and produce a dense mass.

The surface finish of a floor or walk should be obtained by using a wood float. A metal trowel should be used sparingly, if at all, because its use brings to the surface a film of fine material that lacks the wearing quality of the cement and sand combined and may cause the surface to develop hair cracks after the concrete hardens. Although a troweled surface is smoother, it does not wear so well as a floated surface and is likely to be slippery.

If concrete is left exposed to sun and wind before it has properly hardened, much of the water necessary to hardening will evaporate and the concrete will simply dry out. Moisture is needed for the proper hardening of concrete because, as has already been mentioned, this process is due to changes that take place in the cement when it is mixed with the proper amount of water. Concrete floors, walks, pavements, and similar large surfaces can be protected by covering them with sand, moist earth, or some other moisture-retaining material as soon as the concrete has hardened sufficiently to permit doing so without marring the surface. In warm weather this covering should be kept moist by frequent sprinkling for a period of about 10 days. Walls and other sections that cannot be covered in this manner can be protected by hanging moist canvas or burlap over them. During colder weather, protection is equally important, but the concrete need not be kept moist, as evaporation is not so rapid.

**Basements.** Basement floors in houses are generally 4″ thick and are preferably laid in one course directly on the ground or on suitable fill (Fig. 16-1). The floor is laid off in strips 4 or 5 ft wide extending the full length or width of the basement. For forms, 2 × 4's are set on edge and so positioned that they provide grounds or guides for the strike board used in leveling off the surface. Stakes are driven along the 2 × 4's to hold them in place. After the concrete has been placed and leveled off with the strike board, the surface is floated to fill in low places and work down high spots. If the floor is to be pitched toward a drain, this should be provided for in setting the 2 × 4's. After the surface is evened off, the concrete is allowed to stiffen until it is ready for the final troweling. Too much troweling is not advisable.

**Sidewalks.** Sidewalks should always rest on a firm base (Fig. 16-2). If the soil on which they are to be laid is well drained, the concrete can be placed directly on it after all refuse, grass, roots, and similar materials have been removed and the ground has been well compacted. If the soil is not well drained, a 6-in. subbase of well-compacted, clean, coarse gravel or clean cinders should be provided. Sidewalks are generally built by the one-course method, as described for basement floors above, and the surface should be sloped slightly toward one side, to make sure that rain water will drain off. For con-

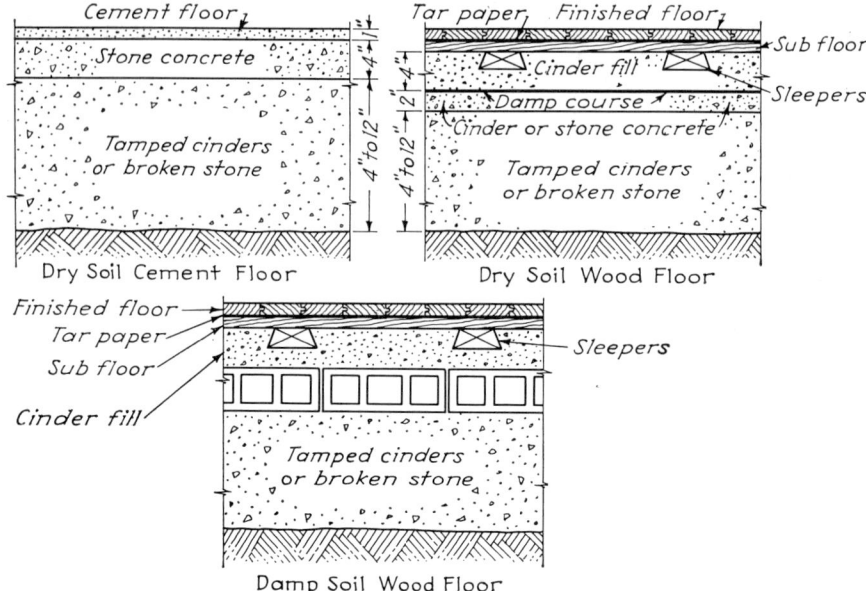

**Fig. 16-1** Floors on earth.

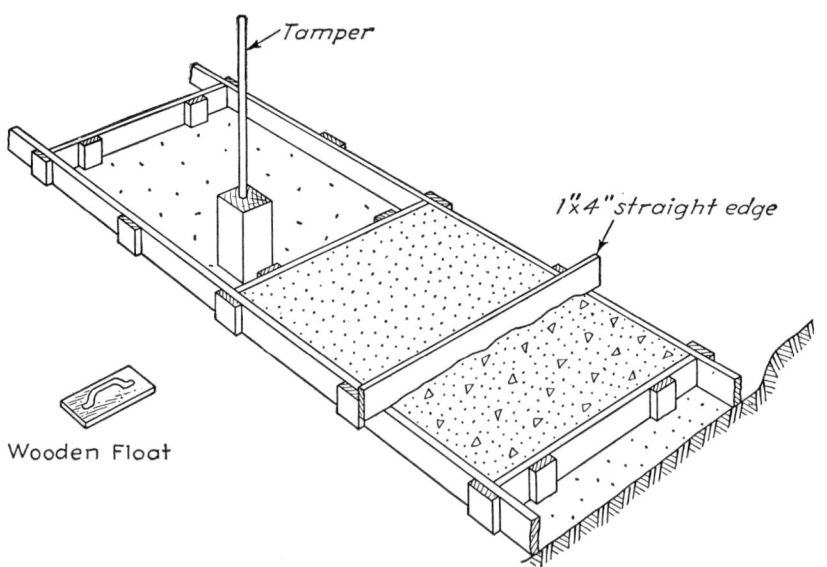

**Fig. 16-2** Sidewalk construction.

venience in building and provision for expansion and contraction joints, walks should be divided at 4- to 6-ft intervals with partition strips placed at right angles to the side forms. Every other section is then concreted. After these have hardened, the cross strips are removed and the remaining slabs are placed.

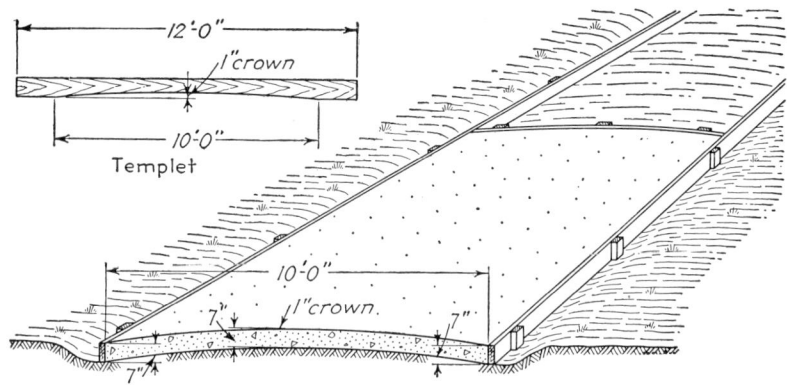

*Fig. 16-3* Concrete driveway.

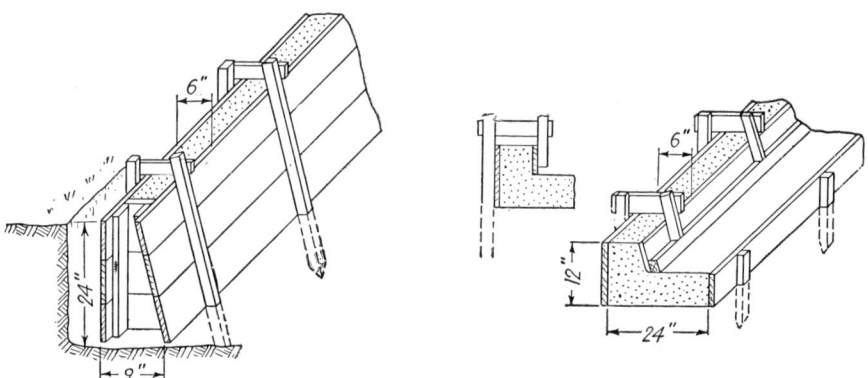

*Fig. 16-4* Curbs.

**Driveways.** Driveways (Fig. 16-3) and curbs (Fig. 16-4) are sometimes included with building work and are listed under Cement Work. Driveways are made not less than 7" thick, in order to carry the load of moving vehicles. The center of the driveway is given a crown of about 1", to ensure rapid drainage. This crown is produced by means of a curved strike board or template, which shapes the surface. The area on which the pavement is to be placed should be brought to the required grade, with such excavation and filling as may be necessary to secure the required uniform slope. All filled-in places should be thoroughly tamped, and the ground should be shaped, with the center 1" higher than the outer edges, so that the finished pavement will have a uniform thickness. Side forms of 2 × 6's or 2 × 8's should be set, with their tops at the right level to serve as guides for the templates. The finishing is done with wood floats attached to long handles. Pave-

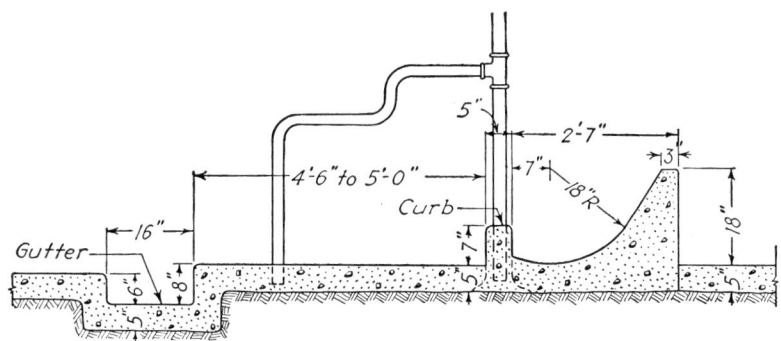

*Fig. 16-5* Cow stall.

ments should be protected from the drying action of the sun and the wind by being covered and kept moist for at least 2 weeks before the driveway is used.

Curbs and similar construction require special formwork, and on large jobs of curbs steel forms are recommended (Fig. 16-5).

**Steps.** Concrete steps (Fig. 16-6) are safe, nonslippery, and so durable as to last indefinitely if they are properly constructed. Stairways and long flights of steps are generally listed under the heading Concrete Work, but small sets of steps are always placed under Cement Work.

When steps are built between side walls, the walls are completed first. When these forms have been removed, the space that will be under the steps is filled in with earth and thoroughly tamped, to provide a firm base on which to pour the concrete for the steps. Steps of this kind resting on well-compacted earth require no reinforcement.

Planks for supporting the riser forms are held firmly in place against the walls by means of cross braces, and pieces of 1 × 4 or 2 × 4 are nailed to the planks for holding the riser forms in place. For a com-

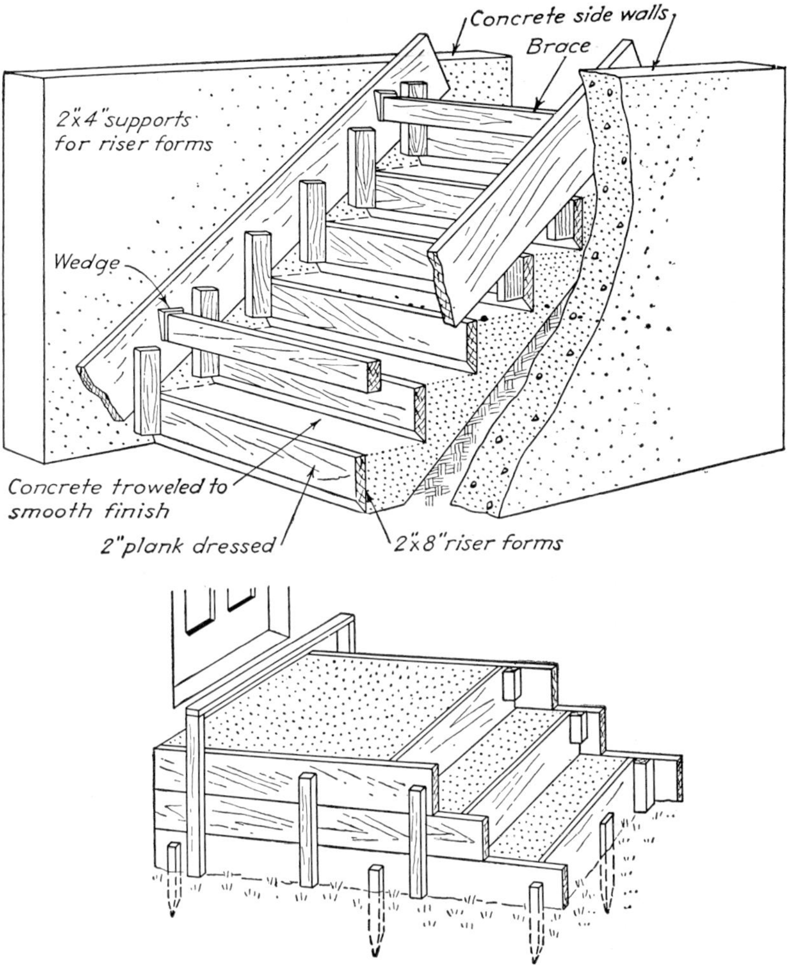

CEMENT WORK 243

Fig. 16-6 Concrete steps.

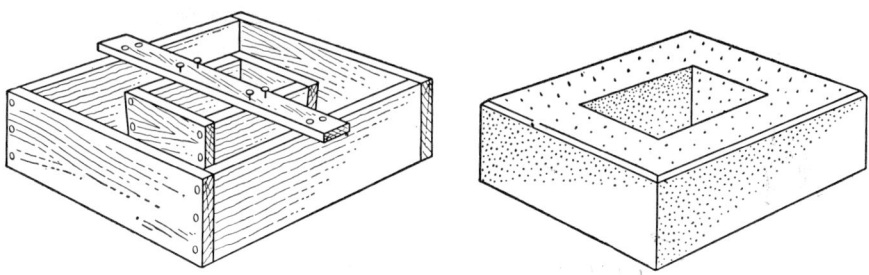

Fig. 16-7 Chimney cap.

fortable flight of steps the vertical distance from one step to the next should not exceed 7½″ and the width of the tread should be at least 10″.

**Chimneys.** Chimney caps (Fig. 16-7) are either cast in moulds, on the ground, and then placed on top of the chimneys, or are cast in place by building forms on the chimney tops. The inner core of the chimney form, which has beveled sides to facilitate removal, is held in place by a strip nailed across the top of both forms.

## ■ EXERCISES

1. Make a sketch of a typical basement floor, in cross section, showing a bed of cinders, with a monolithic-finished concrete slab on top. Mark the thickness of each.
2. Write at least 200 words of notes on cement work, taken from one or two of the books listed in the Bibliography. Name the books used.
3. Make a complete estimate of the cement work required by the house plans shown in Chap. 6. The specifications are stated as follows:

### CEMENT

Cement shall be fresh portland cement. Sand shall be clean, coarse, and sharp. All coarse aggregate shall be either crushed rock or gravel, ranging in size from ¼″ to 1½″ and shall be free from dirt.

All concrete shall be composed of 1 part cement, 3 parts sand, 5 parts coarse aggregate, and ⅒ part hydrated lime. The floor of the basement shall have a 1″ finish coat, consisting of 1 part cement, 2 parts sand, 4 parts crushed rock (pea size), applied immediately after the base slab is laid and before it has taken initial set, floated and troweled to a hard smooth finish.

Construct the playroom and areaway steps and coping with troweled finish, as shown on the plans.

Lay a 3″ concrete bed in the bathroom, vestibule, and foyer floors, ready to receive tile.

CHAPTER 17

# *plastering*

The heading for the plastering division is sometimes Lathing and Plastering. It usually includes all the work done by the lathers and the plasterers.

Plaster is applied directly to brickwork, terra-cotta blocks, gypsum blocks, and other masonry and on wood or metal lath and plasterboards. Stucco, which is just exterior plastering, is included under this heading. Metal corner beads and the iron hangers installed by metal lathers in suspended ceiling construction also come under this heading.

Grounds are wooden strips made to a uniform thickness, which are applied by the carpenters and belong under the heading Carpentry. They are installed to act as guides for the surface of the plaster. They serve as stops against which the plaster is finished and also provide a means of fastening the trim to the walls. Grounds of $7/8''$ thickness are required when three-coat plastering on wood lath is to be used; grounds $3/4''$ thick are required for three-coat work on metal lath; grounds $5/8''$ thick are required for two-coat work on masonry.

Plasterboard work is put under this heading, especially when it is to be coated with plaster. If the boards are simply nailed on and not plastered, the work is sometimes put under the carpentry heading. Each board contains 8 sq ft. The boards are $32'' \times 36''$ and therefore suitable for 12", 16", or 18" nailing on wood studs, etc. Thicknesses obtainable are $1/4''$, $3/8''$, and $1/2''$.

Wood laths are ¼" × 1½" × 48", which allows for 12" and 16" nailing. They are applied ¼" to ⅜" apart. Metal lath comes in many forms of expanded metal sheet and woven wire mesh.

There are many kinds of plaster and stucco materials and finishes. Prepared plasters, already mixed or partly mixed, are in use now to a large extent.

The work included under this heading is measured by the square yard for flat surfaces and by the linear foot for mouldings, cornices, corner beads, etc. Some estimators make no deductions for the openings. Some take half of these "outs." Still others take no "outs" of less than 21 sq ft and, therefore, would measure an average wall or partition as though it had no windows or doors. Although the last of these methods seems to be the one most commonly used, it is best to state at the very beginning of the plastering estimate which method is being used. In this way anybody reading the estimate will note the method adopted and disputes on this will be avoided. Similarly, in bids and letters in which the method of measuring is concerned it is a wise precaution to state which method is assumed.

Care must be taken in listing the work to see that all the items are included. Not only are room walls and ceilings plastered but also walls and ceilings in passageways, halls, and storage rooms and bulkheads. Soffits of stairs, landings, mezzanine platforms, and showrooms often have work under this heading. In addition to the lathing and plastering, there are items of corner beads, arches, and outside plastering, or stucco, to be considered.

The student should not become too much bound by definite rules of procedure in estimating any trade. All the work should be written up, however, so that a complete story is made in the estimate of the proposed building. The test of a good estimate is the ease with which another estimator can understand every item. With plastering work, care should be taken to separate the listing of the quantities into many parts, depending on the location in the building, the kind of plastering, the number of coats involved, the kind of lathing, etc. In this way the estimate may readily be checked or adjusted and, what is more important, it may be priced at proper cost prices to suit each individual item or group of items.

It is recommended that the work for each floor be kept separate from that of other floors. In this way, one large error—that of omitting

| EST. 522 | | | PLASTERING | | | | | | | SHEET 11 | |
|---|---|---|---|---|---|---|---|---|---|---|---|
| | | | (FULL OUTS TAKEN) | | | | | | | | |
| | | | BASEMENT | | | | | | | | |
| | | | Stair walls, on T.C. blks. | | | | | | | | |
| | | | 2 × 25'-0" × 12'-0" | | 600 | | | | | | |
| | | | OUTS  2 × 3-0 × 7-0 | | − 42 | | | | | | |
| | | | | | 9)558 | | | | | | |
| | | | | | 62 yds | 1.70 | | | 105 | | |
| | | | FIRST FLOOR | | | | | | | | |
| | | | Corridor walls, on T.C. & Gyp. blks. | | | | | | | | |
| | N. | | 100'− 0" | | | | | | | | |
| | E. | | 155 − 4 | | | | | | | | |
| | S. | | 101 − 4 | | | | | | | | |
| | W. | | 122 − 0 | | | | | | | | |
| | | | 478 − 8 × 14 − 0 | | 6701 | | | | | | |
| | | | OUTS | | | | | | | | |
| | | | 6 × 6-0 × 8-0 | 288 | | | | | | | |
| | | | 24 × 3-0 × 7-0 | 504 | − 792 | | | | | | |
| | | | | 792 | 9)5909 | | | | | | |
| | | | | | 655 yds | 1.60 | | | 1048 | | |
| | | | Show room walls, on Gyp. blks. | | | | | | | | |
| | | | 84 − 0 | | | | | | | | |
| | | | 55 − 0 | | | | | | | | |
| | | | 117 − 6 | | | | | | | | |
| | | | 97 − 6 | | | | | | | | |
| | | | 354 − 0 × 14 − 0 | | 4956 | | | | | | |
| | | | OUTS | | | | | | | | |
| | | | 3 × 6 − 0 × 8 − 0 | 144 | | | | | | | |
| | | | 22 × 3 − 0 × 7 − 0 | 462 | | | | | | | |
| | | | 1 × 12 − 0 × 8 − 0 | 96 | − 702 | | | | | | |
| | | | | 702 | 9)4254 | | | | | | |
| | | | | | 473 yds | 1.60 | | | 757 | | |
| | | | | | | | | FWD. | 1910 | | |

Fig. 17-1  Plastering estimate.

an entire floor—is less likely to be made. Concentrating thus on one floor, the estimator of little experience will find it easier to visualize the work shown on the plans.

An illustration of part of an estimate in plastering is given in Fig. 17-1. Note the general form of layout and the systematic arrangement of the names of minor headings and the figures. The totals are extended for the quantities and suitable prices are applied.

## Plastering costs

Plain plastering costs the contractor $1.50 to $2 per sq yd for two-coat work on masonry. Three-coat work on lath, including the cost of the lathing, runs from $3 to $4 per sq yd. Hung ceilings, formed of iron hangers, runners, metal lath, and three coats of plaster, cost between $5 and $6 per sq yd. Plain cornices run from 80 cents to $1.50 per lin ft. Corner beads cost 25 to 40 cents per lin ft (Fig. 17-2).

None of these prices includes any general job-expense items, overhead, or profit.

**Labor.** A metal lather applies 100 sq yd of metal lath on wood studs in about 6 hours. To erect metal partition channels and apply metal lath on one side requires 20 hours of lather time, per 100 sq yd, and 27 hours if lathed on both sides. For erecting hanger rods, channel runners, and metal lath for hung ceilings, 22 hours of lather time is required per 100 sq yd, and 27 hours if flat iron hangers, bolted to the channels, are used. To apply plasterboard requires about 9 hours of lather time per 100 sq yd (Figs. 17-3 and 17-4).

To apply three coats of plaster on metal lath, about 22 hours of plasterer time and 11 hours of laborer time are required, per 100 sq yd, on walls, and an additional hour for each on ceilings. On plaster boards or on masonry, applying only the brown and white coats, 16 hours of plasterer time and 8 hours of laborer time are required per 100 sq yd.

**Materials.** Three coats of lime plaster on metal lath requires 2 cu yd of sand, ¾ ton of lump lime (or 1 ton of hydrated lime), 175 lb of plaster, and 2 bu of fibre, per 100 sq yd.

Three coats of gypsum plaster on 100 sq yd of metal lath require 2 cu yd sand, 2,200 lb gypsum cement, 225 lb lump lime or 350 lb hydrated lime (for the white coat), and 175 lb plaster.

Two coats of gypsum plaster on 100 sq yd of masonry require 1¼

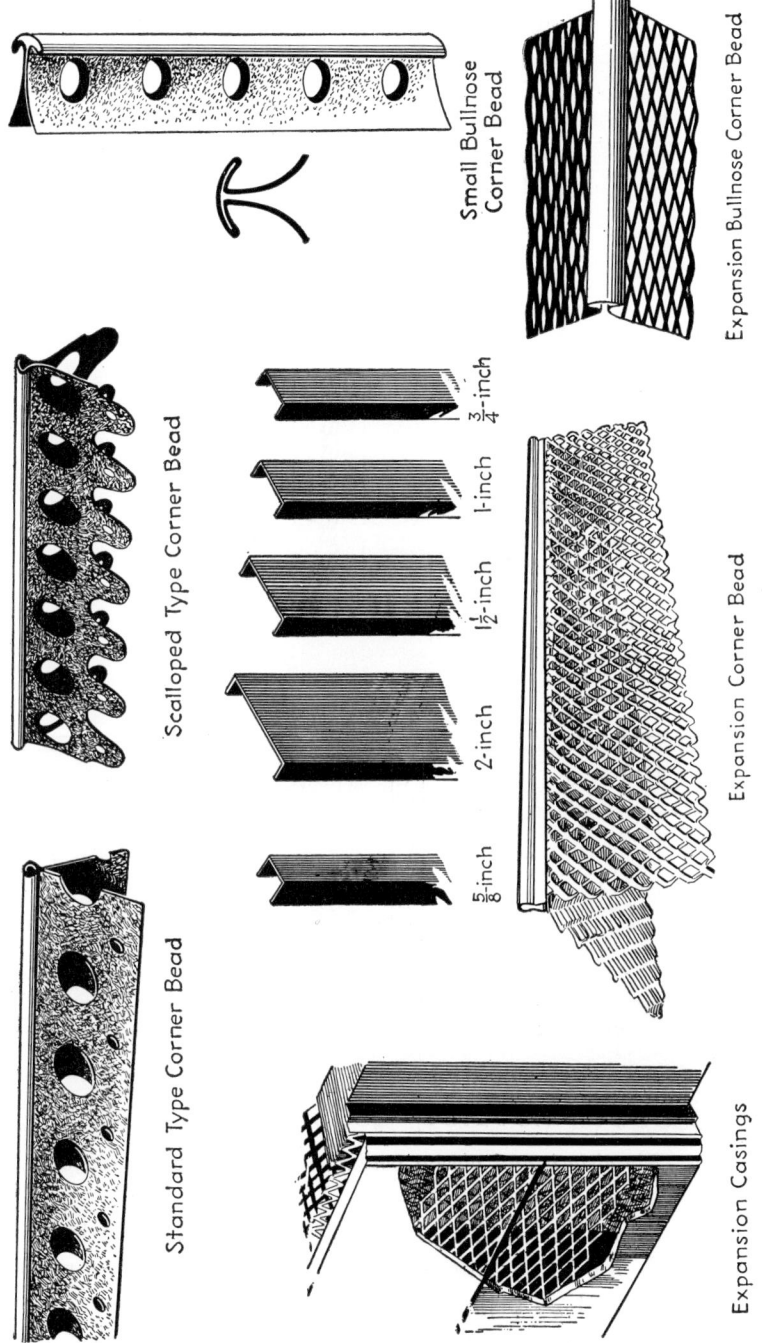

*Fig. 17-2* Corner beads and channels.

cu yd sand, 1,200 lb gypsum cement, 225 lb lump lime or 350 lb hydrated lime, and 175 lb plaster.

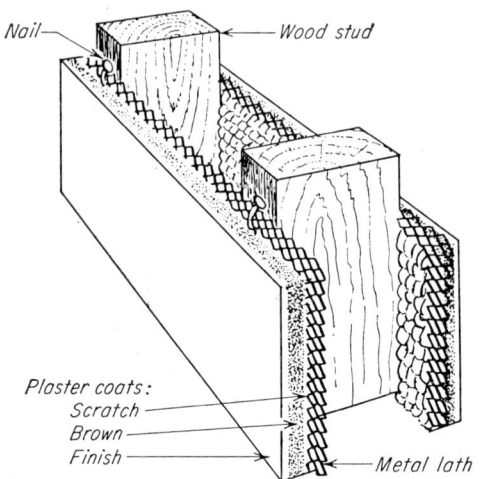

*Fig. 17-3* Metal lath and plaster on wood studs.

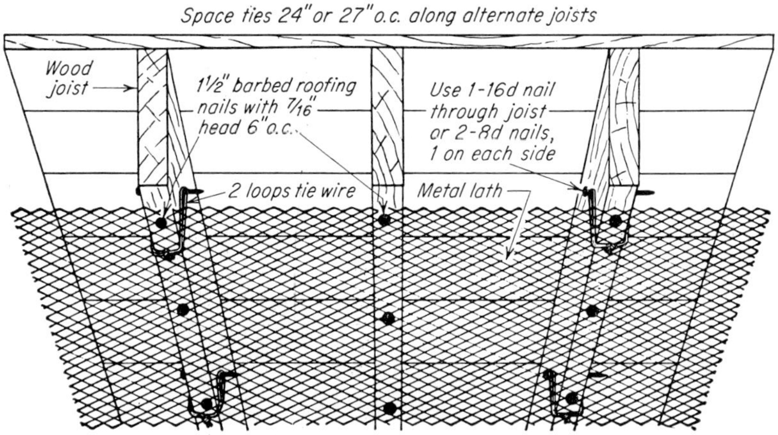

*Fig. 17-4* Metal lath and plaster ceiling, tied-and-nailed wood joists.

Two coats of gypsum plaster on 100 sq yd of plasterboards require one cu yd sand, 900 lb gypsum cement, 225 lb lump lime or 350 lb hydrated lime, and 175 lb plaster.

Gypsum plaster for 100 sq yd of 2″ thick solid plaster partitions requires 4⅔ cu yd sand, 4,700 lb gypsum cement, 450 lb lump lime or 700 lb hydrated lime, and 350 lb plaster.

**Lathing.** Metal lath is made of stamped metal or wire mesh. The stamped or expanded lath comes in sheets 24″ wide and 96″ long. Wire mesh for plastering work comes in rolls 36″ wide and 150′ long. Ribbed lath of many varieties is also obtainable. Metal lath is lapped at least 1″ on abutting edges, and where it finishes against masonry

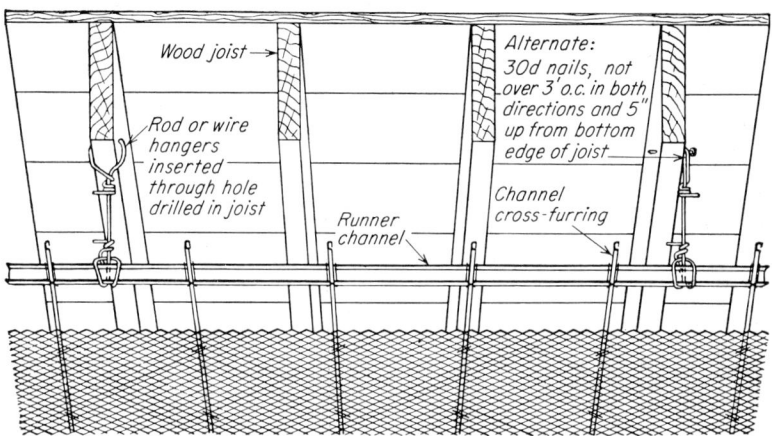

*Fig. 17-5* Metal lath and plaster ceiling suspended from wood construction.

walls, the lath should be extended at least 3″ on the surface of the walls and be securely fastened.

Suspended ceilings are formed of iron hangers and runners on which the lath and plaster are applied (Fig. 17-5). The whole ceiling construction is measured and priced per square yard of plastered surface. Hangers for suspended ceilings must be placed in line in either direction and with maximum spacing of 5′0″ O.C. With floor-arch construction, the hangers extend through the arches and are formed of two pieces of 1″ channels or $\frac{3}{16}″ \times 1″$ flat bars, at least 7″ long, and bolted, riveted, or welded together to form a tee. Purlins of 1½″ channels or angles are bolted to each hanger, with a maximum spacing of 5′0″. Cross furring consists of at least 1″ channels, wired on, with a maximum spacing of 12″ (Figs. 17-6 to 17-12).

Metal corner beads are used on vertical external corners, and sometimes strips of metal lath are applied on internal corners.

## Types of plaster

Plaster is classified as lime plaster, gypsum plaster, or cement plaster, depending upon which of these materials is used in it. Cement plaster is used especially in places where more resistance to fire is desired or in damp places, such as cellars. Gypsum plaster is always

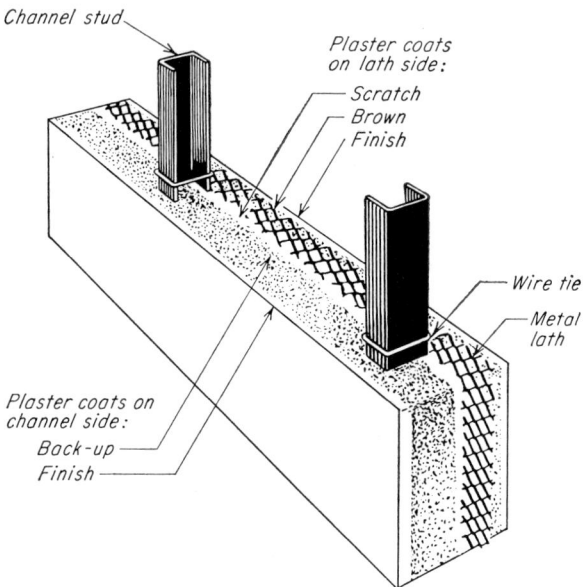

**Fig. 17-6** *Metal lath and plaster solid partition with channel studs.*

used on gypsum blocks and on plasterboards. Any plaster may be used on wood or metal lath.

The first coat of plaster applied to lath is called the *scratch* coat. It consists of lime, sand, and hair and is forced into the lath so that some of the material comes through it to form a key. Before this coat hardens, it is scratched with a scratching tool or with a piece of metal lath, to produce a rough surface for holding the next coat.

The second coat is called the *brown* coat. It is formed of lime and sand only and is applied as the thickest of the three coats and brought to a true surface with a wood trowel called a *float*.

The third coat is called the *white* coat or *finish* coat. It is made of lime and plaster of paris and is only about $\frac{1}{8}''$ thick. It is usually troweled with a steel trowel to a hard, smooth finish, ready to be painted or covered with wallpaper. The plaster of paris gives it the hard, smooth finish. Keene's cement, a very finely ground cement, also is used occasionally for the finishing coat.

Ready-mixed materials, some complete and ready for mixing with water only, and others which require the addition of sand, are obtain-

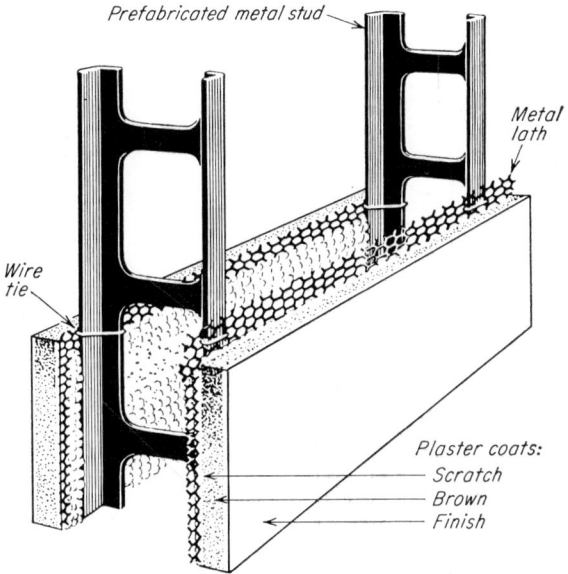

*Fig. 17-7* Metal lath and plaster hollow partition with metal studs.

able in 50- and 100-lb bags. These ready-mixed materials are commonly used in place of plaster mixed at the job.

Stucco, or exterior plasterwork, is generally formed of three coats of cement mortar. Various surface finishes are given by varying the method of floating or troweling employed or by incorporating pebbles or other materials in the finish coat. Ready-mixed stuccoes, some colored, are also obtainable.

Other kinds of plaster, including decorative or colored finishes, are used at times, in addition to the common type of work described above. In any event, the work is measured in square yards of completed surface and the price per yard is applied that will take care of the cost of the entire plasterwork and include the cost of the lathing or other plaster base that may be called for.

Plasterboards are usually estimated under this heading or under the heading Rough Carpentry. Two coats of plaster or just one heavy white coat is usually applied to the plasterboards. If no plaster is to be applied to them, the painters are sometimes called upon to fill the joints before painting. They use a putty knife for filling the joints with spaechel or some other plaster mixture, in order to comply with the

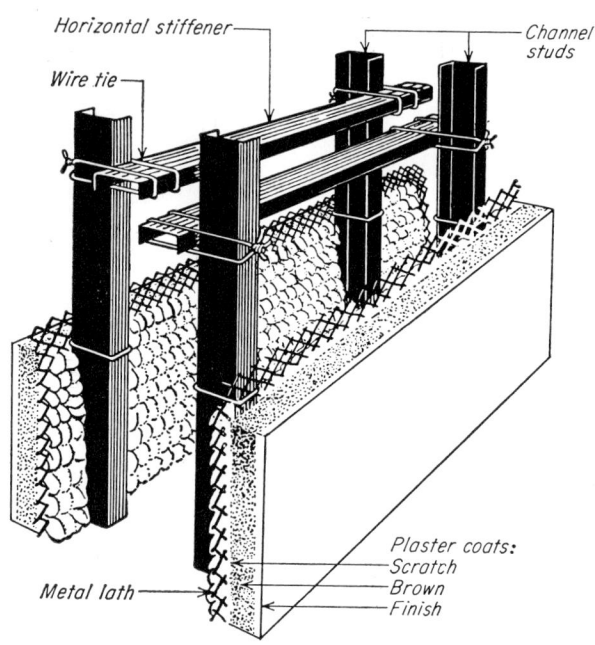

*Fig. 17-8* Metal lath and plaster sound-insulating double partition.

union agreements, which prevent them from using trowels such as the plasterers would use. Plasterboards come in several thicknesses and sizes as has been noted before, some with prefinished surfaces.

**Gypsum Plaster.** This kind of plaster is made by driving off water of crystallization from gypsum rock through calcination, or heat treatment. The addition of water to this product (calcined gypsum rock) forms a plastic mass which unites with the same amount of water as was driven off during calcination. When this plaster is mixed with water it recrystallizes, or sets. Gypsum plaster adheres to gypsum, wood or metal lath and to gypsum blocks, masonry, or other suitable plastering surface.

**Gypsum Ready-sanded Plaster.** This plaster is mixed in the mill and comes ready to use. Nothing but water is added. A small amount is first mixed with the water and then more is added until the required consistency is reached. It is used for the scratch and brown coats. Gypsum neat plaster is composed of calcined gypsum mixed at the

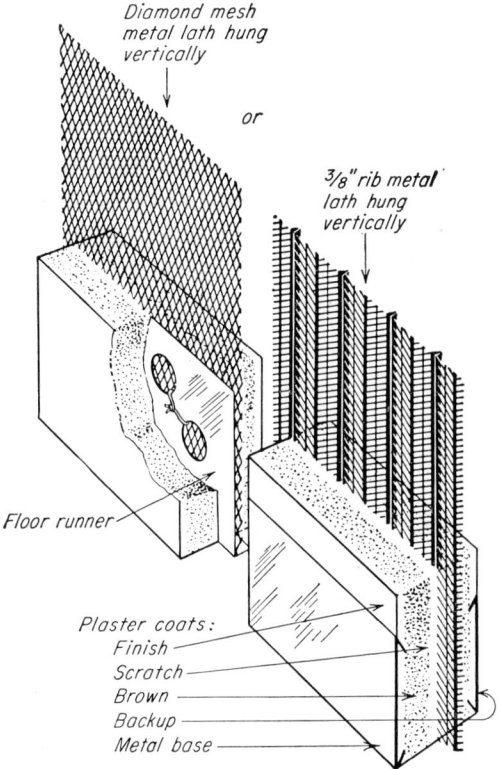

**Fig. 17-9** *Studless lath and plaster partition with diamond mesh or ⅜" rib lath (method of attachment to floor not detailed). Optimum height = 8'4" or less. Maximum height = 10'0".*

mill with other materials to control the working quality, setting time, and the fibring. It is mixed at the job with 3 parts, by weight, of sand.

**Gypsum Wood-fibre Plaster.** This type comes ready to use, requiring only the addition of water. This avoids the use of sand at the job and the dangers of poor sand or oversanding.

**Bondcrete Plaster.** This plaster is specially made for concrete surfaces. Ordinary plaster does not adhere properly to concrete. Bondcrete permits moisture, expelled from concrete, to be carried through and evaporated on the surface, whereas with ordinary plaster there would be a condition of sweating. Finish plaster can be applied over

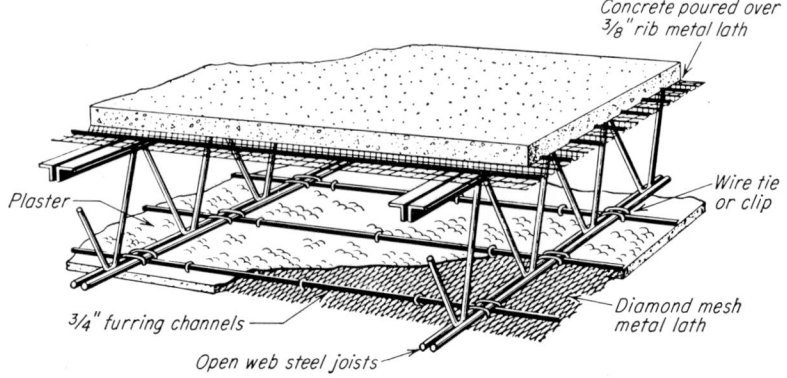

**Fig. 17-10** Metal lath and plaster furred ceiling on the underside of steel joists.

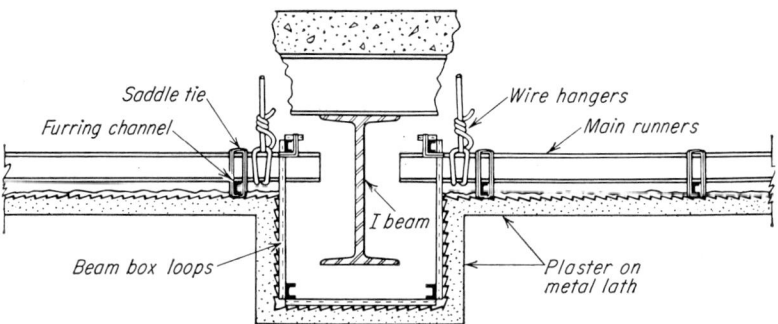

**Fig. 17-11** Metal lath and plaster beam formed as part of a suspended ceiling.

Bondcrete. The Bondcrete is mixed in the water and allowed to soak for 10 minutes and then worked to a creamy consistency. It is applied in two thin coats and broomed before it sets, so as to receive the finish coat.

**Keene's Cement.** This is the hardest and densest gypsum plaster and is made from selected gypsum and excels in smooth, uniform working qualities. It is pure white in color and takes a high, glasslike polish. It

PLASTERING    257

is used neat for extra-hard finish or mixed with lime putty for medium-hard finish.

**Acoustical Plaster.** This is a gypsum mixture which is used, over the standard scratch and brown plaster, as a sound-repellent finish. It

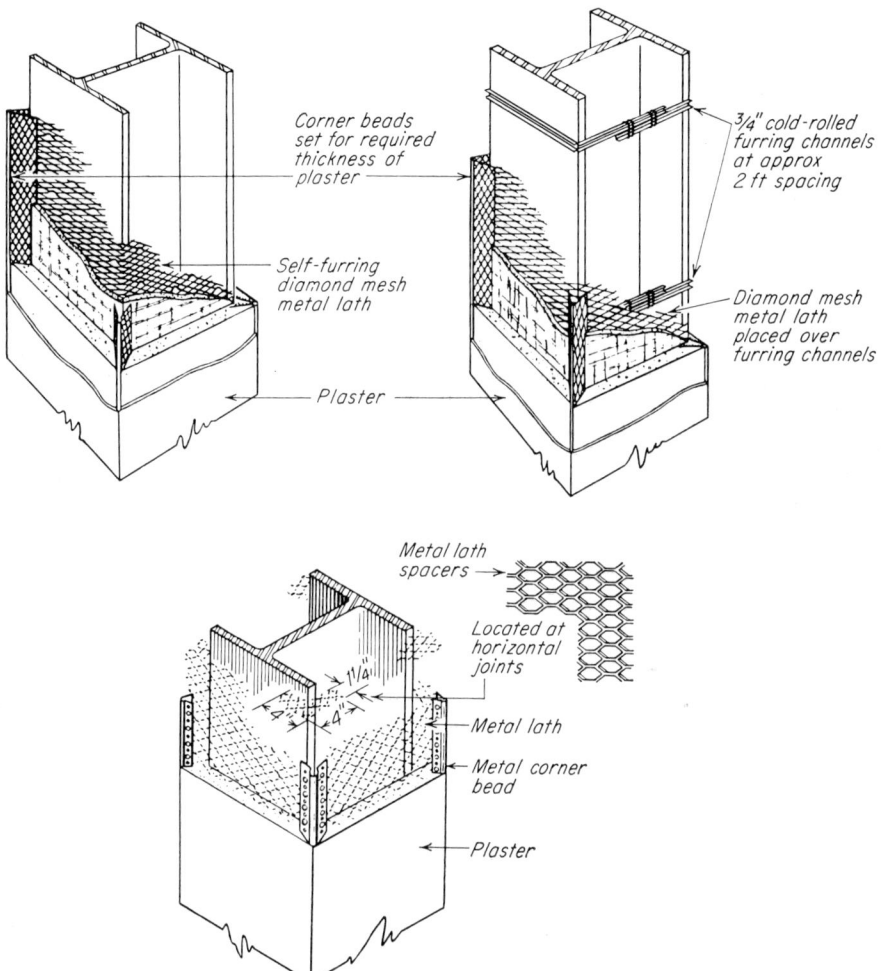

*Fig. 17-12   Optional methods for lathing steel columns.*

absorbs sound because of its multitude of minute passages which have their start in tiny openings at the surface. It is applied ½" thick in two ¼" coats.

## Definitions of terms

The following definitions for metal-lath construction have been established by long usage:

**metal lath.** Metal lath is of three types, designated as diamond mesh (also called flat expanded metal lath), rib, or sheet. Metal lath is slit and expanded, or slit, punched or otherwise formed, with or without partial expansion, from copper alloy or galvanized metal sheets. Metal lath is coated with rust-inhibitive paint after fabrication, or is made from galvanized sheets.

**diamond mesh** or **flat expanded metal lath.** A metal-lath slit and expanded from metal sheets into such a form that there will be no rib in the lath.

**rib.** An unexpanded portion of metal lath which leaves the plane of the lath at a certain angle and returns at the same angle, or a separately attached stiffening member.

**flat rib metal lath.** A combination of expanded metal lath and ribs in which the rib has a total depth of less than ⅛" measured from top inside of the lath to the top side of the rib.

**⅜" rib metal lath.** A combination of expanded metal lath and ribs of a total depth of approximately ⅜" measured from top inside of the lath to the top side of the rib, or other metal lath of equal rigidity.

**¾" rib metal lath.** A combination of expanded metal lath and ribs of a total depth of approximately ¾" measured from the top inside of the lath to the top side of the rib.

**self-furring metal lath.** A metal lath so formed that portions of it extend from the face of the lath so that it is separated at least ¼" from the background to which it is attached.

**paper-backed metal lath** or **expanded metal reinforcing.** A factory-assembled combination of any of the preceding defined types of metal lath or expanded metal reinforcing with paper, fibre, or other backing, the assembly being used as a plaster or stucco base.

**sheet lath.** A metal lath, slit, or punched or otherwise formed from metal sheets (Note: See manufacturers' catalogs for detailed description of various types of sheet lath for special purposes.)

**expanded metal reinforcing (stucco mesh).** A reinforcement similar to metal lath but cut and expanded from heavier metal sheets and especially suitable for exterior stuccowork and for reinforcing cement base for tile and mosaic floors.

**Cornerite** or **corner lath.** A strip of diamond mesh metal lath bent to form a right angle and applied as corner reinforcement at all internal vertical and horizontal angles of any interior surfaces to be plastered, including wood lath, gypsum lath, fibreboard lath, and junctures at wall and ceiling intersections of dissimilar plaster bases; also used in angles where ends of rib or sheet lath abut.

**Stripite** or **strip lath.** A narrow strip of diamond mesh metal lath applied as a reinforcement over joints between sheets of gypsum lath, fibreboard lath and similar material used as plaster bases, or at junctures between such bases.

**main runners** or **carrying channels.** The heaviest horizontal members, supported by hangers, in a suspended ceiling and to which the furring channels or rods are attached; also may be attached direct to the construction above.

**furring channels** or **rods.** The smallest horizontal members of a suspended ceiling, applied at right angles to the underside of carrying channels and to which the metal lath is attached; also the smallest horizontal members in a furred ceiling; also, in general, the separate members used to space metal lath from any surfaces over which it is applied.

**hangers.** The vertical members which carry the steel framework of a suspended ceiling; also the vertical members which support furring under concrete joist construction; also the wires used in attaching lath directly to concrete joist construction.

**contact ceiling.** A ceiling composed of metal lath and plaster which is secured in direct contact with the construction above, without the use of runner channels or furring.

**furred ceiling.** A ceiling composed of metal lath and plaster which is attached by means of steel channels, or rods, or wood furring strips, in direct contact with the construction above.

**suspended ceiling.** A ceiling composed of metal lath and plaster and steel channels which is suspended from and is not in direct contact with floor or roof construction above.

## *Building code*

The following notes are from the New York City Building Code:

**Metal Lath.** Metal lath shall weigh at least three pounds per square yard and shall be galvanized or painted for interior use and either galvanized or of non-corroding metal for exterior use. Welded wire lath shall be galvanized #16 ga. or larger, with a maximum mesh opening $2'' \times 2''$, or equal weight if the mesh is finer, but in any case at least #20 ga. wire. Expanded metal lath shall be of a type suitable to form a key sufficient to retain the plaster firmly. Metal lath shall be lapped at least one inch on abutting edges and where it finishes against masonry walls the lath shall be extended at least $3''$ on such walls and be securely fastened. Metal lath shall be kept at least $3/8''$ away from sheathing or other solid surfaces.

**Suspended Ceilings.** Suspended ceiling hangers shall be placed to line in either direction with a maximum spacing of five feet on centers. Runners, consisting of $1\frac{1}{2}''$ steel channels or $1\frac{1}{2}'' \times 1\frac{1}{2}'' \times \frac{1}{8}''$ steel angles, shall be bolted to each hanger. Cross furring shall consist of at least one inch steel channels with a maximum spacing of $12''$ on centers and secured to each runner with hairpin steel wire clips at least as thick as #8 ga.

**Plaster Board.** Plaster boards shall be nailed directly to wood studding or furring with $1\frac{1}{8}''$ nails with flat $\frac{3}{8}''$ heads, spaced not more than 6" apart on walls and 4" on ceilings. The joints shall be broken at every other board on walls and at right angles to the furring on ceilings. It shall be unlawful to wet plaster boards before plastering.

**Application of Plaster.** Plaster for plastering with hard white finish shall consist of lime, sand, hair or fibre, or a gypsum hard wall plaster, sand or fibre with a finishing coat of lime putty gaged with calcined plaster. The scratch coat shall be mixed in the proportions of one part of lime putty to two parts of sand by volume with sufficient hair or fibre to form a binder. The brown coat shall be mixed in the proportions of one part of lime putty to three parts of sand by volume. The scratch coat shall be applied to all lathed surfaces, and on walls and partitions such coat shall be carried to the floor. The brown coat shall be applied over the scratch coat where used, and on all masonry surfaces, and shall be carried to the floor. Where lime plaster is used, the brown coat shall be applied only after the scratch coat has thoroughly dried. Where gypsum plaster is used, the brown coat shall be applied only after the scratch coat has become thoroughly set. The finishing coat shall be applied after the second or brown coat has become about dry. Where hard wall plaster is used, such plaster shall be received at the structure in the manufacturer's original package and shall be mixed and applied in accordance with his specifications. When plastering is in progress the structure shall be enclosed and heated in freezing weather.

■ **EXERCISES**

1. Write at least 300 words of notes on lathing and plastering from one or two of the books listed in the Bibliography. Name the books used.
2. Why are such strong members and such careful methods required by building codes in suspended-ceiling construction?
3. Make a sketch plan of a plastered bedroom and closet, showing dimensions, and then prepare a quantity survey of all the plastering work involved.
4. Make a complete estimate of the plastering work for the house shown in Chap. 6. The specifications are as follows:

   PLASTERING

   Apply expanded metal lath, weighing at least 3 lb per sq yd, to all the wall and ceiling surfaces in the bathrooms and kitchen, and on the walls and soffit in the stairway leading from the kitchen to the basement.
   Apply Rocklath on the walls and ceilings in the other rooms, closets, and halls on the first and second floors, except in the storage room. Install strips of expanded metal lath, over the Rocklath, in all vertical and hori-

zontal internal corners and extending at least 3" on each of the surfaces. Install galvanized metal corner beads on all external vertical corners throughout, extending the full height of the plaster.

Apply a scratch coat of cement for the tilework in the bathrooms. Apply three coats of lime plaster, with Keene's cement mixed half and half in the finish coat, on all the metal lath elsewhere.

Apply two coats of gypsum plaster on all the Rocklath.

Finish all plaster to a hard, troweled surface.

Apply three coats of cement stucco on the outside of the foundation walls above grade and on the walls in the rear areaway, with smooth, wood-float finish.

All work shall be done by skilled mechanics and in strict conformity with the building code.

CHAPTER 18

# finish carpentry

Finish carpentry includes all the items of carpentry except what is required for concrete forms and rough-framing work. The formwork is always included under the concrete headings. The rough framing, rough flooring, wall and roof sheathing, and other items of carpentry that are put in place before the lathing is started, are put under the heading Rough Carpentry. On very small jobs, however, all the carpentry may be included under a single heading. On very large jobs it is usually advisable to set up separate headings for the special items, such as stairwork, finish flooring, kalamein work, overhead-type doors, etc. The estimator uses his own judgment about arranging his estimate. The main thing is to see that every item is properly estimated under some suitable heading and that nothing is estimated more than once.

## Millwork

Most contractors take estimates from millwork manufacturers for furnishing all the millwork required for a job. Then they either add a certain percentage of the material figure, to cover the labor, or list the items and price them at prices to suit each one. Others list all the items in detail, just as the millwork dealer would do, and then send copies of their list to several dealers for prices.

It is dangerous to use the percentage method of pricing the labor unless the work is of a type that can be judged accurately. The labor cost, not including insurance or general expense, runs usually between 90 and 120 per cent of the material cost, for ordinary work.

It is a good idea to make small sketches of each piece of millwork listed, especially those that cannot be described in a few words. A sketch often gives more information than even a lengthy description. The sizes of the various members shown in each sketch should, of course, be given in the sketch if they are not listed clearly.

In listing windows, care must be taken to state that the dimensions are sash sizes or mason-opening sizes. Sash sizes are usually given for windows in frame buildings and mason-opening sizes for windows in masonry buildings.

The thickness of doors and of window sashes makes a decided difference in the cost, both for the material and for the labor of installing. The two most common thicknesses are 1⅜″ and 1¾″. Sizes of doors and windows are always given by stating the width first, then the height, and finally the thickness. Thus, a door might be 2′10″ × 6′8″ × 1⅜″, and a window might be 2′9″ × 5′2″ × 1¾″. Note that the thickness of a door refers to the thickness of the stiles and rails and not of the panels, and the thickness of a window means the thickness of the sash and not of the frame. All other essential information should, of course, also be given.

Wood doors are made in many stock sizes and stock patterns. White pine doors are generally used for exterior openings. White pine, fir, and birch are commonly used for interior doors, sometimes one material being used for the stiles and rails and another for the panels. The common thickness for exterior doors is 1¾″, and the thickness most used for interior doors is 1⅜″. These, as has been noted, are the thicknesses of the stiles and rails. Unless raised panels are called for, the panels are usually very thin. Cupboard doors are only about 1″ thick.

A single door consists of one leaf. Double doors, or pairs of doors, have two leaves. Double-acting doors are those that swing into both rooms on double-acting hinges.

Overhead doors, which swing up above the door opening, are provided with special fittings and hardware for this purpose.

Sliding doors slide sidewise or vertically and are provided with special tracks, rollers, guides, etc.

Fireproof doors may be of kalamein, hollow metal, or all steel. Kalamein doors are those made of wood and covered over with metal.

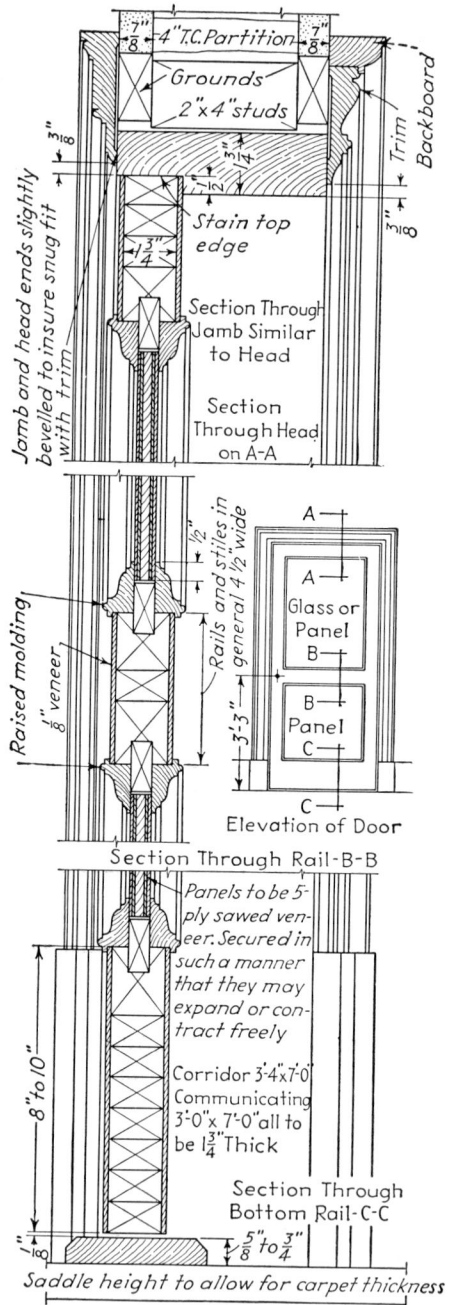

**Fig. 18-1** Vertical sections through door.

# FINISH CARPENTRY 265

Figure 18-1 shows vertical sections through a typical interior door opening. These illustrate the construction of a well-made door. A rabbeted solid frame, rough buck, grounds, two types of trim, and a backband are also shown.

Figure 18-2 shows horizontal sections through two types of doors, one with a solid rabbeted frame and the other with jambs and stops.

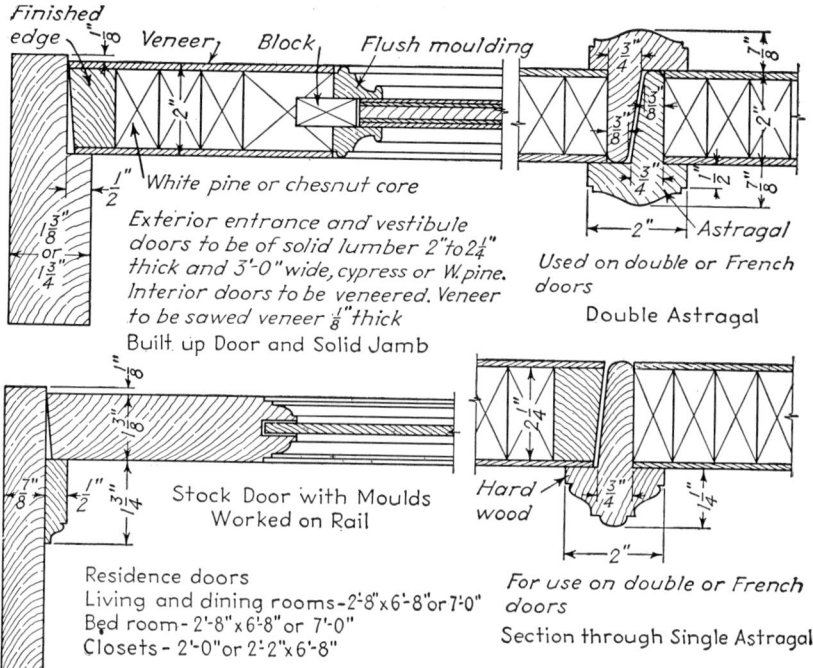

*Fig. 18-2* Horizontal sections through doors.

Figure 18-4 shows the designation of doors as to the swing. These designations are used also in describing the hardware requirements.

Figures 18-5 to 18-8 show various window details. They will also serve to illustrate the entire construction around window openings in both masonry and frame buildings. The student should go over all of these carefully, noting every member and its relation to all the other members.

Exterior trim for water tables, eaves, etc., is carefully listed and priced to suit the work called for on the plans or in the specifications. Simple items may be merely named and the size and number of linear feet given; but elaborate cornices, etc., will require sketches and description.

## Siding

Siding is made in a number of woods. It is usually about ¾" thick and is termed, by the shape of the pieces, as tongue-and-groove, shiplap, or beveled. It is further described with terms such as plain, rabbeted, drop siding, horizontal, and vertical. Figure 18-3 shows some shapes.

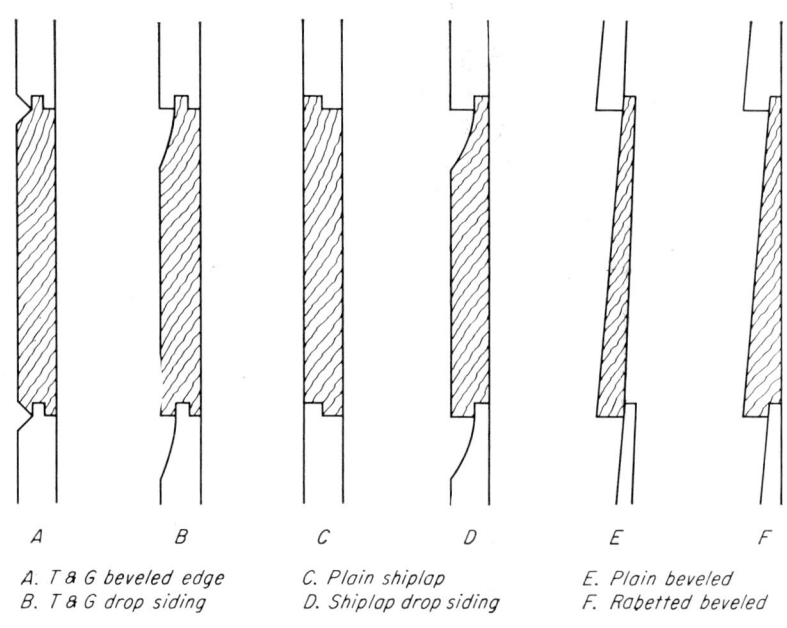

A. T & G beveled edge
B. T & G drop siding
C. Plain shiplap
D. Shiplap drop siding
E. Plain beveled
F. Rabetted beveled

Fig. 18-3  Siding.

The outside of a house, which is subjected to sun, wind, rain, and snow, must withstand severe usage. Exterior wood should be of a decay-resisting species that will hold tight at the joints and will take and hold paint. It should be thoroughly seasoned material. Siding, cut in various patterns, is made in sizes ranging from ½" × 4" to ¾" × 12" and is available in western red cedar, Idaho white pine, northern white pine, and ponderosa pine. In applying siding, all joints around window frames and corner boards should be carefully fitted. Spliced joints should be tight to prevent the infiltration of moisture. No spliced joints in three consecutive boards should be directly over one another. Proper nailing with the right size of nails is important. The nailheads should be set into the wood and the holes filled with putty after the first coat of paint is applied. All exposed woodwork should be given a coat of priming paint as soon as it is in place.

## Trim

Interior trim, such as baseboards, chair rails, wall mouldings, and plain handrails, can generally be listed by stating the number of linear

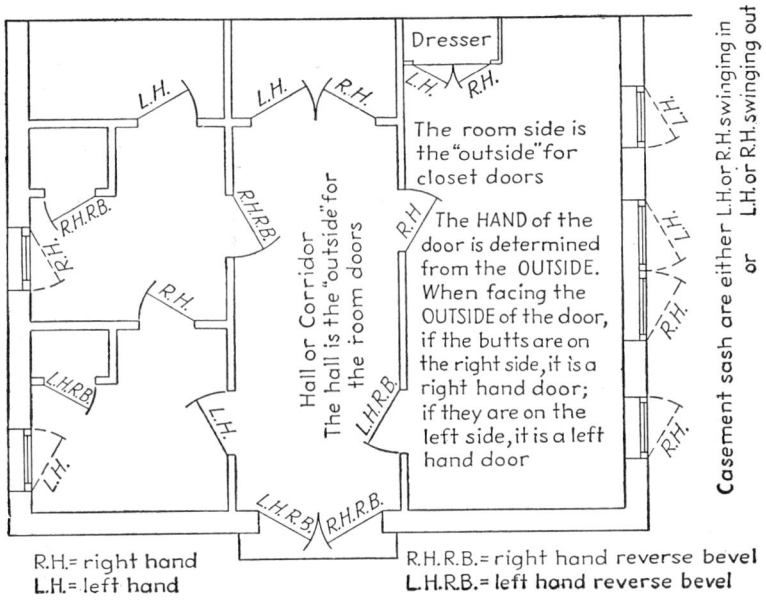

*Fig. 18-4   Designation of doors.*

feet required and the sizes. Careful measurements should be taken of these items, as well as those of more elaborate nature, with particular care that all the items are included and all the rooms are taken into account.

The trim for doors and windows is often included with the pricing of the doors and windows. If not, these items must be listed and priced, either by the linear foot or by the side of trim. A side of trim means the trim required for one side of the door or window. Elaborate trim must, of course, be described and sketched with care.

Handrails, although they may be priced per linear foot for straight work or for straight portions, also require pricing by the piece for any easements, twists, etc., in addition, and pricing per piece for handrail

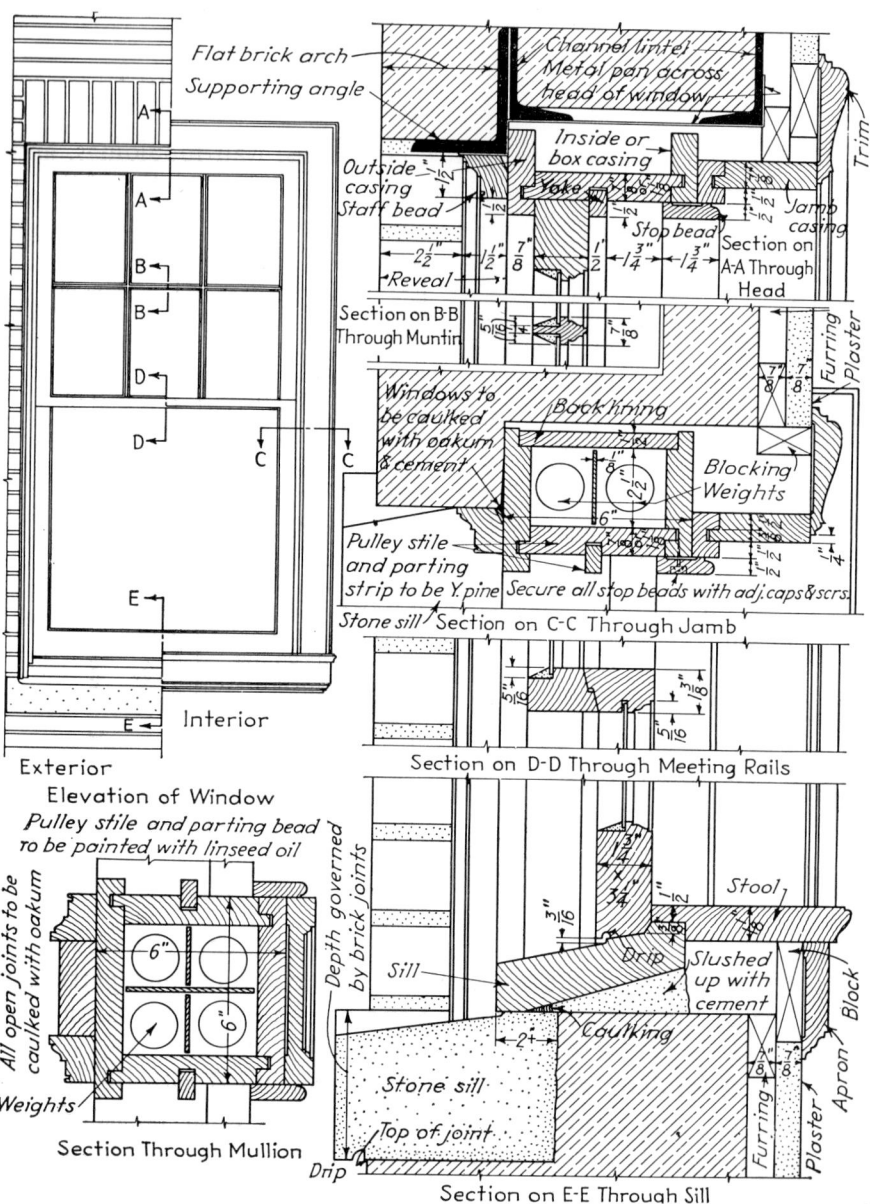

**Fig. 18-5** *Double-hung window in masonry.*

# FINISH CARPENTRY 269

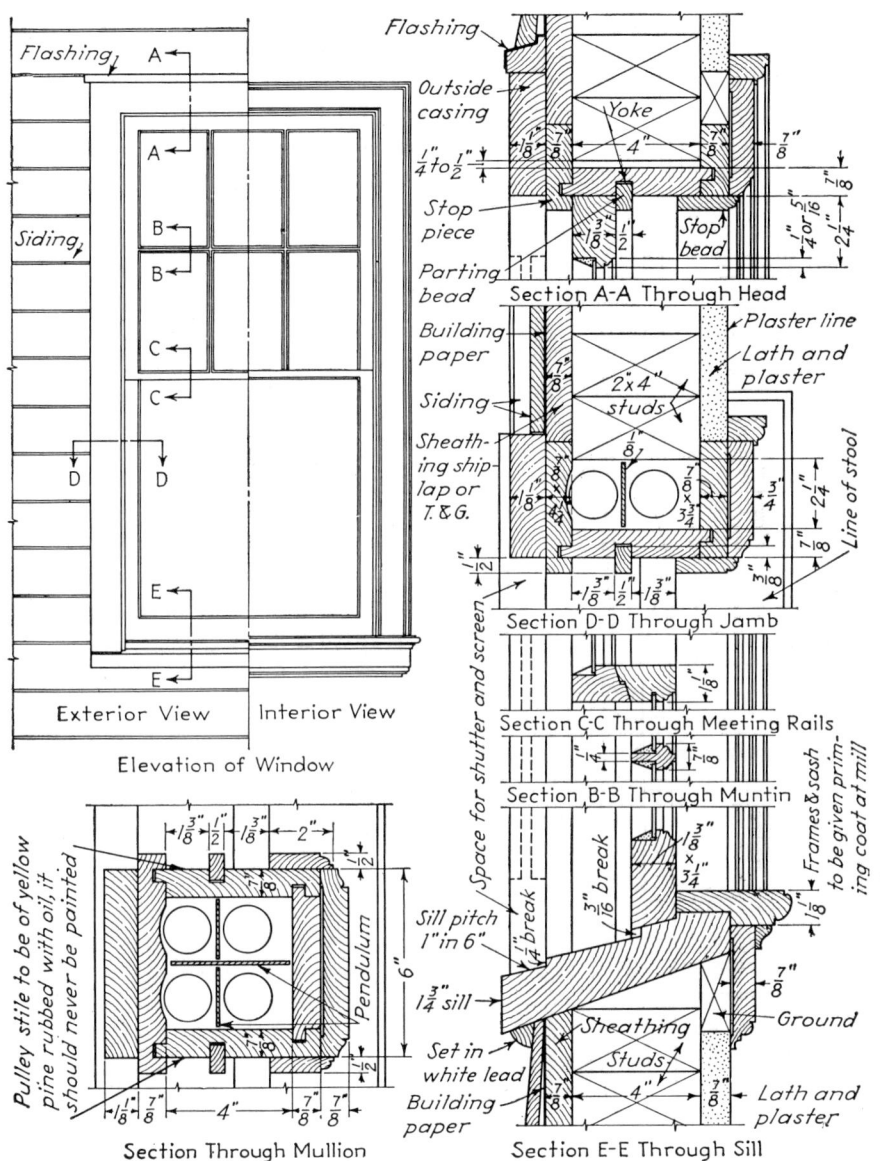

*Fig. 18-6* Double-hung windows in frame.

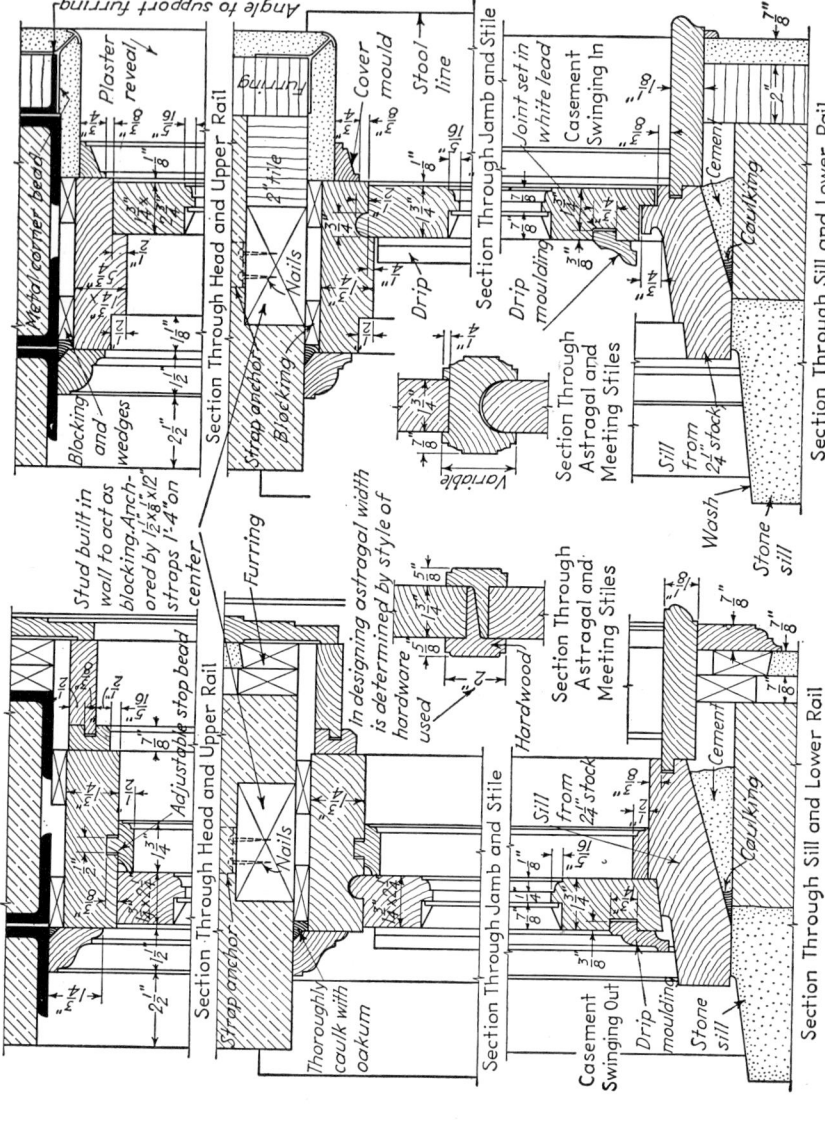

Fig. 18-7 Casement windows in masonry.

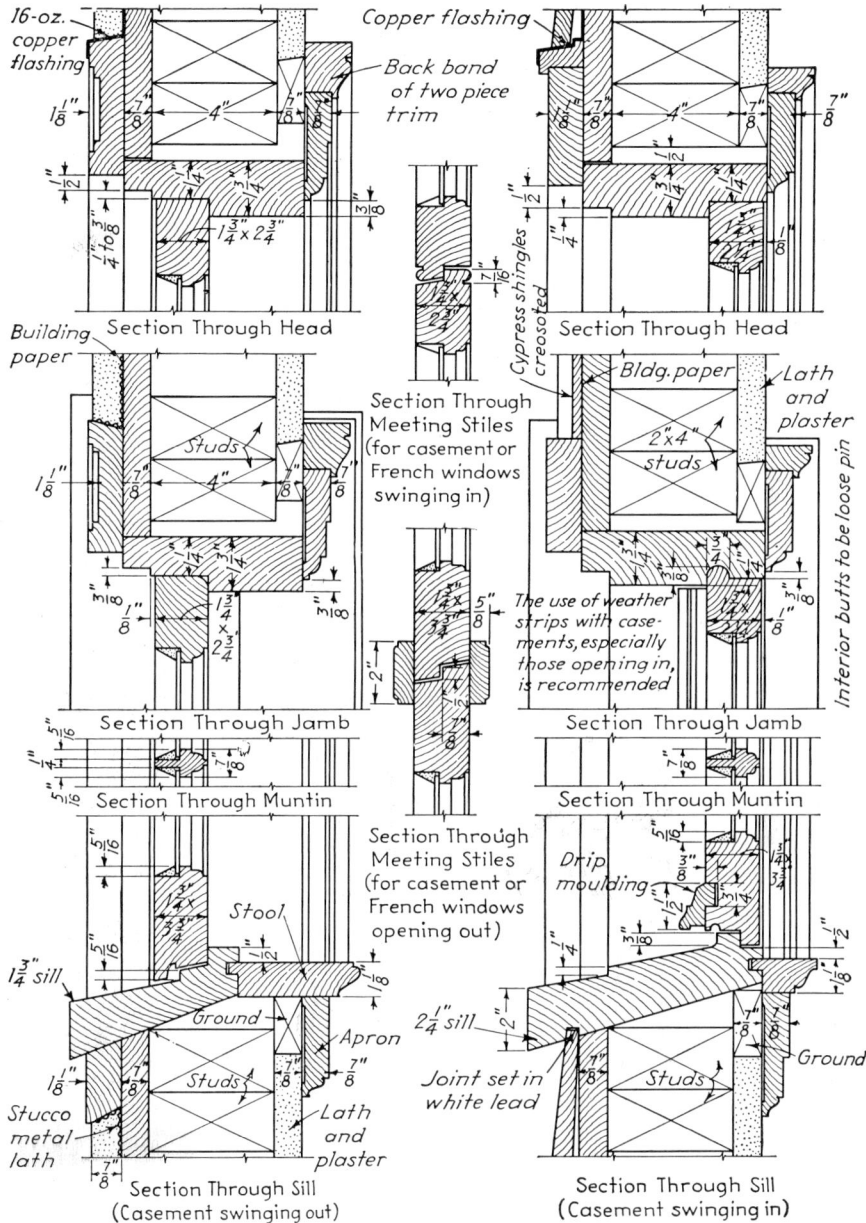

Fig. 18-8 Casement windows in frame.

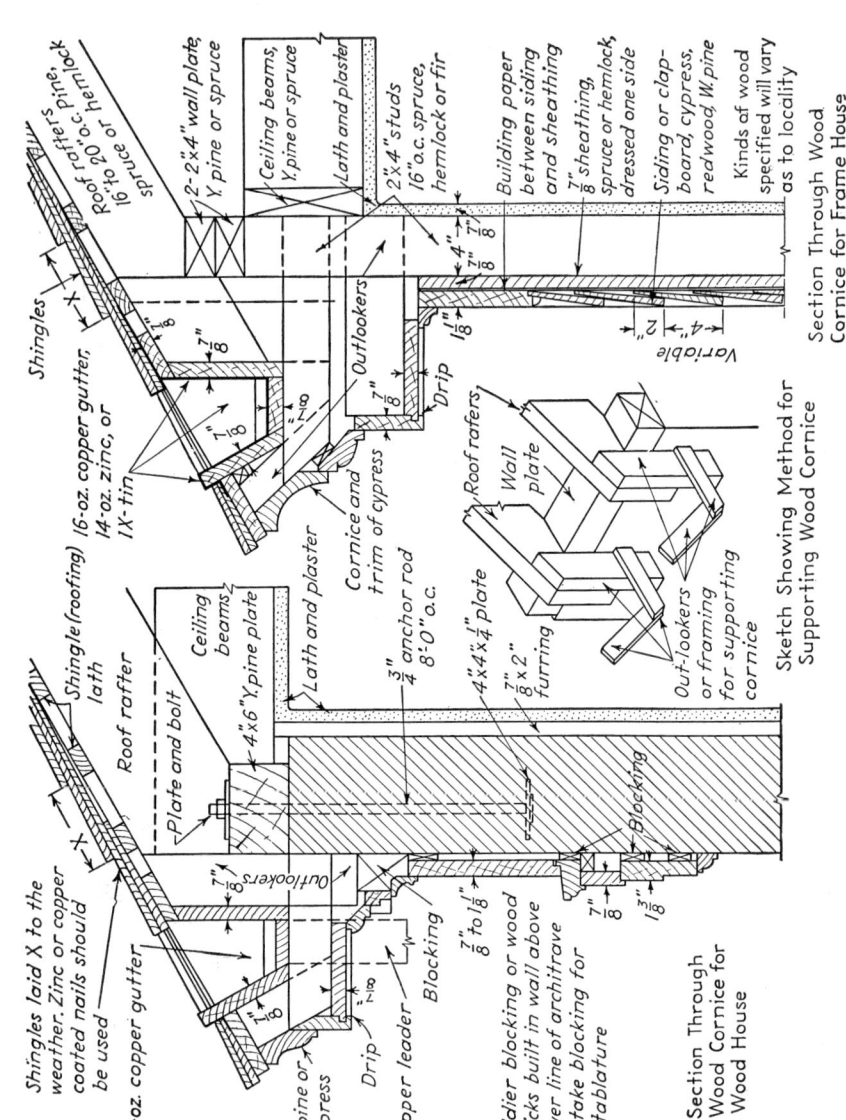

Fig. 18-9 Wooden cornices.

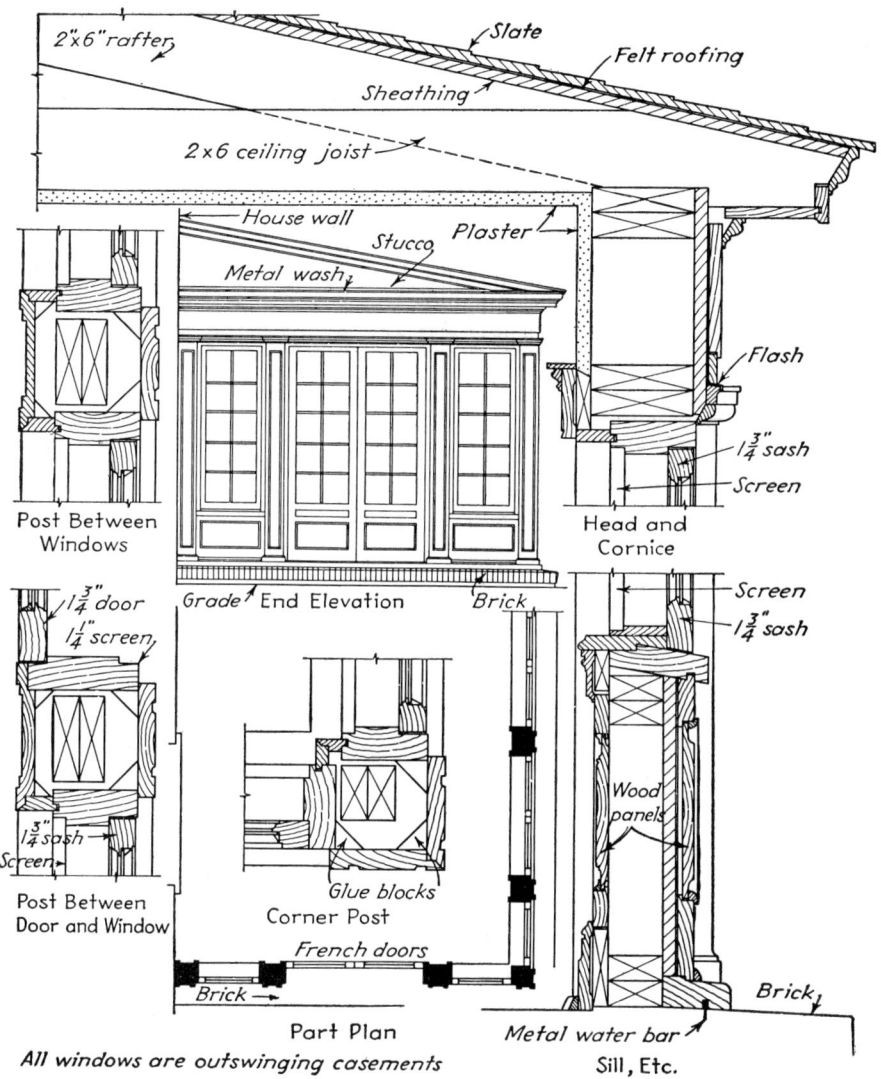

Fig. 18-10  A sun porch.

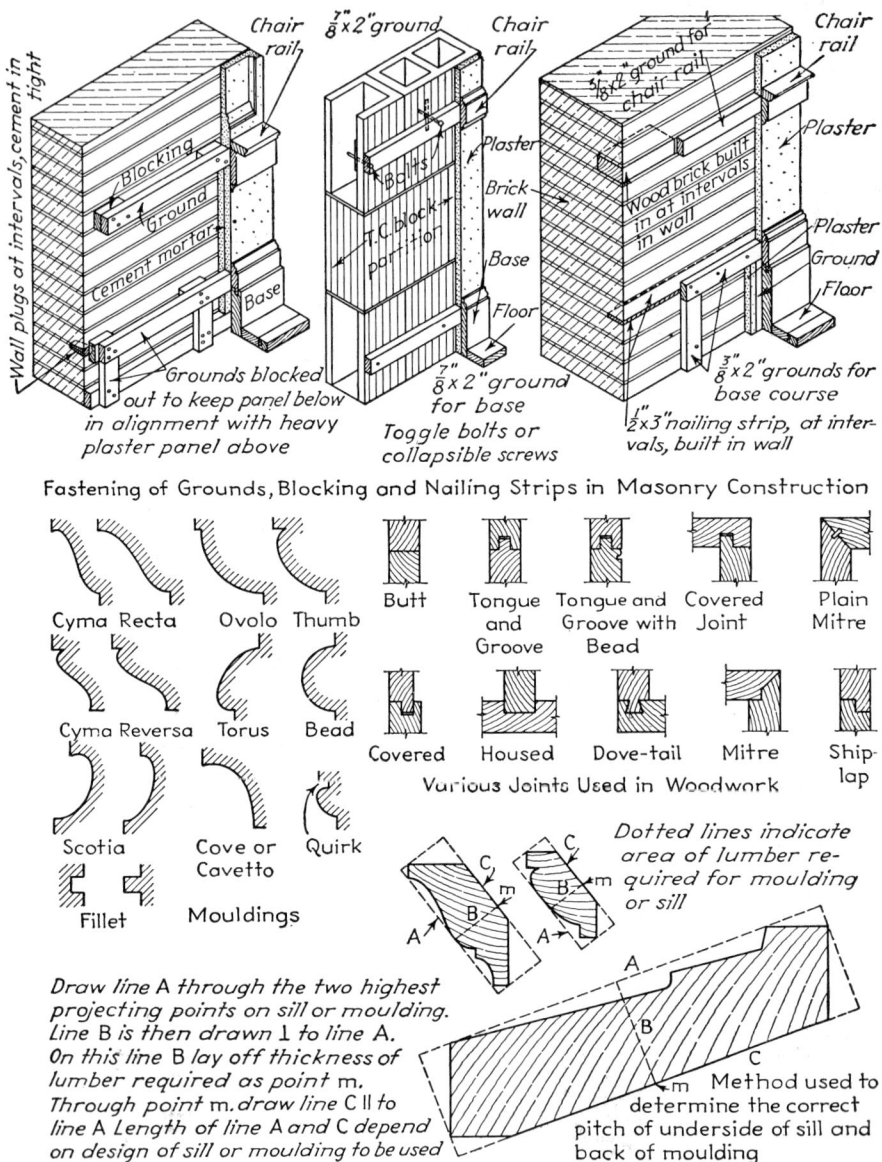

Fig. 18-11 Grounds, joinery, and mouldings.

brackets that may be required. Easements and twists may cost $8 to $25 each.

Figure 18-9 shows details of exterior cornices on masonry and frame buildings. While these are more elaborate than those found on ordinary buildings, they will serve to illustrate many features of construction with which the estimator should be familiar. Go over these details, item by item. Note the work involved in putting the various parts together. Note the amount of blocking required for work of this character. Realize the substantial scaffolding that must be provided for doing the work safely.

Figure 18-10 shows the construction of a typical porch of the sunroom type. Note here the great amount of blocking and the large number of members required. The hardware alone for a porch with many doors and windows, such as this one has, is an item of considerable expense to be estimated, especially if it is solid bronze as it should be.

Figure 18-11 shows various items of interior trim and some of the blocking that must be considered in connection with it, if the blocking is not provided for under the heading of Rough Carpentry.

## Labor

Table 18-1 gives the approximate time required to install some items of finish carpentry.

## Flooring

Wood flooring is generally $25/32''$ thick and tongued-and-grooved, or "matched." The most common face widths are $2\frac{1}{4}''$ and $3\frac{1}{4}''$. These widths are made from 3" and 4" wide stock and are figured on the basis of 3" and 4" widths. As several grades of flooring are made in each kind, the estimator must be careful to price the work to suit the grade specified.

Oak flooring is classified first as white or red oak. It is further classified as plain-sawed or quartersawed. Still further it is graded as clear, select, No. 1 or No. 2, if it is plain-sawed material, or as clear, sap clear, or select, if it is quartersawed. Maple flooring is classified as white clear, red clear, first-grade, second-grade, or third-grade. Pine flooring is first classified as flat grain or comb grain (sometimes called edge grain), and is further graded as Grade A, Grade B, or Grade C.

As will be seen from the stock required to make the flooring, there is considerable loss of surface, owing to the manufacturing. There is also waste in cutting at the job, owing to pieces being rejected. These losses make it necessary to add between one-third and one-half to the

TABLE 18-1

### Interior

| | |
|---|---|
| Plain window casings | 1¼ hr per side |
| Backband window casings | 1½ hr per side |
| Moulded window casings | 2 hr per side |
| Plain door casings | 1 hr per side |
| Backband door casings | 1¼ hr per side |
| Moulded door casings | 1½ hr per side |
| Jambs and stops | 1½ hr per set |
| Oak doors and hardware | 2½ hr each |
| Pine doors and hardware | 1½ hr each |
| Hardwood base | 12 hr per 100 lin ft |
| Pine base | 9 hr per 100 lin ft |
| Base floor mould | 4 hr per 100 lin ft |
| Picture moulding | 4½ hr per 100 lin ft |
| Cove cornice | 13 hr per 100 lin ft |
| Two-member cornice | 26 hr per 100 lin ft |
| Chair rail | 10 hr per 100 lin ft |

### Exterior

| | |
|---|---|
| Plain outlookers | ¼ hr each |
| Facia | 8 hr per 100 lin ft |
| Plancier | 8 hr per 100 lin ft |
| Verge boards | 13 hr per 100 lin ft |
| Crown or bed mould | 5 hr per 100 lin ft |
| Cove or quarter round | 3 hr per 100 lin ft |
| Single window frame | ¾ hr each |
| Mullion window frame | 1 hr each |
| Triple window frame | 1¼ hr each |
| Window sash | 1 hr each |
| Doorframe | 1 hr each |
| Front door | 3 hr each |
| Rear door | 2 hr each |
| Screen or storm door | ½ hr each |

floor area, in order to obtain the number of board feet required. If the work is priced on the basis of the square foot of floor surface, this fact must be carefully borne in mind and the unit price must be increased to suit the case.

Wood flooring is generally laid on top of wood underflooring. The underflooring is of ordinary 1″ boards, usually T&G and usually laid diagonally across the floor beams; 1" × 6" boards are most commonly used for this purpose. These are actually only about ¾" thick and have a face width, when laid, of only 5¼". This rough flooring is listed under Rough Carpentry.

Wood flooring is generally scraped and then finished in one of several ways. When these items are called for, they are priced in addition to the flooring itself, on the basis of the number of square feet of floor area. Scraping and finishing are sometimes placed under the heading Painting, instead of Carpentry. The additional cost runs between 4 and 7 cents per sq ft for scraping, and between 6 and 9 cents for the finishing.

Flooring costs between $180 and $300 per MBM for the material and about the same range for the labor for regular kinds. This total range of $360 to $600 equals about $500 to $950 per thousand sq ft, or 50 to 95 cents per sq ft. As has been stated before, the narrower the board, the greater will be the percentage of waste as a result of forming the tongues and grooves, etc.

**Oak Flooring.** Oak is heavy, hard, and strong. Commercial white oak averages 47 lb per cu ft at 12 per cent moisture content. Commercial red oak averages 44 lb per cu ft at 12 per cent moisture content. Pieces or boards are considered as quartersawed when 80 per cent of the surface shows the radial grain at an angle of 45° or less. Figure 18-12 shows a quartersawed board (the one taken from the center of the log) and its position in the log. The radiating lines of the medullary rays (shown at the end of the log) lie nearly parallel to the plane of the board, while the growth rings cut across through the board at nearly right angles. As the sawing continues and the quartered boards are cut at distances away from the plane through the center of the log, the angle made by the rays with the plane of the boards increases and, after a 45° angle is obtained, the boards are classed as plain-sawed and have a similar appearance to the board shown coming from the outer portion of the log.

In addition to having desirable beauty, quartersawed material shrinks and swells less than plain-sawed lumber, and it does not surface-check or split as much as the plain-sawed kind. It takes about 10 operations with specialized machinery to produce good flooring. After the oak lumber comes from the dry kilns, it is fed through the ripping machines, which cut it into widths suitable for flooring. It then travels on endless chains along the platform and through a planer and a flooring machine. The flooring machine tongues and grooves, hollows the back, and stamps the maker's name and trademark on the strips. The strips are next conveyed through an end-matcher machine and are then bundled.

**Maple Flooring.** Maple, beech, and birch are heavy, strong, and hard woods. They are very much alike, and sometimes one is substi-

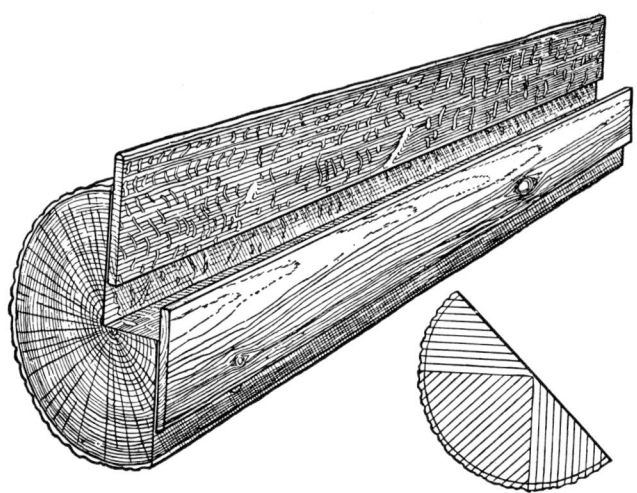

*Fig. 18-12   Quartersawed and plain-sawed boards.*

tuted for another. All three are close-grained, hard-fibred, and free from slivering and splintering.

Grades 1, 2, and 3 are the standard commercial grades. The clear grades are special grades selected for color. First grade is the highest standard grade and is the most durable and desirable for fine houses, churches, dance floors, gymnasiums, public buildings, and other places where appearance as well as durability must be considered. Second grade is practically as serviceable as first grade and has only slight imperfections. Third grade is of poorer quality and shorter lengths but is suitable for factories, workshops, and other places requiring a strong, hard floor.

# Wood

Softwoods used in building work are fir, yellow pine, white pine, spruce, redwood, and occasionally cypress and cedar. Hardwoods used are oak, maple, birch, walnut, gum, mahogany, and occasionally others.

**Plywood.** Plywood in general use is composed of three to five layers of wood glued together, under pressure, in the mill. Waterproof glue is used in plywood intended for exterior use. Thicknesses of ¼″ to ¾″ are commonly used. Other plies and thicknesses are also available. The sheets, or panels, are 48″ wide and some types are also made in 30″, 36″, and 42″ widths. Lengths of 8′, 9′, 10′, and 12′ are obtainable in most types, and some come in other lengths. Plywood is made of many kinds and grades of wood. The grading refers to the appearance of the outer surface and to the quality of the wood used for the inner layers.

**Wallboards.** Wallboards for interior use are composed of compressed gypsum compositions, wood fibres, or insulating materials, in various thicknesses. They are 48″ wide and come in 6′ to 12′ lengths.

Gypsum board is fireproof. Some types are made with simulated wood finishes on the paper facing. Plaster may be applied to the plain types, if desired. Gypsum board is made in ¼″, ⅜″, ½″, and ⅝″ thicknesses.

Fibreboards, or hardboards, are made in standard types and also in tempered types with a smooth, hard finish on one side. Some are scored on one side in imitation of tile. Some have a finish coating. Fibreboard is made in ⅛″, 3/16″, ¼″, and 5/16″ thicknesses.

Insulating board is made in plain types and also with various finishes, in thicknesses of ½″, ¾″, and 1″. Tiles of 12″ × 12″, 12″ × 24″, and other sizes, and also planks, are made in some types of insulating board.

■ EXERCISES

1. Name 10 items of finish carpentry.
2. Write at least 300 words of notes on finish carpentry from one or two of the books listed in the Bibliography. Name the books used.
3. Make a list of all the members shown in Fig. 18-6 and give a brief description of the meaning and use of each of these terms.
4. Make a complete estimate of the finish carpentry for the house shown in Chap. 6. The specifications are as follows:

FINISH CARPENTRY

**Materials.** All materials shall be thoroughly seasoned and protected from the weather until placed in the building. All interior trim and doors shall be sandpapered. All tool and erection marks shall be carefully removed from finished surfaces. All workmanship must be perfect. All joints, where possible, shall be tongued and rabbeted together to conceal shrinkage. Exterior wood finish shall be selected stock white pine and all interior finish southern poplar, except where otherwise noted below.

**Doors.** Exterior doors shall be 1¾" thick of stock design as selected by the architect. The interior door in the garage shall be a flush kalamein door with kalamein frame. All other interior doors shall be white pine 1⅜" thick, six-panel colonial design, with raised panels. Exterior door frames shall be solid rabbeted 1⅛" thick white pine. Interior door jambs shall be ⅞" thick. Install hardware as specified under the hardware heading. Mirrored doors shall have full-length plate-glass mirrors on one side as indicated on the plans.

**Windows and Screens.** Provide a hinged steel window in the basement wall near the boiler, and aluminum louvered frames in the gables. All other windows shall be aluminum of approved good quality with approved hardware and adjusters. The double-hung windows shall have sash balances, locks, handles, and stainless steel weather strips. Provide full-size aluminum insect screens, made by the window manufacturer, on all windows that open.

**Exterior Trim.** Provide all exterior trim required, including 1⅛" × 2¾" plain window frames and casings, 1¾" × 7¾" oak window sills, 1⅛" × 3¾" moulded door casings, as well as the bay-window and living-room-window woodwork, cornices, cornice returns, rakes, brackets, mouldings, overhead-door frame, shutters, flower boxes, arbor, and gate, etc., all of substantial construction and as directed by the architect.

**Wall Covering.** Cover all exterior walls, except in the gable areas, with No. 1 red cedar 16" shingles set with 10" exposed to the weather. Cover all gables with redwood boards ½" × 10", set vertically and 1 in. apart on ¾" × 4" redwood battens, battens behind the joints. Provide black waterproof building paper, well lapped, behind all siding and casings and wall trim.

**Porches.** Provide front- and rear-porch posts and railings, step and screen at rear porch, and seat on front porch, all of white pine substantially constructed. Lay yellow pine or cypress ¾" × 2½" flooring and nosings on the rear porch and on the sun deck. Provide ½" × 2½" T&G white pine or cypress ceilings in the front and rear porches.

**Interior Trim.** Provide ¾" × 3¾" moulded trim around all doors and windows. Provide 1⅛" stools rabbeted over ¾" × 3¾" moulded aprons at all window openings. Provide ¾" × 3¾" moulded chair rail in the dining room. Provide ¾" × 5¼" baseboard, with two mouldings, and ¾" × 4¾" cornice mould throughout, except in the basement, bathrooms, and closets. Provide one shelf, a chrome-plated hanger pole, and ¾" × 3¼" baseboard in all closets, with six shelves in the linen closet. Provide substantial shelves in the basement as indicated on the plans, consisting of three 16" shelves under the stairs, 16" workbench and two 12" shelves in the laundry, 16" counter and ten 12" shelves in the pantry, and one shelf in the garage. Provide plywood cabinet with two doors, and folding ironing board with cabinet, in the laundry. Construct cedar closet in the storage room.

**Cabinetwork.** Provide plywood and cabinets in the foyer and vestibule and at the fireplace in the living room, as indicated on the plans, including door trim and trimmed openings, all of oak in high-grade cabinet construction. All the kitchen and dining-nook cabinets, counters, and other fittings will be provided by others under a separate contract.

**Stairs.** The stairs, from basement to second floor, shall have $2'' \times 12''$ carriages, $1\frac{1}{8}''$ oak treads, $\frac{7}{8}''$ white pine risers, $1\frac{1}{8}''$ sq balusters set three to a tread, and $3\frac{1}{4}'' \times 3\frac{1}{4}''$ moulded oak newel posts. Treads and risers shall be mortised together, wedged, and glued. Provide moulded oak $2\frac{1}{4}'' \times 2\frac{1}{4}''$ handrails.

**Flooring.** Lay $\frac{3}{4}''$ thick Douglas fir plywood flooring, 4-ft sheets in long lengths, in the living room and kitchen, well nailed down at all bearings. Lay No. 1 quartered oak flooring, $\frac{25}{32}'' \times 2\frac{1}{4}''$ face, in the dining room. Lay No. 2 plain-sawed oak flooring, $\frac{25}{32}'' \times 2\frac{1}{4}''$ face, in all the other rooms, halls, and closets on the first and second floors, except in the bathrooms. Lay $\frac{1}{2}''$ thick plywood in the storage room. Scrape all ridges from the plywood flooring. Scrape all oak flooring and sand it to a smooth surface ready for finishing.

CHAPTER 19

# steel and iron

The subcontractors for steelwork include all the steel columns, girders, and beams, and the steel plates and connections that go with these items. They also include the heavier types of steel lintels and any special steel framing that may be required for tank and machinery supports and other purposes, in addition to that for the general support of the building. If there are steel trusses, these also would be figured by the steel contractors' estimators (Figs. 19-1 to 19-9).

The subcontractors for ornamental and miscellaneous ironwork include all the steel and iron items called for on the plans and in the specifications, except those that are put under the Structural Steel heading. Sometimes there is a question as to the heading under which certain items should be entered. Small items of steel, especially those that are plain members or that do not frame into other steel, are examples. It is part of the general contractor's work of coordinating the trades to see that questionable items are taken care of by one or the other of steel and iron contractors.

Occasionally, one subcontractor will include all the structural steel and the miscellaneous and ornamental iron. This happens especially when there is a considerable amount of ironwork and only a few items of structural steel.

**Structural Steel.** To compute the cost of the structural steel, the steel estimators first list all the pieces of steel shown on the plans. The cross-section type and size, the length, and the weight per linear foot are generally taken. The extensions are made in tons, and the work is priced on that basis. Some of the factors considered in making up the estimated cost are shop drawings, erection drawings, time for delivery, mill or stock material, fabrication, shop painting, freight, trucking, steel members, rivets, welding, bolts, connections, bearing plates, unloading, erection, equipment, scaffolding, planking, field painting, foremen, insurance, general job expense, overhead expense, and profit.

If inspection of the structural steel is called for, either at the steel plant or fabricating shop or at the job, the expense in connection with this must be considered also. Perhaps the specification makes a lump-sum allowance for this purpose or specifies the name of an inspection concern to do the inspecting. In addition to this element of cost, which may have to be included by the general contractor's estimator instead of by the steel contractor, other items require careful consideration. The shop or erection drawings are sometimes so specified that the general contractor has to judge the cost and make a suitable entry in the general estimate, instead of having the steel contractors include such cost in their estimates to him.

Steel specifications often call for the grouting or bedding of the steel members that bear on concrete or masonry. This work is practically never included by the steel contractors, and the general contractor's estimator must therefore make sure that the cost is covered in his estimates for the concrete or masonry work.

The student should become aware of the fact that architects assume no responsibility for the correctness of shop drawings and schedules sent to them for approval, in any line of work, even if they sign the drawings or schedules as meeting with their approval. The approval is always understood to cover the general design of the work only and not the correctness of the detail members or measurements, or the fitting together of the work in the job. Too many contractors rely on the architect's approval of the drawings submitted by themselves or by their subcontractors. This approval does not constitute a thorough check of the drawings. Neither does an architect guarantee that various drawings sent in by subcontractors in different trades coordinate with one another. This all means that the estimator has to consider carefully any cost items of this sort that come to his mind in estimating. He may have to make the drawings himself for some of the subcontractors, or he may have a lot of trouble checking and correcting drawings. It may even be necessary for him to reconsider the type of subcontractors that his firm plans to use on a particular job.

The contractor assumes practically all the responsibility for the job, once he has signed the contract. He may pass along a considerable amount of this responsibility to his subcontractors and, of course, this is what he should do, because they are the experts and specialists in their own fields of work and may, therefore, be expected to undertake

**Fig. 19-1** *A steel-frame building.*

the responsibilities that their work entails. The estimator must see that subcontracts are so worded that the subcontractors take the whole of their fair share of responsibility. The estimator should also endeavor to have his firm's interests properly covered in the general contract with the owner. Contractors are too prone to sign the contract form that is given them by the architect (and usually double-checked beforehand by the owner or the owner's lawyer) without considering whether changes or additional clauses might not reasonably be requested by himself, the better to protect his own interests.

The steel framework of buildings is made up mainly of H-columns, I-beams, channels, and angles. This is all highly standardized. Reference lists like Tables 19-1 to 19-5 are used by architects and engineers when they design the structure and by steel contractors when they prepare the material lists and schedules. The tables give only the more commonly used sizes and the weights per linear foot. The wide-flange I-beams are used also for columns.

TABLE 19-1 WIDE-FLANGE I-BEAMS

| Size, in. | Lb per ft | Size, in. | Lb per ft |
|---|---|---|---|
| 27 | 163 | 14 | 74 |
|  | 145 |  | 68 |
|  | 106 |  | 61 |
|  | 98 |  | 53 |
|  | 91 |  | 48 |
| 24 | 150 |  | 43 |
|  | 130 |  | 38 |
|  | 120 |  | 34 |
|  | 100 |  | 30 |
|  | 87 | 12 | 72 |
|  | 80 |  | 65 |
|  | 74 |  | 58 |
| 21 | 132 |  | 53 |
|  | 112 |  | 45 |
|  | 96 |  | 40 |
|  | 82 |  | 36 |
|  | 73 |  | 28 |
|  | 68 |  | 25 |
|  | 63 |  | 22 |
|  | 59 |  | 19 |
| 18 | 114 | 10 | 66 |
|  | 105 |  | 54 |
|  | 96 |  | 49 |
|  | 85 |  | 41 |
|  | 77 |  | 33 |
|  | 70 |  | 26 |
|  | 64 |  | 21 |
|  | 55 |  | 19 |
|  | 50 |  | 17 |
| 16 | 96 |  | 15 |
|  | 88 | 8 | 40 |
|  | 78 |  | 35 |
|  | 71 |  | 31 |
|  | 64 |  | 27 |
|  | 58 |  | 24 |
|  | 50 |  | 21 |
|  | 45 |  | 17 |
|  | 40 |  | 15 |
|  | 36 |  | 13 |
| 14 | 84 | 6 | 16 |
|  | 78 |  | 12 |

TABLE 19-2 STANDARD I-BEAMS

| Size, in. | Lb per ft |
|---|---|
| 24 | 105.9 |
|  | 100 |
|  | 90 |
|  | 79.9 |
| 20 | 95.0 |
|  | 85 |
|  | 75 |
|  | 65.4 |
| 18 | 54.7 |
| 15 | 50.0 |
|  | 42.9 |
| 12 | 50.0 |
|  | 40.8 |
|  | 35.0 |
|  | 31.8 |
| 10 | 35.0 |
|  | 25.4 |
| 8 | 23.0 |
|  | 18.4 |
| 7 | 15.3 |
| 6 | 17.25 |
|  | 12.5 |
| 5 | 10.0 |
| 4 | 9.5 |
|  | 7.7 |
| 3 | 7.5 |
|  | 5.7 |

TABLE 19-3 STANDARD CHANNELS

| Size, in. | Lb per ft |
|---|---|
| 18 | 58 |
|  | 51.9 |
|  | 45.8 |
|  | 42.7 |
| 15 | 55 |
|  | 50 |
|  | 40 |
|  | 33.9 |
| 12 | 30 |
|  | 25 |
|  | 20.7 |
| 10 | 30 |
|  | 25 |
|  | 20 |
|  | 15.3 |
| 9 | 20 |
|  | 15 |
|  | 13.4 |
| 8 | 18.75 |
|  | 13.75 |
|  | 11.5 |
| 7 | 14.75 |
|  | 12.25 |
|  | 9.8 |
| 6 | 15.5 |
|  | 10.5 |
|  | 8.2 |
| 5 | 9.0 |
|  | 6.7 |
| 4 | 7.25 |
|  | 6.25 |
|  | 5.4 |
| 3 | 6.0 |
|  | 5.0 |
|  | 4.1 |

**Steel Costs.** Structural steelwork usually costs the general contractor between $180 and $220 per ton in place. About 4 per cent for rivets and bolts is added to the total weight of the members listed from the plans. Between 15 and 20 per cent is added for connections and other details. For every ton of steel about $3 is added for the cost of the many drawings required. The shop painting of the steel costs between $12 and $15 per ton of steel. Before any of the steel can be erected,

TABLE 19-4 EQUAL-LEG ANGLES

| *Size, in.* | *Lb per ft* |
|---|---|
| 8 × 8 × 1 | 51.0 |
| 7/8 | 45.0 |
| 3/4 | 38.9 |
| 5/8 | 32.7 |
| 1/2 | 26.4 |
| 6 × 6 × 7/8 | 33.1 |
| 3/4 | 28.7 |
| 5/8 | 24.2 |
| 1/2 | 19.6 |
| 3/8 | 14.9 |
| 5 × 5 × 3/4 | 23.6 |
| 5/8 | 20.0 |
| 1/2 | 16.2 |
| 3/8 | 12.3 |
| 4 × 4 × 5/8 | 15.7 |
| 1/2 | 12.8 |
| 3/8 | 9.8 |
| 5/16 | 8.2 |
| 3½ × 3½ × 1/2 | 11.1 |
| 3/8 | 8.5 |
| 5/16 | 7.2 |
| 1/4 | 5.8 |
| 3 × 3 × 3/8 | 7.2 |
| 5/16 | 6.1 |
| 1/4 | 4.9 |
| 2½ × 2½ × 3/8 | 5.9 |
| 5/16 | 5.0 |
| 1/4 | 4.1 |
| 3/16 | 3.07 |
| 2 × 2 × 5/16 | 3.92 |
| 1/4 | 3.19 |
| 3/16 | 2.44 |

TABLE 19-5 UNEQUAL-LEG ANGLES

| Size, in. | *Lb per ft* |
|---|---|
| 8 × 6 × 7/8 | 39.1 |
| 3/4 | 33.8 |
| 5/8 | 28.5 |
| 1/2 | 23.0 |
| 8 × 4 × 3/4 | 28.7 |
| 5/8 | 24.2 |
| 1/2 | 19.6 |
| 7 × 4 × 5/8 | 22.1 |
| 1/2 | 17.9 |
| 7/16 | 15.8 |
| 3/8 | 13.6 |
| 6 × 4 × 5/8 | 20.0 |
| 1/2 | 16.2 |
| 7/16 | 14.3 |
| 3/8 | 12.3 |
| 5 × 3½ × 5/8 | 16.8 |
| 1/2 | 13.6 |
| 3/8 | 10.4 |
| 5/16 | 8.7 |
| 4 × 3½ × 1/2 | 11.9 |
| 3/8 | 9.1 |
| 5/16 | 7.7 |
| 4 × 3 × 1/2 | 11.1 |
| 3/8 | 8.5 |
| 5/16 | 7.2 |
| 1/4 | 5.8 |
| 3½ × 3 × 1/2 | 10.2 |
| 3/8 | 7.9 |
| 5/16 | 6.6 |
| 1/4 | 5.4 |
| 3½ × 2½ × 3/8 | 7.2 |
| 3/16 | 6.1 |
| 1/4 | 4.9 |
| 3 × 2½ × 3/8 | 6.6 |
| 5/16 | 5.6 |
| 1/4 | 4.5 |
| 3 × 2 × 3/8 | 5.9 |
| 5/16 | 5.0 |
| 1/4 | 4.1 |
| 2½ × 2 × 3/8 | 5.3 |
| 5/16 | 4.5 |
| 1/4 | 3.62 |
| 3/16 | 2.75 |
| 2½ × 1½ × 5/16 | 3.92 |
| 1/4 | 3.19 |
| 3/16 | 2.44 |

288  BUILDING CONSTRUCTION ESTIMATING

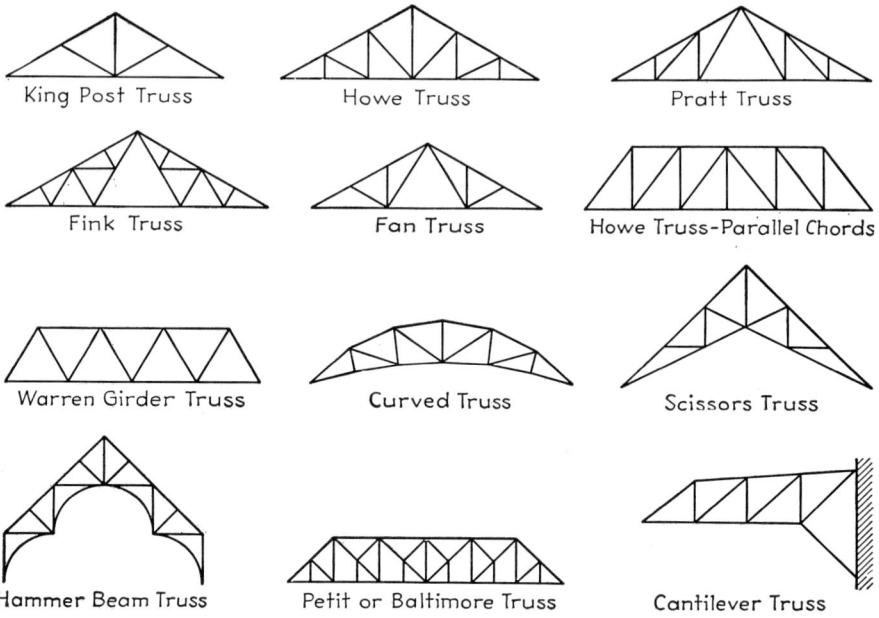

*Fig. 19-2*  *Types of steel trusses.*

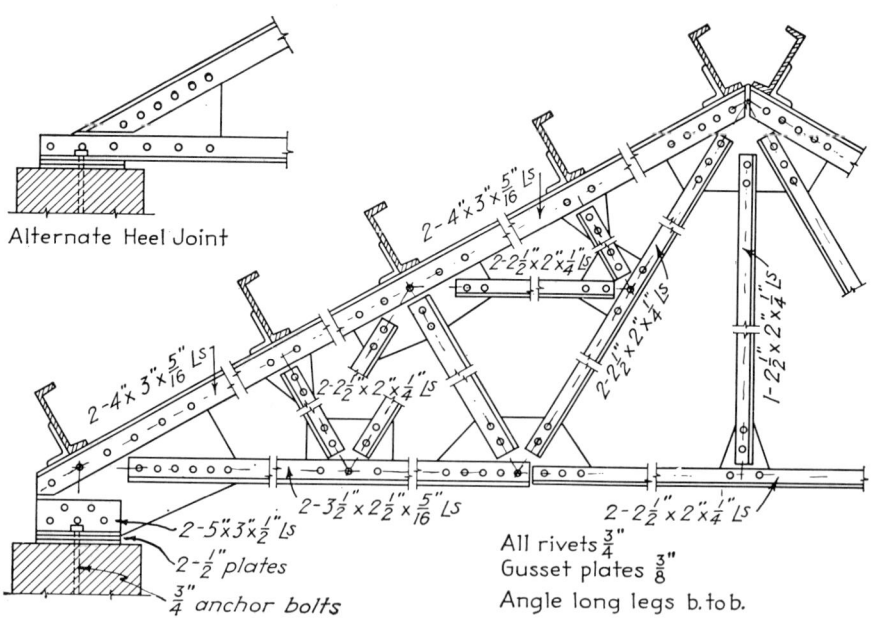

*Fig. 19-3*  *Truss details.*

the contractor sends a considerable amount of equipment to the job. This may include derricks, cranes, hoisting engines, tackle, scaffold planks, a shanty, and an assortment of small tools. Getting all this to the job, installing it in readiness to start operations, maintaining it in good order while the job is under way, dismantling it, and finally removing it from the job are all necessary elements of cost. Setting up a large derrick alone may cost $200. The labor of unloading and erecting the steel will depend upon the type of building, the accessibility of the work, and the amount of steel involved. Riveting or welding, field painting, and other items of direct job cost require consideration also, and the insurance, general expense, and overhead expense must be considered before adding the profit. The cost of workmen's compensation insurance for erecting structural steel amounts to about 42 per cent of the job payroll. Thus, if a small building requires a steel gang for about 2 weeks and the payroll amounts to $1,200, an amount of $504 must be added to cover this one form of insurance alone.

**Miscellaneous Iron.** To compute the cost of the miscellaneous and ornamental ironwork, the iron subcontractors' estimators list their work much as the general contractor's estimator lists millwork. Some of the items, like stairs, fire escapes, door bucks, lintels, and other commonly used items, have become standardized and require only simple listing of the amounts and sizes. Special ornamental railings, grilles, curved features, and other decorative work, however, require careful study of the plans and specifications and careful analysis of the costs involved. Stairs cost between $140 and $180 per story for plain work. Ordinary fire escapes cost between $100 and $150 per story. Door bucks, plain grilles, plain window guards, and other simple items of ironwork cost between $40 and $80 each.

The specification division dealing with miscellaneous ironwork is one that demands careful scrutiny by estimators. Architects use this as a division into which they may throw many items of a special nature that should have special headings, and also some that are already covered under Masonry, Carpentry, or other headings, besides the ironwork heading.

Subcontract bids may read "For all the Miscellaneous Ironwork" or "For all the Ornamental Ironwork" or other wording that does not mean to include everything that is in the architect's specification division referred to. It is the estimator's task to make a list of possible items that he thinks will need checking. Wirework, bronze and other nonferrous items, safety treads, masonry anchors, flagpoles, metal chimneys, coal doors, ash hoists, awning boxes, steel sash, metal toilet partitions, and fittings of various kinds are some of the things that are often thrown into the iron specifications by the architect or the specification writer. The estimator usually has to pull these items out and make special headings in his estimate for them or include them in other divisions of his estimate.

*Fig. 19-4* Steel-truss construction.

Generally speaking, the miscellaneous ironwork includes area gratings, covers and frames, coal chutes, door bucks, ladders, sidewalk doors, steel stairs, wheel and corner guards, curb angles, etc.

Ornamental iron includes grilles, balconies, lamp standards, ornamental brackets, ornamental railings, iron canopies, and special iron construction for decorative purposes.

The iron subcontractors usually suit themselves as to which items on a job they will include in their bids. Some submit, along with their bids, a list of the items included. It frequently happens that a few items are left hanging, being covered by neither sub-bids nor material quotations. The general contractor's estimator then has to scout around and get prices on them or price them himself. Loose lintels, intended to be set by the masons, are often omitted both by the steel contractors and by the iron contractors.

Many items of steel and iron are handled and set by the concrete workers or by the masons, and the cost of this handling and setting must be provided for under appropriate headings in the estimate. Anchors, inserts, sockets, safety treads, gratings, door frames, guards, curb angles, bearing plates, and small steel lintels are some of these items.

| Approx. fraction of inch | No. and size of wire | Decimals of inch |
|---|---|---|
| 1/4 | No. 3 | .2437 |
| 7/32 | 4 | .2253 |
| 13/64 | 5 | .2070 |
| 3/16 | 6 | .1920 |
| 11/64 | 7 | .1770 |
| 5/32 | 8 | .1620 |
| 5/32 | 9 | .1483 |
| 9/64 | 10 | .1350 |
| 1/8 | 11 | .1205 |
| 7/64 | 12 | .1055 |
| 3/32 | 13 | .0915 |
| 5/64 | 14 | .0800 |

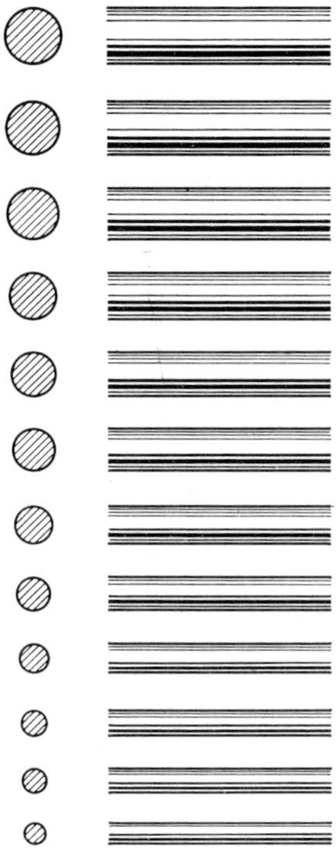

All sizes are A.S. & W. or W. & M. wire gage

*Fig. 19-5  Steel wire.*

**Steel Joists.** Open-web steel joists are used for floor and roof supports. They are generally spaced not more than 24" on centers in floors and not more than 30" in roofs. The ends of the joists must bear at least 4" on masonry walls and at least 2½" on steel supports unless the ends butt and are securely fastened by welding, bolting, or riveting. The ends in walls are built into the walls and every third joist is anchored to the wall with steel anchors. All steel joists must be fastened in place and the permanent bridging installed before any construction loads are placed on them.

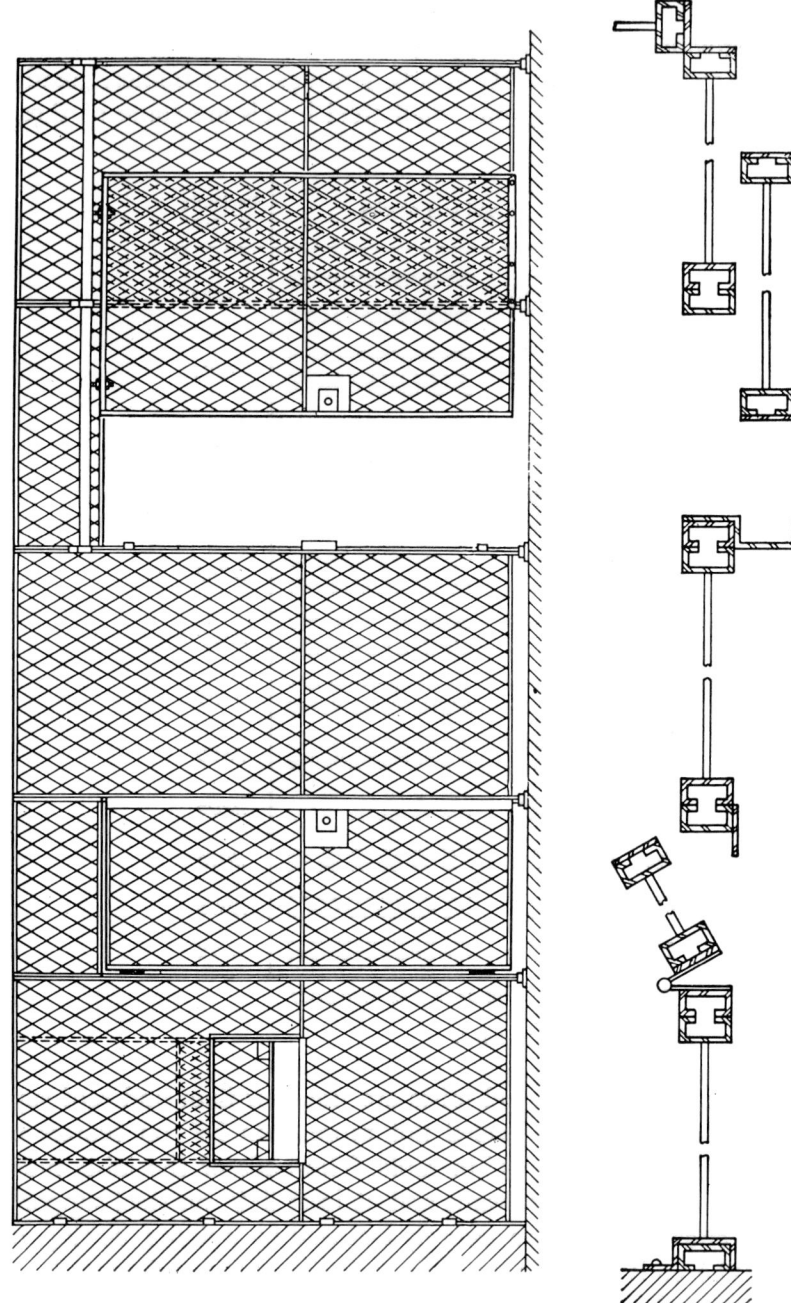

Fig. 19·6 Wire partition.

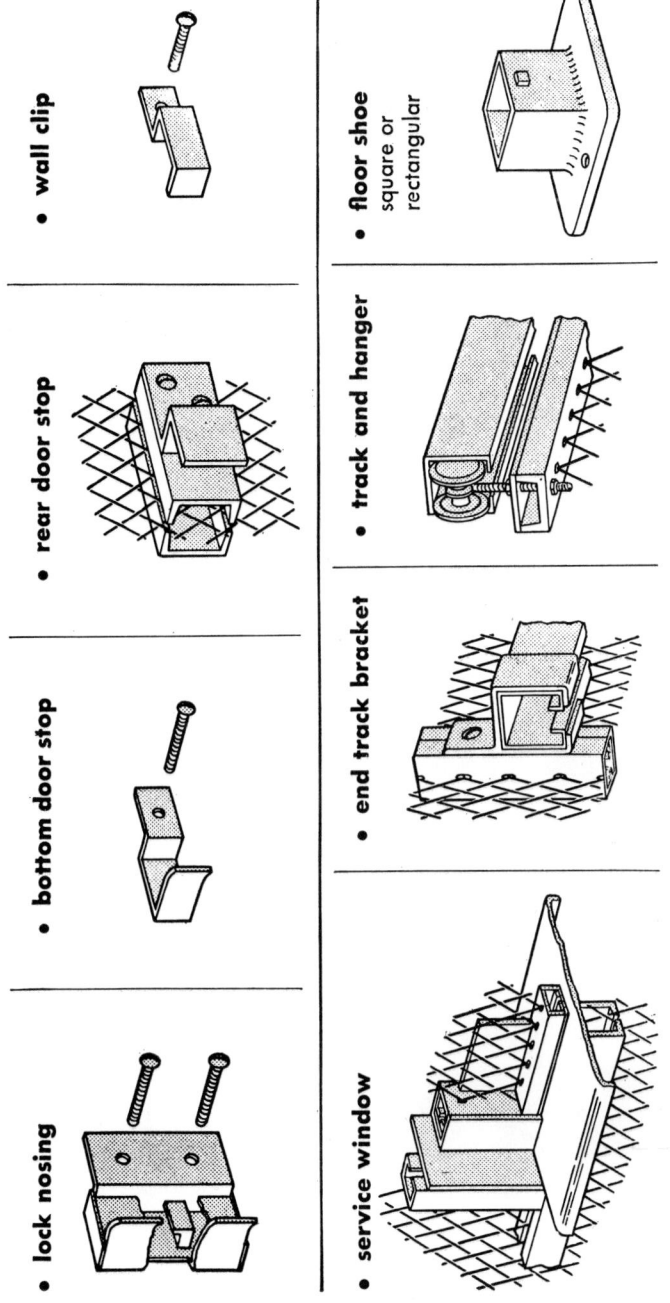

Fig. 19-7 Wire partition details. Specifications: No. 10 wire, 1½" diamond mesh, frames to be 1¼" × ⅝" "C" section channel for vertical and 1" × ½" channels for horizontals. Center reinforcing stiffeners of two ¾" channels riveted or bolted together. Finish: one shop coat of paint. Hinged doors to be fitted with fixtures and padlocks, cylinder rim latches, or mortise-type cylinder locks. Sliding doors to have bronze mortise-type cylinder locks operated by key from outside and recessed knob inside.

293

Fig. 19-8   Steel frame and steel joists.

Fig. 19-9   Steel girders and tapered columns.

■ EXERCISES

1. Make sketches of four typical steel sections. Name them and place dimensions on them.
2. List the items shown on the house plans in Chap. 6 that would be included under the Steel and Iron heading in an estimate.
3. List only the names of the items shown on the plans in Chap. 29 that would be included under the heading of Structural Steel.
4. List only the names of the items shown on the plans in Chap. 29 that would be included under the heading of Miscellaneous and Ornamental Iron.

CHAPTER 20

# roofing and sheet metal

These two lines, roofing and sheet metalwork, are practically always handled as one subcontract. The work involves the roof coverings and flashings, metal leaders and gutters, skylights, wire guard screens, ventilating ducts, registers, etc.

**Roofing.** Built-up, or composition, roofing (Fig. 20-1) consists of several layers of felt and asphalt (or tar). The asphalt or tar is applied hot, with mops. Asphalt is refined from a natural substance found in large deposits in Trinidad and Venezuela. On top of the several layers, a surface coating of fine gravel or slag or a layer of smooth- or mineral-surfaced roofing is applied. The number of layers and the surface finish are, of course, given in the specifications. Slate roofing is made in shingle form, usually measuring $12'' \times 16''$ or $14'' \times 20''$, and the thickness of commercial grades runs between $3/16''$ and $1/4''$. Other forms of roofing also are used, such as copper, zinc, and asbestos. Asbestos shingles, imitating the appearance of wood shingles, are available. All roofing is measured by the square foot or by the square containing 100 sq ft.

Composition roofing costs the general contractor between 20 and 35 cents per sq ft. Asbestos, clay tile, slate, and other heavy types of roofing run between 60 and 90 cents per sq ft, depending upon the quality, shapes, colors, and amounts involved.

**Flashings.** The joints between the roof covering and the adjoining higher walls, chimneys, etc., are sealed with flashings. Strips of copper, galvanized-iron or composition sheets, or plastic materials are used for this purpose. The valleys in roofs are similarly flashed. These are measured in linear feet of each kind, and the material and width are noted (Figs. 20-2 and 20-3).

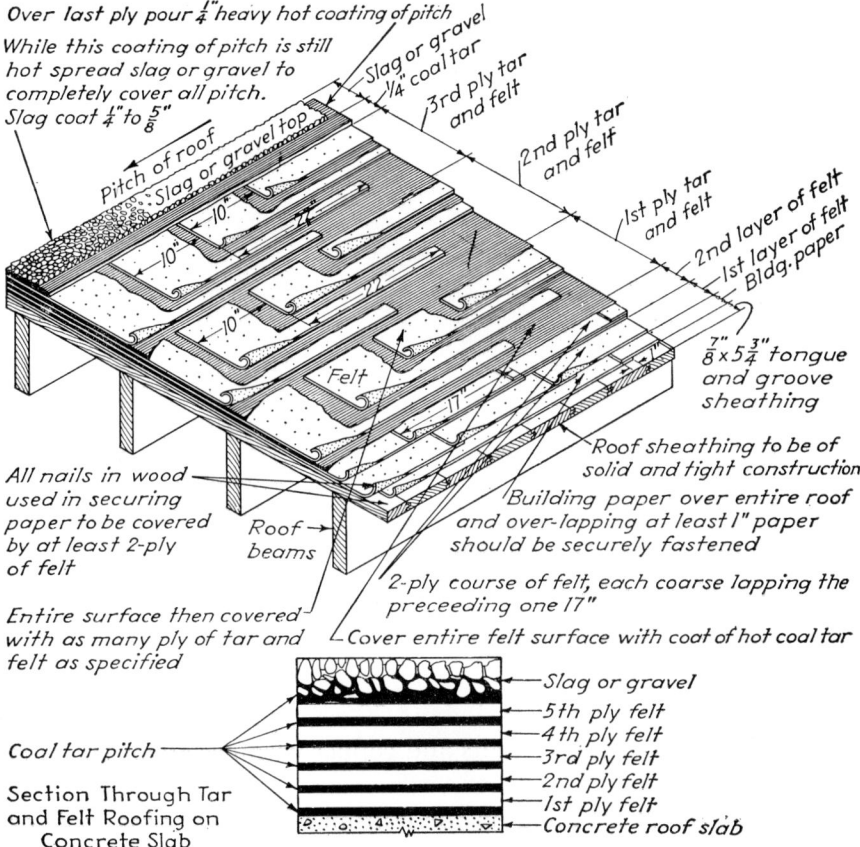

*Fig. 20-1* Built-up roofing. Composition (slag or gravel) roof over concrete is similar to that over wood except for the use of the building paper and untarred felt, which are eliminated. The hot coal tar is applied directly to the concrete and then followed with as many ply as required. The average good roof is 5 ply.

Reglets formed or cut in the walls for cap flashings are sometimes specified under Roofing, and may not be specified or shown elsewhere. Inasmuch as the roofers will probably not include these in their bids, the estimator must provide for the expense involved under Masonry or some other appropriate heading. If, however, a patented type of reglet is called for, it will probably be included by the roofers. No definite general rules can be laid down covering items that may or may not be included in bids from subcontractors. It is essential to realize the importance of taking care in analyzing these bids. Experience along this line is necessary.

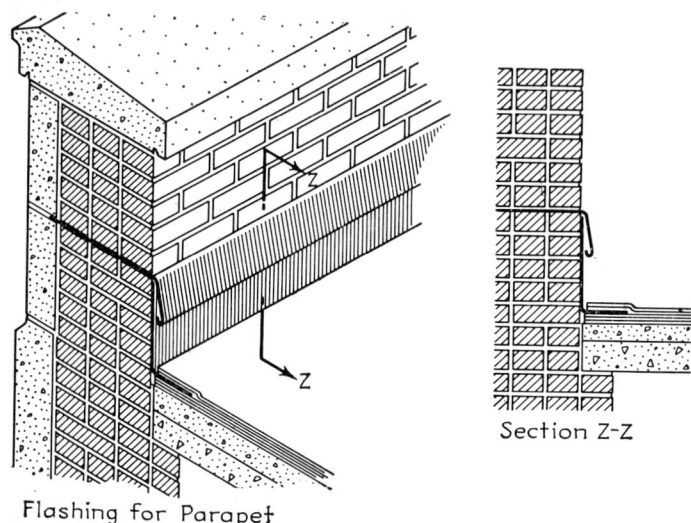

Flashing for Parapet

*Fig. 20-2* Wall flashing.

Ordinary plastic flashings cost about 20 cents per lin ft. Galvanized-iron flashings cost about 40 cents and copper flashings about 80 cents per lin ft.

**Gutters.** Gutters are hung on the bottom edges of sloping roofs or are formed in the roofs themselves (Fig. 20-4). Copper or aluminum are the materials most used for gutters and gutter linings. For estimating purposes, a sketch of the gutter giving the shape and size is required, together with the number of linear feet and a notation as to the kind of material called for.

Metal cornices, which are included under this heading, often require special analysis. Plain cornices of metal are measured in linear feet and the full width of the metal is noted, together with a sketch showing the profile.

Copper gutters and leaders cost the general contractor about 70 to

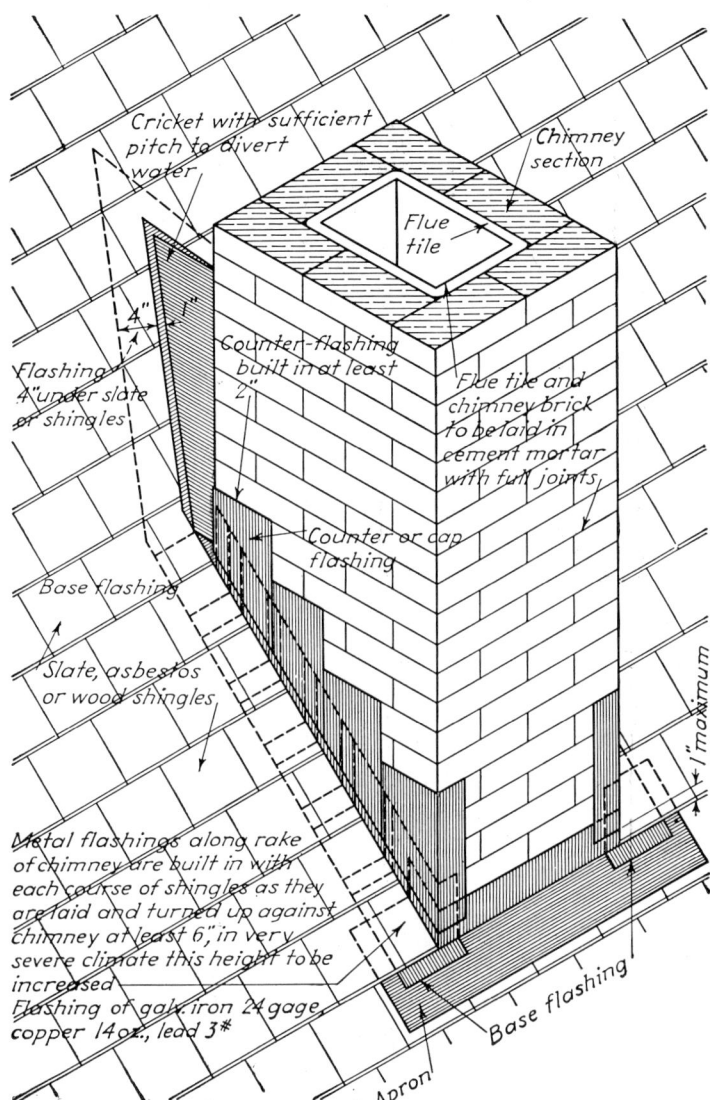

**Fig. 20-3** *Chimney flashing.*

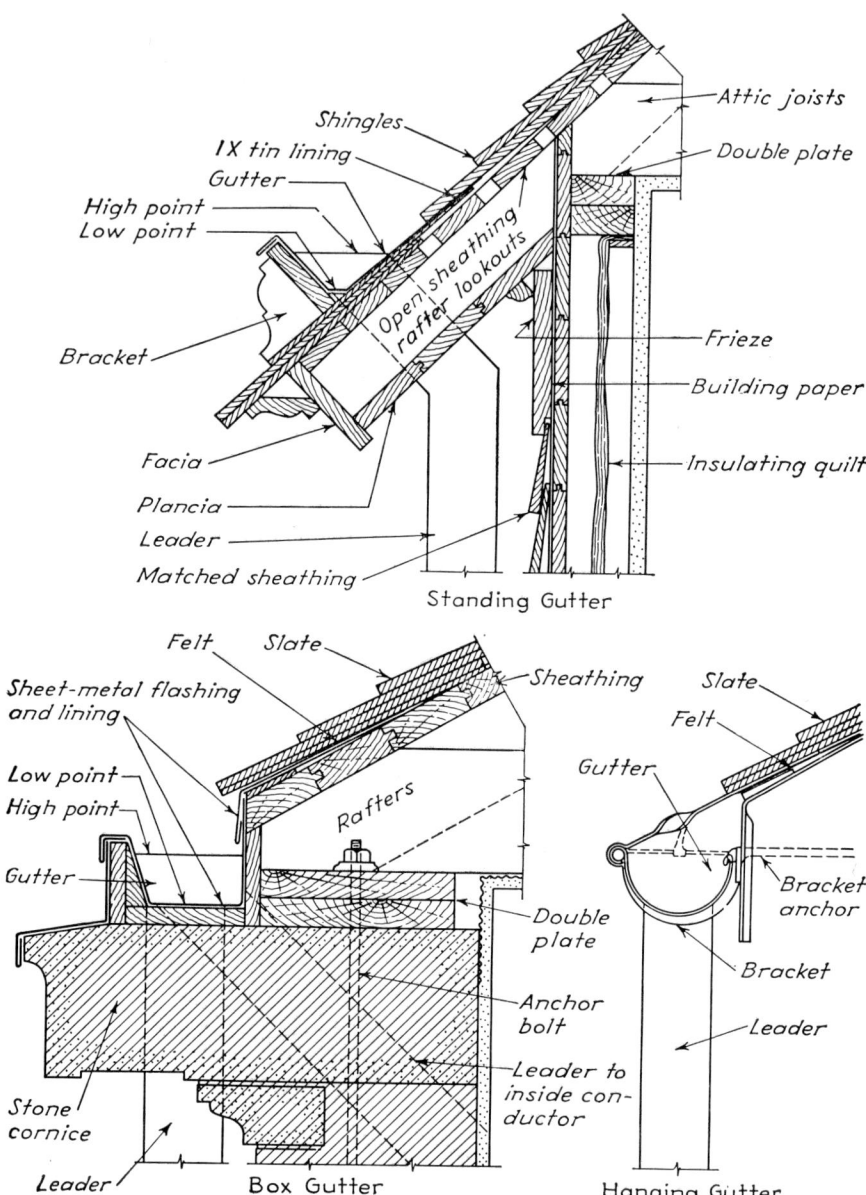

**Fig. 20-4** Gutters and leaders.

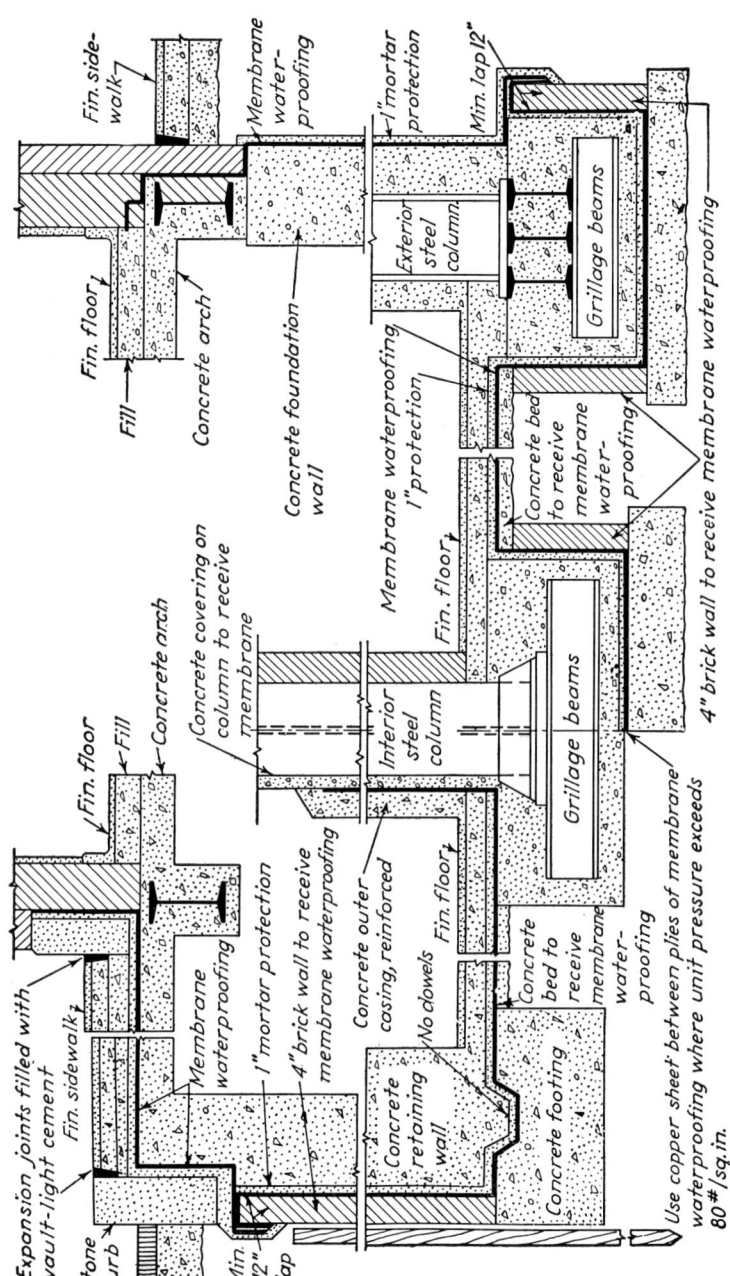

Fig. 20-5 Cellar waterproofing.

90 cents per lin ft. Plain cornices of galvanized iron cost about $2.50 per lin ft for small sizes and those of copper about $4 per lin ft. Large and decorated cornices easily cost twice as much or more.

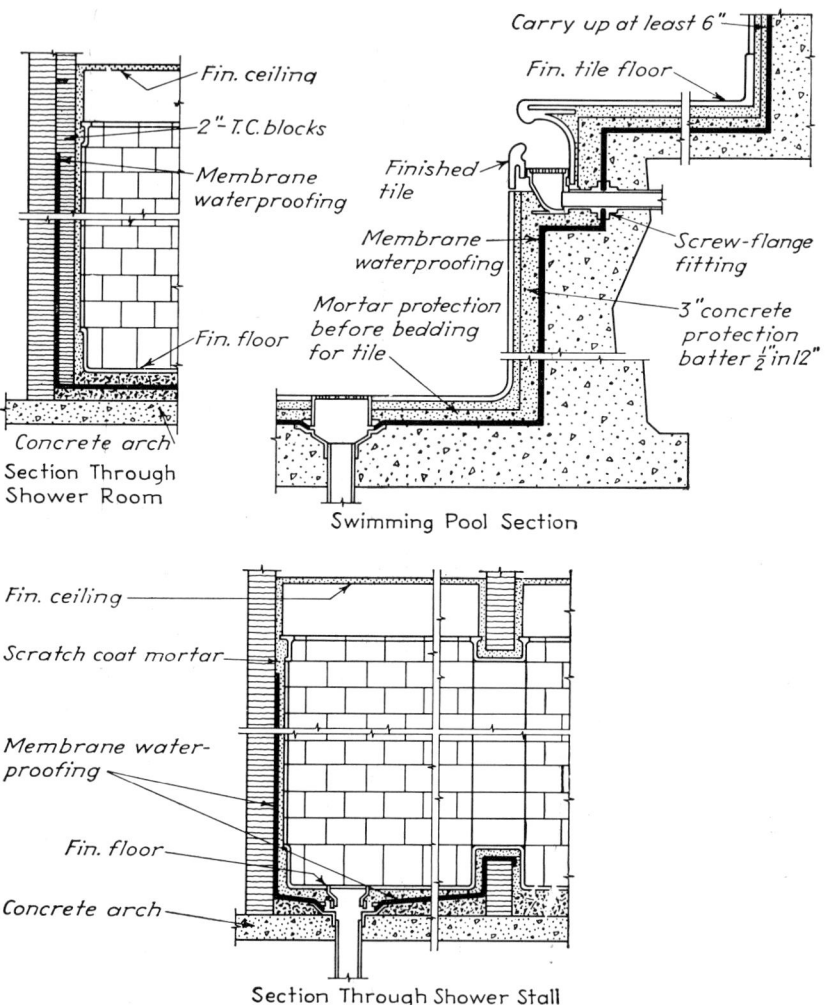

Fig. 20-6  *Shower rooms and swimming pool.*

**Skylights.** Skylights, which are of many types, require careful listing as to size, shape, kind of metal, glass, ventilating openings, etc. Galvanized-iron skylights cost about $40 apiece for small ones 2 or 3 ft square, $100 apiece for those of about 5 ft. Copper skylights are much more expensive. A large one, say 8'0" × 15'0", with ventilators, might cost $500 to $600. The glass used in skylights is always included with the skylights and is not listed under the regular glass heading.

**Ducts.** Plain ventilating-duct work is listed in linear feet. The cross-section size is given. All bends and special shapes are carefully counted and described separately. Items of registers, louvers, and other fittings and supplies are also listed and described separately. Straight duct work in galvanized iron, 12" × 12" in cross section, will cost about $2 per lin ft, and each intersection or bend in it will cost $6 to $8 extra.

**Waterproofing.** Membrane waterproofing is similar to composition-roofing work and is done by the roofers. It consists of several layers of felt or burlap, each embedded in hot asphalt or tar, and is applied on foundation walls and in bathrooms, showers, and other locations requiring such protection. This work is measured in square feet, and the cost is generally between 20 and 30 cents per sq ft (Figs. 20-5 and 20-6).

### ■ EXERCISES

1. List the items of work generally included under Roofing and Sheet Metal, and give opposite each one the unit of measure used and an average price per unit.
2. Make an approximate estimate of all the roofing and sheet metalwork for the plans shown in Chap. 29.
3. Make a good estimate of all the roofing and sheet metal for the house plans shown in Chap. 6. The specifications are as follows:

    ROOFING AND SHEET METAL

    **Roofing Felt.** Lay 15-lb asphalt roofing felt horizontally on all roof surfaces, with all joints lapped 4", nailed down with copper roofing nails. Lay an extra ply, 36" wide, lengthwise, over all ridges and along higher walls, stuck to the first ply with roofing mastic.

    **Flashings.** Flash where required with 16-oz copper. Install base flashing at all walls and chimney, at least 6" high and projecting at least 3" on the roofs, and cap flashings turned down not less than 4" over the base flashings. Install proper step flashings at the chimney and where vertical surfaces meet sloping roofs. Flash all exterior door and window heads, extending the flashing at least 3" above the frames.

    **Gutters and Leaders.** Install heavy moulded aluminum gutters, 5" wide and 4" deep, and corrugated rectangular aluminum leaders, 4" × 3", where indicated on the plans. Provide all necessary heavy aluminum leader heads, hangers, and straps.

**Asphalt Shingles.** Lay standard asphalt strip shingles, 36" × 10", with 4" exposure, on all roof surfaces except the two window roofs and the sun deck. Butts shall be ⅜" thick, color as selected. Nail with three large-head copper roofing nails per strip. This work shall be done in strict accord with the recommendations of the shingle manufacturer.

**Canvas Roofing.** Cover the sun deck with 12-oz cotton duck, properly laid over a coat of white lead and linseed oil, stretched, and tacked at all edges with large-head copper nails ¾" apart. When the canvas is thoroughly dry, paint it with one coat of lead and oil and two coats of approved gray deck paint.

**Copper Roofing.** Cover the living-room window roof and the dining-room window roof with 16-oz soft copper, coated on the back with 15-lb lead per square. Use sheets not larger than 14" × 20" and lay with flat lock seams, soldered.

**Flower Boxes.** Provide flower boxes of 16-gauge copper, with top edges turned and reinforced, in the wooden flower boxes. Provide three boxes in the living room flower box and one each in the other boxes, a total of five copper boxes.

**Dampproofing.** Apply a full, heavy coat of hot asphalt on the outside of all exterior walls, below finish grade.

CHAPTER 21

# stonework

Stonework is a general term for several divisions that may be required in the estimate. Some subcontractors handle several kinds of stone, others only one (Figs. 21-1 to 21-4).

Plain cellar walls or other walls of rubble stone are measured in cubic feet. However, rubble stonework is more often included with

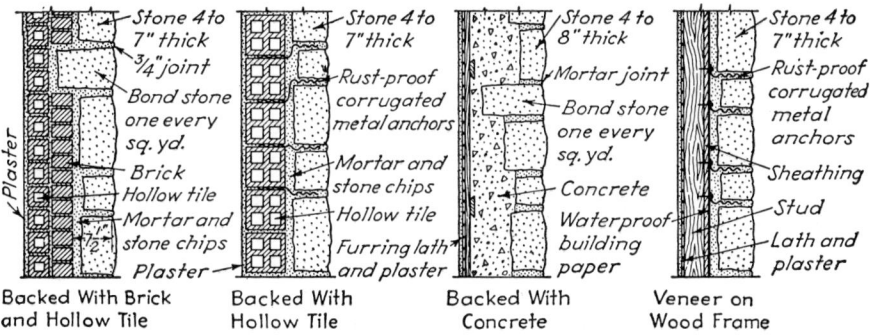

Fig. 21-1  Rock-face ashlar.

the regular masonwork instead of being treated as a subcontract item.

Granite and limestone are the two kinds of stone most commonly used for facing purposes; but bluestone also is used, especially for sills, copings, lintels, and curbing.

Cast stone is an artificial stone which, by the right combination of

305

the proper ingredients, can be made to imitate any of the natural stones.

All the different kinds of stone are measured in approximately the same manner for estimating. The plain work of ashlar facing is measured in square feet. Curbs and copings are measured in linear feet. Sills, lintels, keystones, and other small units are counted per piece. Special shapes are analyzed and, in the case of cut stone, the size of the block from which each piece will be cut is taken into account.

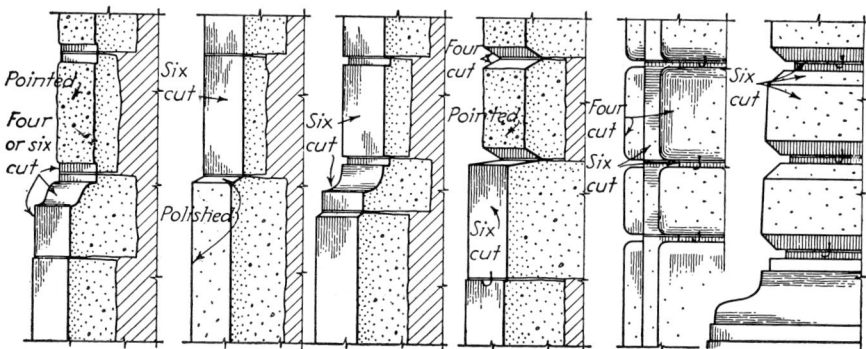

*Fig. 21-2  Stone base courses and walls.*

Random ashlar differs from rubble stone in that it is available in exact heights and, with a ½" mortar joint, can be coordinated with brick. Random ashlar is shipped to the job in lengths that enable the mason to set it in a practical manner. Using a masonry saw, he can break the stones to the lengths required. Rubble stone, often referred to as native stone because it can be found in practically every state, does not have standard course heights. This type of stone permits a more random or rustic effect. Rubble stone is found in a variety of colors ranging from white to dark pink or red. For the most part, random ashlar and rubble stonework is confined to residential construction. It is also adaptable, however, to commercial and religious buildings when used with a cut-stone trim which provides a contrast in color and texture.

Stonework used for buildings has changed considerably in type of usage with the changes in construction in recent years. Before the advent of steel-frame construction, cut stone formed the basis for the solid walls required. The exterior walls supported the entire structure and cut granite or limestone provided the ideal material for these walls, which generally were several feet thick at the base of the structure and diminished in thickness in the upper stories. Present-day construction, with steel, reinforced concrete, or prestressed concrete frames, relieves the exterior walls of the necessity of supporting the weight of the structure. This makes it possible to use the stone as a veneer facing, supported and anchored on the frame at each floor level.

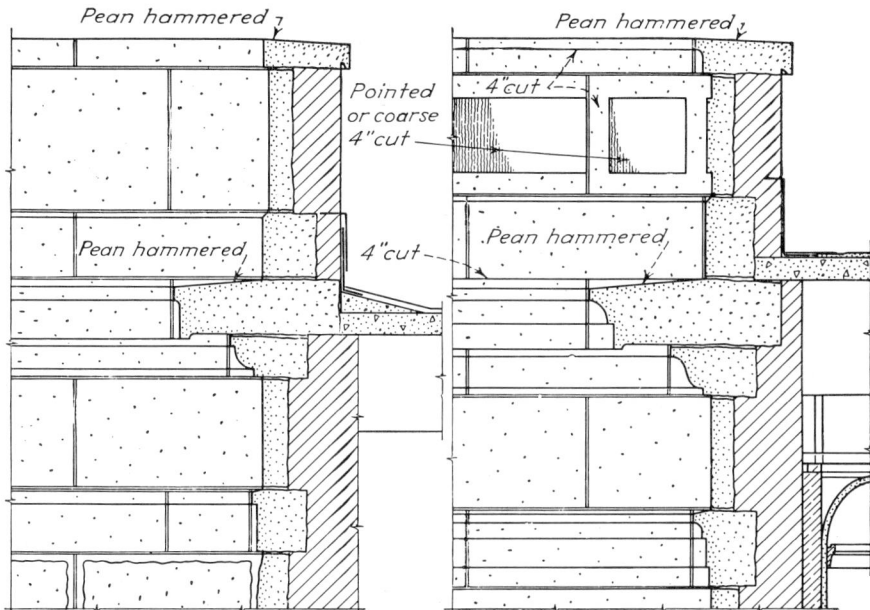

*Fig. 21-3* Stone cornices and parapets.

Stone used as a veneer, 2″ and sometimes less in thickness, is priced per square foot, with the price depending on the type of material and finish. A 2″ thick granite with a polished finish can be fabricated and delivered to the job for approximately $7 per sq ft. Erection will cost an additional $2.25 per sq ft.

Cut stone in thicknesses greater than 2″ is priced per cubic foot, with the type of material, the color, and the finish being determining factors. Granite, with a bushhammered finish, is delivered to the job for approximately $18 per cu ft, and another $6 per cu ft is required for erection.

Stonework, other than plain ashlar surfaces, such as sills with raised lugs, carved work, moulded trim, lintels, etc., must be analyzed, and the amount of handwork, machine work, type of finish, and the size of the rough block must be taken into consideration in determining the cost.

Limestone, marble, bluestone, and cast stone run slightly less in price than granite because the fabricating costs are slightly less. The erection costs are the same for all types of exterior stone.

In most cities the cut stone is set by stone setters. In other areas it may be set by the bricklayers. Local regulations should be investigated and taken into account when making up the estimate.

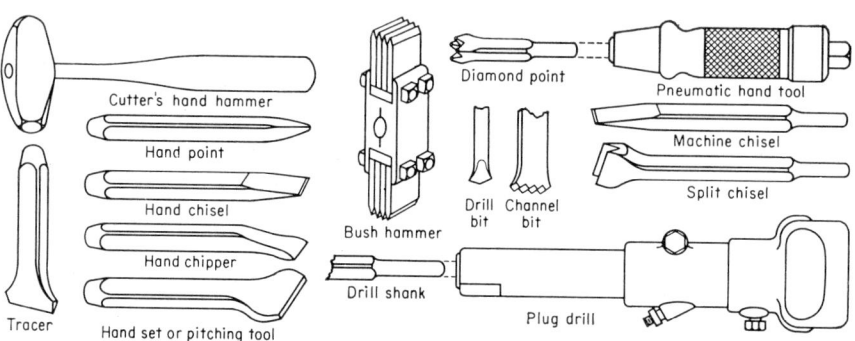

*Fig. 21-4   Hand cutting tools.*

In some cases, prices on the stone will be furnished to the general contractor by firms that do only quarrying and fabricating. These prices are based on delivery of the fabricated stone to the job; the contractor will have to add the cost of setting in order to establish a complete price. Sometimes the small isolated pieces of stone are delivered by the stone contractor but are set in the brick walls by the bricklayers. In all cases the stone contractor will exclude the cost of mortar, anchors, scaffolding, and the protection of the stonework at the job.

Occasionally, as in the case of certain types of stone arches, timber braces or supports, called *centers*, must be provided while the stone is being erected. These temporary supports are removed after the stonework is completed or, in the case of a stone arch, after the keystone has been set in place. The erection of these centers is handled by the carpenters on the job, and this cost must also be taken into consideration.

There may be many contingent parts that go together to make up the complete cost of the stonework, particularly if there are several

### STONEWORK 309

types of stone involved. The estimator must make a careful study of the plans and specifications in order to be familiar with the requirements in each case. He must also make sure that any subcontractor's prices he anticipates using are complete in all respects and are based on the use of materials acceptable to the architect.

### ■ EXERCISES

1. Name four kinds of stone and state what each is generally suitable for.
2. List the items from the plans and specifications in Chap. 29 that would be included under stonework.
3. Prepare an estimate for the stonework shown for the house in Chap. 6. The specifications are as follows:

STONEWORK

**Front Porch.** Lay 1½" thick bluestone flags, honed finish, on the floor of the front porch. Only large flags shall be used and they shall be set in cement mortar with tight joints, in pattern as directed by the architect.

**Walks.** Lay standard 1" thick slate flagstone walks as shown on the plans. The flags shall be laid on a bed of compacted sand 1" thick. The joints between the flags shall not exceed ¼" in width. The flags for the walks shall be one piece in width of the walks and of plain slate color, except those for the front walk which shall be of colors as selected by the architect.

**Driveway Wall.** Construct the wall at the driveway, above the grade, 16" thick, of selected rough local stone set neatly in random-ashlar formation in cement mortar. The stones shall be properly bonded together and no small stones shall appear on the exposed surface. Provide three one-piece, 1½" thick copings of bluestone with honed finish on top and all edges.

**Front Steps.** Construct straight and radial steps as shown on the plans, set solidly in cement mortar, on the rough concrete steps. The risers shall be of large pieces of rough local stone. The treads shall be of 1½" thick bluestone with honed top and front edge, each tread in one piece and set to project out about ½" over the risers. Care shall be taken to make all the steps uniform in height and, as far as practicable, uniform otherwise.

CHAPTER 22

# fireproof doors and windows

This chapter deals with kalamein doors and windows, windows and doors of steel or hollow metal, show-window construction, and special types of fireproof doors. On a large job each would probably have a separate division in the estimate, while on a small job some or all of them might be included under Carpentry.

Fireproof doors and windows are required by law in certain locations, as in fireproof buildings and in the boiler rooms, public halls, and elevator shafts of nonfireproof buildings of some types. Insurance requirements also call for fireproof protection of this nature if the buildings are to be insured. The tinclad or other types of fire doors are especially required by insurance regulations.

Combination frames and trim for doors are obtainable and are popular in use with wood doors, as well as with fireproof doors. These are made of sheet steel bent to the shape of frame and trim. They come all made up as a unit, ready to be installed in place, and have suitable notches in them for the hinges and for the lock strikes (Fig. 22-1).

**Kalamein.** Kalamein work refers to that type of construction in which sheet metal is built on a wood core. This is done in the shop. The ordinary type of fireproof door, such as entrance doors to individual apartments in a modern apartment house, is a kalamein door.

Ordinary kalamein doors cost about $28 each for the door, delivered, and the combination pressed-steel frames that are generally used with them cost about $14, delivered. A complete assembly of frame,

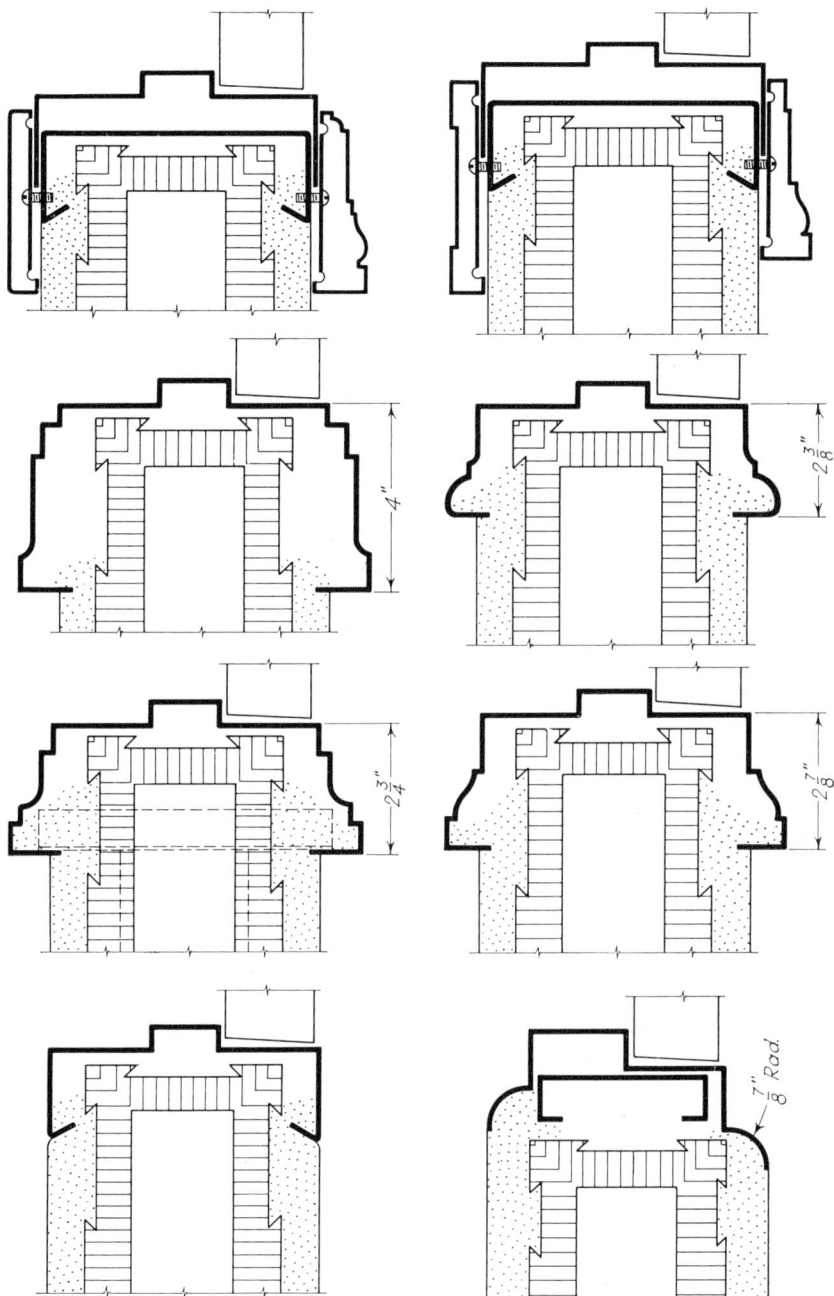

*Fig. 22-1* Sections through door frames.

312 BUILDING CONSTRUCTION ESTIMATING

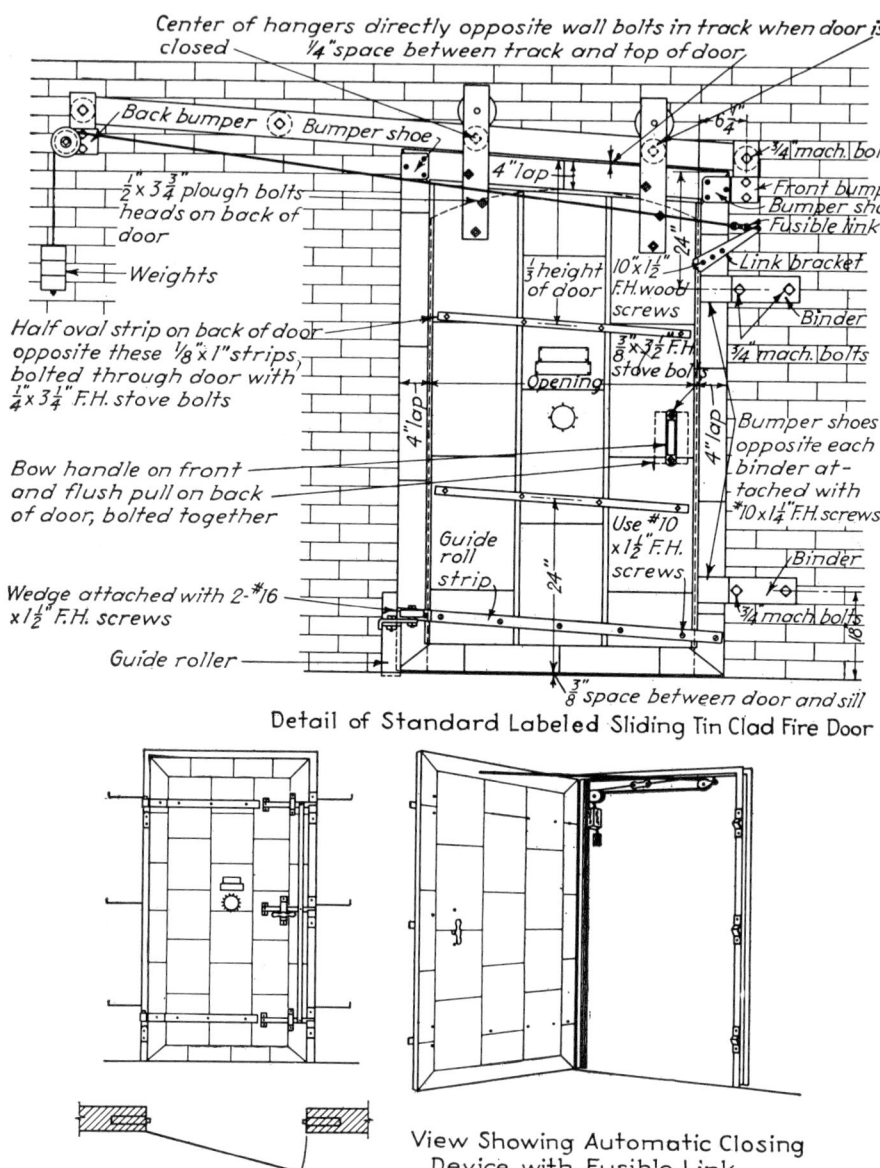

Fig. 22-2  Tinclad doors.

door, and hardware with spring hinges with a cylinder-lock set, will cost between $55 and $65, installed in place.

Tinclad fire doors are made of a wood core covered with one or more layers of tin, and are especially resistant to high temperatures. This is a type of kalamein work, but it is of such special nature that the doors are always called *fire doors* or *tinclad doors* or *tinclad Fire*

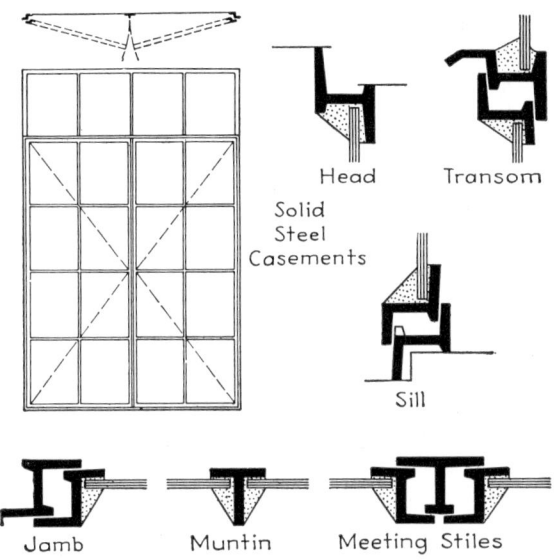

Fig. 22-3  Steel casements.

*Underwriters' doors,* to distinguish them from the ordinary kalamein door (Fig. 22-2).

A regulation fire door, set, costs between $150 and $250 for an opening of ordinary size. The cost depends mainly upon whether a single door is suitable or whether a door on each side of the opening is required.

Fireproof windows are made of kalamein, hollow-metal, and steel construction. Steel factory sash and residence steel casements are of the solid-steel types. Kalamein and hollow-metal double-hung windows cost around $50 each, and the rough wire glass that is generally used in them costs about $15 additional. If clear wire glass is used, the glass cost would be $20 to $30, instead of $15. The glass is generally estimated under the heading of Glazing.

Kalamein contractors commonly furnish all the kalamein doors and windows and the pressed-steel frames for the doors, and the tinclad fire doors. They always install the tinclad doors, but more often the regular carpenters on the job install the other items.

Kalamein store-front construction is a specialty that is handled by separate subcontractors and is, therefore, placed in a division of the estimate by itself.

**Steel Sash.** Steel windows are popular for all types of buildings. They may be of the casement type (Fig. 22-3), which swing on side

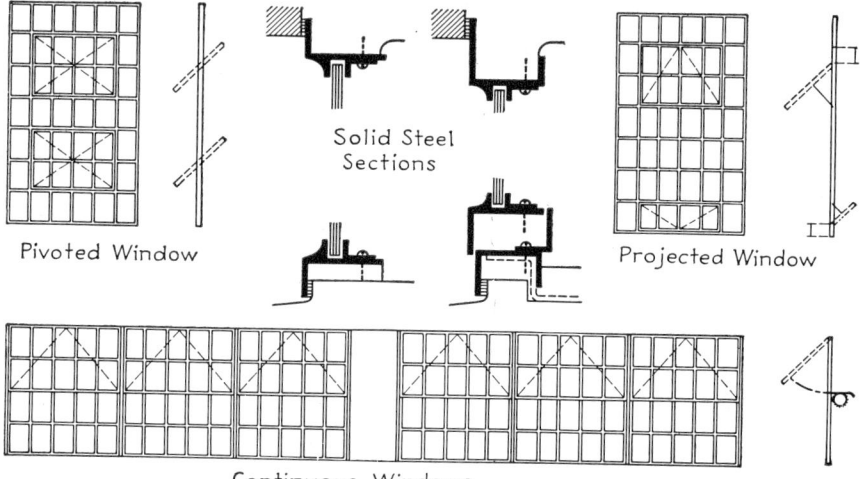

*Fig. 22-4  Pivoted and projected types.*

hinges inward or outward; the pivoted or projected types (Fig. 22-4), which are common in factories and public buildings; or the double-hung type. Many grades of these are made, and the estimator must take care to price the exact type and grade called for in the plans and specifications.

Steel sash is often given a separate division in the estimate, if many such items are required. Sometimes, however, they are specified under Masonry or Carpentry in the specifications and are then likely to be carried under one of those headings also in the estimate. If they are specified in the Iron division of the specifications, they are invariably omitted by the iron contractors, because this is a special iron item with which they do not deal. Steel sash is a stock item sold by the manufacturers directly to contractors. Some of these manufacturers quote on the installation of their windows, also.

**Rolling Steel Doors.** These doors coil up into head frames and are sold and installed by the manufacturers. Some of the features are patented. Both manually operated and motor-operated doors are obtainable (Fig. 22-5). Figure 22-6 shows elevator doors.

*Fig. 22-5  Rolling steel door.*

**Store Fronts.** Store-front construction, consisting of the frames of show windows and the surrounding trim, etc., is usually made either of kalamein (metal on wood cores) or of solid-metal members. Many grades are made and many combinations of mouldings, etc., are possible. The work requires care in pricing. Generally, the mouldings are priced by the linear foot and an additional price per piece is applied for the corners, decorations, and special items. The flat surfaces of base under the show windows and the sign spaces over the windows are priced per square foot. Marble and other materials are often employed in store-front work, and these are estimated under the appropriate heading for the material involved.

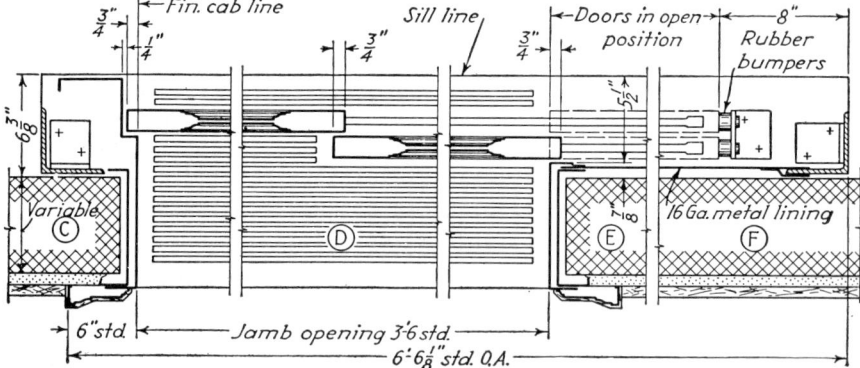

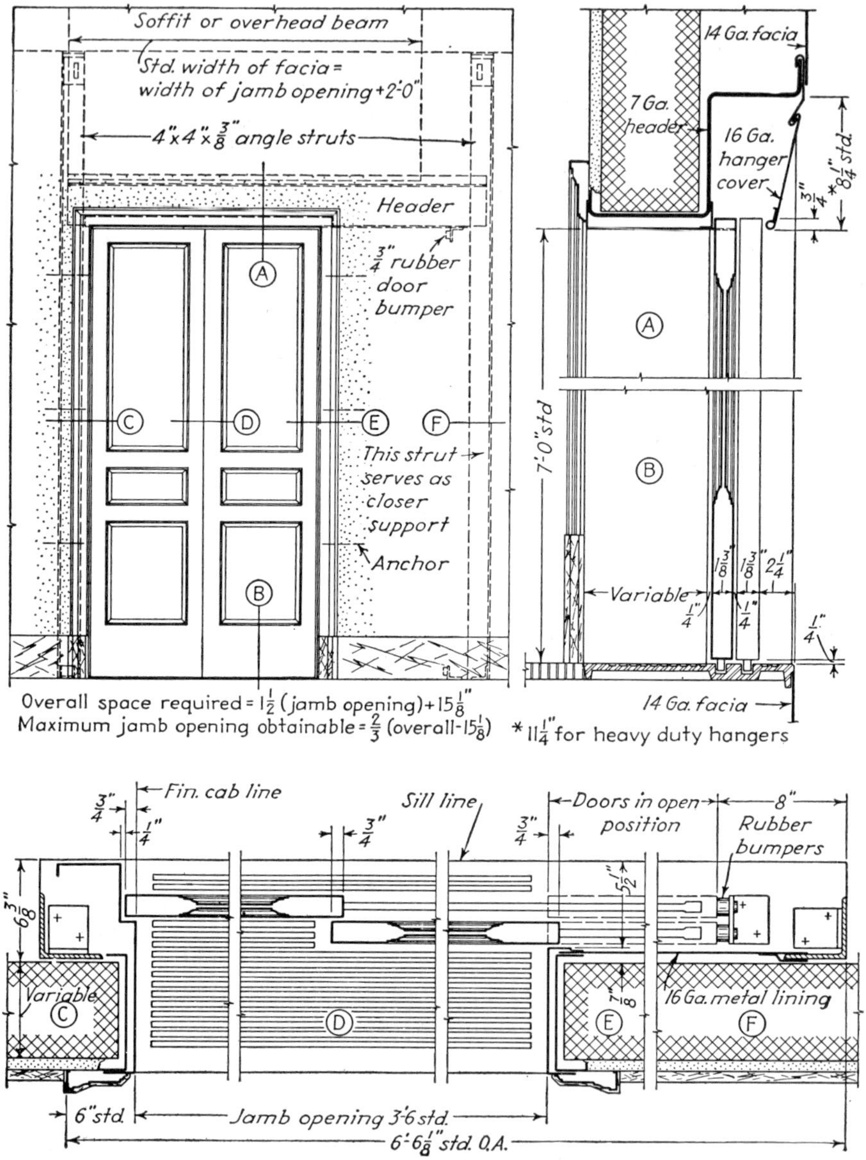

Fig. 22-6  Hollow-metal elevator doors.

Store-front show-window frames cost about $100 for a small single store and about $200 for a regulation double-window job in aluminum or copper kalamein. Bronze kalamein runs a little higher. Stainless-steel fronts and fronts of elaborate design, as well as hollow-metal or extruded or cast-metal fronts, cost thousands of dollars.

■ EXERCISES

1. What is kalamein work?
2. List the items on the plans in Chap. 29 that would be included under the heading of Fireproof Doors and Windows.

CHAPTER 23

# tile, terrazzo, and marble

Tile, terrazzo, and marble are three separate lines of work. Some subcontractors handle two or all three of them, others only one. Slate also is handled by marble subcontractors.

Plain floors and walls are measured by the square foot. Stair treads, door saddles, sills, and other such members are counted and priced per piece. Special items and special features require careful analysis of the shopwork and field labor required.

Bathroom-wall accessories are generally included under the tile heading. For estimating their cost the type, size, and number of each are listed.

Plain bathroom-type tile flooring costs the general contractor about $1 per sq ft, and tile wainscoting about $2 per sq ft. Accessories cost $2 to $3 each. Medicine cabinets and other large fittings cost $10 to $30 each for the usual sizes and types, but these are often included under separate headings or may be listed under the Carpentry heading.

Marble floor and wall covering runs about $3 to $5 per sq ft. Matched or moulded marble may run to $8 per sq ft.

Terrazzo flooring costs about $1.75 per sq ft, but if only a small area is involved, a minimum of about $150 is charged because of the expense of transporting the equipment and organizing the work of installation.

## Regulations

Regulations of subcontractors' associations call for certain items to be supplied by the general contractor. Some of these for this work are the following:

**Light.** Wherever necessary, sufficient electric light shall be installed and maintained.

**Heat.** Whenever necessary, sufficient heat to prevent freezing of work shall be supplied.

**Water.** Water connections shall be furnished on every floor.

**Elevator.** An elevator or hod hoist with operator shall be furnished free of charge to the subcontractor.

**Power.** 110- and 220-volt electric current with connections and maintenance shall be supplied on each floor adjacent to terrazzo work.

**Debris.** All dirt and debris shall be piled on each floor by the subcontractor and shall be removed by the general contractor.

**Foundations.** Concrete foundations shall be prepared by the general contractor to within 2" below the finish-floor level.

**Cleaning.** The surfaces will be cleaned only once. Plasterers and painters must take proper precautions to protect the surfaces and must remove any dropped plaster or paint while it is still wet. All rooms must be properly cleaned out to the satisfaction of the subcontractor before he commences work.

## Tile

Tile is made in many different materials, colors, and sizes. Ceramic tile is a clay composition with glazed or unglazed surfaces. Metal tile is made of stainless steel, copper, aluminum, or porcelain-surfaced metal. Plastic tile is also available. All are made in individual tiles and in sheets of tiles or in sheets scored to look like tiles. Special tile shapes are made for corners, bases, caps, etc., and these are made of metal or plastic for use with either plastic or metal tile. Sometimes these metal items cost as much as the tile itself. Figures 23-1 and 23-2 show shapes and sizes of ceramic tile as recommended by the Tile Council of America.

## Terrazzo

Terrazzo is a marblelike composition used for flooring, base, and stair treads, and occasionally for wainscoting and other uses. It is a plastic mixture of cement, sand, and marble chips, about 1″ thick, mixed almost dry, and poured over a layer of concrete or cement mortar. It is then rolled with a heavy roller, and the surface is finally honed and polished with machines. The quality and color of the surface are determined by the selection of marble chips used.

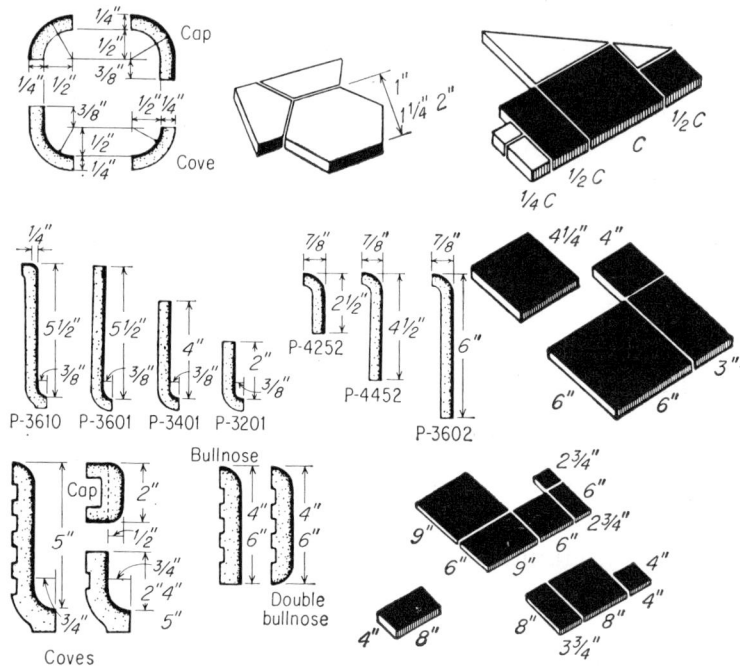

Fig. 23-1  *Unglazed ceramic tile.*

Terrazzo floors are generally divided into panels, bands, or design patterns by the use of brass, zinc, or other strips which are set in place before the first layer is poured. The strips extend up to the finish-floor level. Wire mesh is often incorporated in the first course or laid under it. Precast terrazzo tiles, treads, base, and other members are also available.

## Marble

Marble and slate are made in thin slabs and special shapes and serve for flooring, base, door saddles, hearths, counter tops, shelves, toilet-stall partitions, and many other uses. Thick marble is handled by stone setters and is therefore estimated under the Stonework heading.

## ■ EXERCISES

1. What is terrazzo work?
2. List the items specified and shown on the plans in Chap. 29 which would be included under the heading of Interior Marble and Terrazzo.

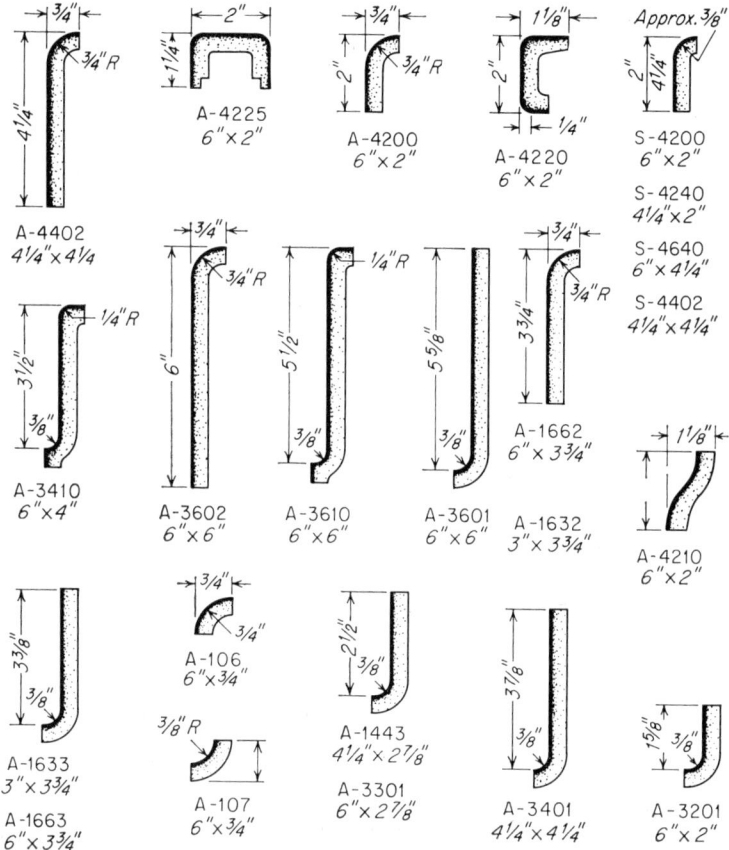

*Fig. 23-2   Glazed ceramic tile.*

3. Make an estimate of the tilework shown on the house plans in Chap. 6. The specifications are as follows:

   TILEWORK

   This work shall be done in strict accord with the recommendations of the Tile Manufacturers Association. The surfaces shall be cleaned and polished and left in perfect condition.

Lay 6" × 6" semivitreous unglazed quarry-tile flooring in the entrance vestibule and foyer.

Install 4¼" × 4¼ semivitreous matte-glazed tile flooring and 42" high wainscoting, including 50" high above the bathtubs, in the two bathrooms. Provide coved base, cap, and corners. The floor border and wainscot base and cap shall be one color and the balance of the tile another color, which will be selected by the architect.

Install glass towel bars in chrome-plated brackets, chrome-plated shower-curtain rods, and chrome-plated soap dish and combination tumbler and toothbrush holder in each bathroom. Provide recessed tile toilet-paper holders and combination soap holder and grab bars.

CHAPTER 24

# painting and glazing

The painting subcontract generally takes in painting and papering done at the building only. However, owing to union regulations especially, it is sometimes necessary to have the painting subcontractor include other items, besides. These include shop coats in some cases, priming in the shops or at the job, painting of structural steel, dampproof painting, etc. It is part of the general contractor's estimator's duty to see that all painting is properly provided for under one or more divisions of the estimate.

Some painting estimators go over the plans and gauge the amount of labor involved in each room or floor and then add items in their estimate for the amount of white lead, oil, color, and other materials. Other painting estimators measure all the wall and ceiling areas in square feet and count the doors, windows, radiators, and other features with great care. Strange as it may seem, the first method seems to give as good results as the second. However, only a man with thorough experience can use the shorter method. It would probably be good practice to adopt both methods on every job and check the resulting figures of one against the other.

Painting estimates vary greatly, probably because many painting estimators have careless methods or because they expect to attempt cheating on the job. It is well for the general contractor's estimator to check the lowest price carefully before using it in his summary.

Finished floors usually require protection. This may sometimes be accomplished by merely locking the doors of the finished rooms. In other cases, it is necessary to cover the floors with building paper or something better, and this represents expense that the estimator must provide for under General Job Expense, if it is not included under the headings of Floor Work or Painting.

In checking the painting estimates, it may be found that some items have been omitted in them. Floor finishing, for example, may be included by some of the painting estimates and not by others. Papering or other treatments that do not involve paint may likewise be omitted in some of the estimates, or the bids may state that the special materials thus involved are to be furnished by others or are to cost not more than a certain amount.

Painting of new work generally involves three or more coats, and old surfaces, two coats. Each coat costs between 3 and 6 cents per sq ft, depending upon the quality of the material used and the workmanship called for, and also upon the amount of work involved, the kind of surface, the amount of scaffolding required, etc. If the work is out of the ordinary, these prices may be doubled. An average room costs about $50 for painting the walls and calcimining the ceiling. A new store of average size may cost $300 for the interior alone. A new building of good grade may cost about $1,000 per story for the painting.

Of the many kinds of paint that are made, the most common for exterior work is linseed-oil paint. This is used on interiors, also, especially for high-grade work, but water paints are more frequently used inside. Varnishes, enamels, and stains are needed, as well. All are generally measured in square feet of surface to be covered, and a price is applied to suit the material, number of coats, etc. Wallpapering also comes under the heading of Painting.

## *Painting*

The success of a painting job depends principally on four factors: the condition of the surface to be painted, the condition of the preceding paint coat in a repainting job, the prevailing atmospheric conditions, and the quality of the paint and its suitability for the service expected. It is probable that the disregard of these factors is the chief cause of paint failures with all materials.

Painting should be done only in clear, dry weather. The surface should be free from dew or frost and should not be painted until it feels dry to the hand. Painting should not be done when the temperature is below 50°. In a falling temperature the work should be stopped early enough in the afternoon to allow the paint to set before the temperature drops below 40°. If lumber is green when placed or has been made very wet for some other reason, at least a week of dry, sunny weather should precede the painting.

**Exterior Wood.** Oil paint applied over knots and sappy wood is likely to alligator and peel and with pine woods to discolor. There is apt to be more alligatoring of paint over knots on southern exposure, because of more sunshine, and more yellow discoloration of the paint over knots on the north side where sunshine is weakest. The knots may be spot-painted with orange shellac varnish, or a high-grade aluminum paint applied a day or two before applying the priming coat of paint. An effective sealer for knots consists of pure shellac varnish plasticized with 6 oz of blown castor oil to the gallon of shellac varnish. However, experience in painting No. 2 common southern yellow pine containing many knots indicates that ordinary shellac varnish or aluminum paint is of little value in sealing knots. When shellac varnish is applied too liberally over the knots, the paint tends to peel off on weathering. When the shellac varnish is applied thin, the paint may not peel, but yellow discoloration may show through from the pitch. All pitch should be scraped or burned from knots and other places.

The usual system of painting houses in white and light tints includes the use of raw linseed-oil paints and white-lead paints mixed on the job. These paints have the advantage of easy-brushing properties and can be spread over large areas of 600 to 700 sq ft per gal. The first coat should always be carefully brushed on and crossbrushed, if practicable, over all parts of the surfaces so that cracks and nail holes receive enough paint to wet the surfaces. The older, unbodied linseed-oil first coat should be brushed out to cover about 600 sq ft per gal, the newer type priming paints and the white-lead mixed-on-job paints used on two-coat work should be applied at about 450 sq ft per gal. Nail holes and cracks should be properly filled with putty as soon as the priming coat is dry. After about a week of clear, warm weather, the putty should be hard enough to receive the second coat.

Window sills, outside doors, and other parts are sometimes varnished. For this work no wood filler is necessary if the wood is close-grained, but several coats of spar varnish should be applied. A high-grade paint job will last at least 4 years before repainting is necessary, but exposed varnished woodwork will probably need to be revarnished annually.

Shingle stains are thin, fluid paints in which shingles are dipped before being laid. Suitable stains properly applied to shingles or rough siding give a durable finish on exterior woodwork. Ready-mixed stains are used more often than those mixed on the job. The manufacturer's directions for application should be followed.

**Exterior Masonry.** As moisture back of the paint film will seriously impair the life of oil- or varnish-base paints when applied to concrete, stucco, or masonry, walls made of these materials should be thoroughly dry before being painted. The drying may require a long time and will depend upon weather conditions and upon the thickness and porosity of the walls. It is equally important to prevent water from entering the wall after painting.

To seal open-textured walls, it is advisable to apply a cement-sand grout as the base coat. This is composed of equal parts of cement and fine sand and water. This is scrubbed in with a stiff fibre brush, thereby forcing the grout into the openings and producing a surface of uniform porosity. This should be sprayed with a fine spray from a garden hose about every 6 hours for 48 hours. A minimum of 60 days should elapse before oil paint is applied over the grouted surface. Exterior masonry can be successfully coated with water paints if the surface is properly prepared and directions for applying the paint are carefully followed.

**Interior Wood.** The directions for preparing the surfaces of new exterior wood for painting apply also to interior woodwork. In addition to having the surface clean and dry, it is advisable to roughen very smooth wood slightly with fine sandpaper.

Linseed-oil paints used for exterior wood may not be suitable for interior work because of slow drying and a tendency to turn yellow.

Varnishes applied to interior surfaces darken with age. Consequently, wood coated with interior varnish and kept indoors may darken if not exposed to direct sunlight but otherwise remain in good condition for many years. Wood stains are used either to change or to modify the color of interior wood without obscuring the natural grain.

The usual system of staining and varnishing interior wood is to apply the coats in the following order:

1. Stain coat.
2. Filler coat, for open-grain woods only.
3. Sealer coat, bleached shellac varnish—this coat needed only if the stain contains aniline dye (mahogany stain, for example).
4. First-coat varnish, thinned with ½ pt of turpentine per gal. The dried surface should be sanded lightly when dry with No. 4/0 sandpaper.
5. Second-coat varnish, applied without thinning; should be sanded same as first coat.
6. Third-coat varnish. When a dull, rubbed finish is wanted, let the last coat of varnish dry hard for 72 hours and then rub with powdered pumice and water to a uniform dull finish.

The best results will be obtained if varnish is used according to the following instructions:

Smooth, thin coats should be applied, preferably at a temperature between 70 and 80°, in a well-ventilated room.

Varnish should be brushed on with the grain of the wood, cross-brushed, then brushed again with the grain of the wood.

The minimum number of coats necessary to attain the desired finish should be applied.

Ample time should be allowed for the coats to dry.

Freshly varnished work should be kept clean and dust-free.

Each coat should be rubbed lightly with fine sandpaper to a dull finish before the next coat is applied.

Brushes and varnish cans should be kept free from dust.

In painting interior woodwork for flat finish, the first coat may be primer and sealer followed by one or two coats of eggshell flat paint. The primer does not dry hard enough to sand and should not be considered as an enamel undercoater. This material can be made by adding ½ gal of primer to 1 gal of paint. It could be used directly on the wood, omitting the primer, or as an enamel undercoat over the primer.

A semigloss finish on interior wood may be produced by using two coats of paint composed of equal parts of paint and enamel. The first coat should be thinned with 1 pt of turpentine to 1 gal of paint.

A full-gloss finish on interior wood can be obtained by the application of two coats of enamel. The first coat should be thinned with 1 pt of turpentine to each gallon of enamel. For better hiding power, as an alternate gloss finish, two coats of enamel undercoater may be used, followed by a single coat of enamel.

**Interior Masonry.** Interior surfaces of masonry, concrete, and plaster should be prepared for painting in the same way as similar exterior surfaces. Finish coats may be either oil paints or water paints.

In flat-finish oil painting on new walls of plaster or concrete, the first coat should be a primer and sealer, applied without thinning. The second coat should be eggshell flat paint applied without thinning.

With water-thinned paint, the plaster or concrete should be free from grease, wax, dirt, or calcimine. Cracks and holes should be filled with patching plaster and these areas spot-primed with the water paint.

For semigloss-finish oil paint, the first coat should be a primer and sealer. If suction spots appear after this paint is dry, they should be spot-primed with the primer. The second coat should be semigloss wall paint. This may be prepared by mixing equal parts of flat and enamel.

For full-gloss-finish oil paint, the first coat should be a primer and sealer, allowed to dry 24 hours. The second coat should be enamel undercoater, allowed to dry 24 hours. The third coat should be enamel.

**Floors.** Varnish-base paint, known as floor-and-deck paint, is intended for interior and exterior use on wood and concrete floors. Concrete basement floors should age thoroughly before painting and they should be painted when dry and when the humidity in the basement is very low. The paint should be applied in a thin layer and be brushed out well. Two coats of paint are generally sufficient. On wood floors it is suggested that for the first coat of paint a quart of thinner, composed of 2 parts of boiled linseed oil and 1 part of turpentine, be added to each gallon of paint. For the second coat, a pint of boiled linseed oil per gallon may be used. On concrete floors, for the first coat 1 qt of liquid in the proportion of 2 parts of spar varnish and 1 part of turpentine is added to each gallon of paint, and the second coat can be applied as furnished.

Rubber-base, ready-mixed paint is used on concrete flooring. The first coat should be diluted with 1 qt of thinner, composed of mineral spirits, to each gallon of paint.

Lacquer-type floor sealer is intended for sealing oil-treated flooring and oiled wood floors that have been sanded and cleaned. Sealing the oil within the flooring prevents subsequent marring of the finished surface. This type of sealer also smooths and integrates the surface to which it is applied, thus producing a finished appearance and furnishing a base for wax and varnish. Open-grained flooring should be treated with wood filler before the application of sealer.

## Glazing

Glazing covers an ever-widening field. Glass is now used not only in windows and doors but in many other places as well. Structural glass, which is rapidly coming into common use, offers all sorts of complications for the glazing estimator and for the general contractor's estimator. In this, as in all modern construction, the young estimator will do well to analyze all the possible incidental work and expense, as well as the actual glasswork. Any work out of the ordinary run of established materials and methods should be thought of in this way.

There are many types, thicknesses, and grades of glass. Large plate glass is measured very carefully, because of the rapid increase in square foot prices due to the special handling necessary for large pieces of such fragile material. Ordinary glass for windows and doors is counted in number of pieces, or lights, of the different sizes required. Mirrors are part of the glazing division. Leaded glass also is handled here, unless it is of a highly elaborate nature, in which case a separate special heading is usually set up in the estimate.

Union regulations generally require that all glazing shall be done at the job and that no windows or doors shall be sent to the job with the glass in them.

Large plate glass is usually not installed until the very last, and even then it is immediately protected against being scratched or broken. This protection is to be provided for under the heading General Job Expense.

Ordinary glass is subject to considerable breakage, which the contractor will generally accept rather than stand the expense of protecting the glass. The judgment of the estimator must make allowance for this in the estimate. The amount to allow will depend upon the type of building, the type of neighborhood it is in, the time of year, and the amount of glass involved. Ordinary glass is installed before the end of the job if it is necessary to have this protection against the weather.

Small sizes of double-thick glass cost the general contractor about 60 cents per sq ft, and plate glass, in small sizes, about $1.25 per sq ft. The glass in a show window 8'0" × 10'0" may cost between $100 and $120. Pyramid glass, prism glass, and other types not in general use must be checked closely, since the names are tricky and often misleading, and there are usually several thicknesses to choose from. Bent glass, leaded glass, and other special forms require careful analysis also. There are already several types of structural glass on the market and new ones may be expected. Some of these that do not cost much in themselves may involve a great amount of labor on the job, in order to satisfy the architect or the owner as to the effect they are looking for.

**Plate Glass.** Plate glass is a transparent glass which has been cast, rolled, ground, and polished on both sides. The two surfaces are true and parallel and provide clear, undistorted vision. Wired plate glass is also made.

Standard plate glass is graded in three qualities: silvering quality—the highest grade and best quality; mirror-glazing quality—free from defects but not of the perfection attained in silvering quality; and glazing quality—the usual type suitable for general purposes. Standard plate glass is made in three thicknesses and the maximum sizes usually obtainable are as follows: 6'0" × 10'3" in $\frac{1}{8}$" thickness; 10'3" × 18'0" in $\frac{13}{64}$" thickness; and 13'4" × 18'4" in $\frac{1}{4}$" thickness.

Heavy plate glass ranges in thickness from $\frac{3}{8}$" to $1\frac{1}{4}$". It is suitable for table tops, shelves, shower doors, partitions, etc. Heat-tempered plate glass is made by a process which produces a glass highly resistant to breakage. Heat-resisting or heat-absorbing plate glass is made by a process which gives it the capacity for absorbing a high percentage of the infrared rays of the sun without interfering with the transmission of light.

**Sheet Glass.** Flat sheet glass is graded and labeled in accordance with Federal government standards in the following grades: AA quality—the best select quality of window glass available for high-grade work and made only on special order; A quality—select glass of superior glazing quality suitable for commercial purposes, contains no imperfections that will appreciably interfere with straight vision; B quality—suitable for general glazing purposes where visual qualities need not be quite perfect.

Single-strength sheet glass is about $\frac{3}{32}$" thick and double-strength is about $\frac{1}{8}''$ thick. Heavy sheet glass $\frac{3}{16}''$ and $\frac{7}{32}''$ thick is also available.

## ■ EXERCISES

1. Make a quantity survey of the glass required by the plans and specifications in Chap. 29.
2. Make an estimate for the glazing required for the house shown in Chap. 6. The specifications are as follows:

### GLAZING

Glaze the stationary windows in the living room and dining room with $\frac{13}{64}''$ thick glazing quality plate glass.

Glaze all the other windows in the front wall and the stationary window in the kitchen with A-quality, double-strength sheet glass.

Glaze all the other windows with B-quality, single-strength glass.

Glaze the rear basement door, the two rear porch doors and the sundeck door with A-quality, double-strength glass.

Provide a $34'' \times 60''$ polished-edge mirror, of silvering quality, $\frac{1}{4}''$ thick plate glass with copper-coated back, on the wall in the entrance vestibule, held in place where directed by six glass rosettes.

The door mirrors are specified under Finish Carpentry.

CHAPTER 25

# *hardware*

Preparing a hardware specification or schedule is an involved matter. Writing a complete list of the hardware required for a job requires close attention to many details, and only a man of considerable experience is able to do this properly.

Architects sometimes call in the representative of a hardware manufacturer or dealer for this service on larger jobs and to help in selecting the hardware. This usually means that the order for the hardware is given to the person who prepares the list. On small jobs the contractor is expected to submit a list and catalog cuts to the architect for his approval.

Architects frequently specify that a certain specified amount of money for the hardware shall be included by the contractor in his bid. This is generally understood to be his cost price for the hardware, delivered to the job. He would later have to show his hardware bills and adjust the total contract amount to suit any savings or any additional cost. This method is not used on public work, however, and for such jobs the hardware is usually specified in detail.

## *Expenses*

Expense for the contractor, in addition to the first cost, is involved under the Hardware heading. He or his estimator may have to prepare the detail lists and schedules or have them prepared by others. Samples must be procured and submitted to the architect. Revised lists, and more samples, may be required before the hardware can be ordered. Shop drawings must be checked and coordinated, and templates for some items may be required. The types of screws and bolts for the hardware must be indicated.

Finally, when the hardware is received at the job, great care must be taken to store it and the keys safely, to sort it out, and to distribute the many items to the locations throughout the building where they are to be installed. In some cases it is necessary to send certain items of hardware to material manufacturers, subcontractors, and others, in order that proper provisions for the hardware may be made in doors, frames, and other parts that are affected by the hardware. The additional cost of all this varies with each job but must be included in the estimate under Hardware or General Job Expense or other appropriate headings. A good estimator will not hesitate to put in $100 to $500 to cover this expense on a fair-sized job.

Key racks or cabinets are sometimes specified. In addition to the cost of these, the labor of installing them, and probably also the labor of preparing key tags and schedules to accompany them, must be provided for in the estimate. Master keys and extra keys, especially for hotels, offices, and institutional buildings, also involve extra expense which must be considered. The use of master keys for important doors should be avoided, if possible, as cylinders without master keying are harder to pick.

## Omissions

The hardware division of an estimate does not provide for all the hardware required on a job. It applies only to the finish hardware and seldom includes even all the items of this nature. It is necessary for the estimator to read all the specifications and to examine all the drawings carefully, in order to determine what to put under this heading. Other divisions of the specifications will undoubtedly demand provision for some hardware expense.

Many types of doors, such as fire doors, all-glass doors, overhead doors, some sliding doors and folding doors, and some metal doors, come equipped with hardware, and this would therefore be included under the appropriate headings with these doors. Most steel windows and some wood windows also come equipped with all or some of the hardware for them.

Cabinets, wardrobes, and lockers sometimes come with the hardware on them and sometimes they do not.

Hardware allowances and hardware lists may or may not include such items as coat hooks, clothes-hanger rods, shoe racks, hat racks, shelf supports, bathroom-wall fittings, shower-curtain rods, drapery fittings, mirrors, medicine cabinets, door chimes, coal chutes, package receivers, grilles, dampers, clean-out doors, display racks, handrail brackets, thresholds, weather strips, etc. Such items as are not placed under this heading must, of course, be put elsewhere in the estimate.

Rough hardware should not be put under the Hardware heading. Items such as nails, spikes, bolts, bridle irons, etc., belong under the carpentry headings. Iron railings and other large iron items should be put under the Iron heading.

## Hand of doors

The "hand" of a door is established from the outside of exterior doors and from the hall side of interior doors. If the butts are on the right, the door is a right-hand door. If the butts are on the left, it is a left-hand door. See Fig. 18-4 for a diagram illustrating the hand designation of doors.

**Bevel of Doors.** If you are standing outside a door and the door opens away from you, the door is a regular-bevel door. If it opens toward you, it is a reverse-bevel door. Some locks are reversible and therefore may be used without regard to the hand or bevel of door.

## Finishes

**Metals.** High-grade hardware, especially for exterior use, is generally of solid bronze, brass, stainless steel, or other noncorroding metal. Other hardware may be of these metals or of plain steel, plated steel, or plastic. Obviously, the grade of material specified will greatly affect the cost of the hardware. On fine work, matching hardware and matching finishes are required. Sometimes, to obtain this, the specifications may state that all the hardware is to be of one manufacture. This, of course, limits the contractor by prohibiting any splitting up of the hardware order that he might otherwise wish to do. Some hardware finishes, as established by the National Bureau of Standards, are as follows:

| | | |
|---|---|---|
| U.S. P | Prime coat for painting |
| U.S. 1B | Bright japanned |
| U.S. 1D | Dead-black japanned |
| U.S. 2C | Cadmium plated |
| U.S. 2G | Electroplated zinc |
| U.S. 2H | Hot-dipped zinc |
| U.S. 3 | Polished brass |
| U.S. 4 | Dull brass |
| U.S. 10 | Dull bronze |
| U.S. 14 | Polished nickel-plated |
| U.S. 26 | Polished chrome-plated |
| U.S. 26D | Dull chrome-plated |
| U.S. 27 | Satin aluminum lacquered |
| U.S. 28 | Satin aluminum anodized |

## Doors

The common types of door hinges are called butts or butt hinges. Some have fixed pins and some loose pins. Some are ball-bearing.

The lock sets most commonly used are mortise locks, tubular locks, and unit locks. They fit into mortises or into holes cut in the edge of the door. The bolts of these locks extend into strike plates screwed to the jamb of the opening.

The hardware on doors is applied before any weather stripping is started. Hinges, and lock sets or latches, are used for front and rear doors, interior doors of many uses, and screen and storm doors. The front-door hardware may be quite elaborate and expensive and may also include a letter box and door knocker. Some doors require door checks, holders, bumpers, panic bolts, pull handles, push plates or bars, kick plates, and even saddles, grilles, numerals, and other fittings. Sliding doors require tracks, hangers, guides, and pulls, and may also have locks. Pairs of doors require top and bottom bolts.

Skilled carpenters are needed for properly installing hardware, especially lock sets and other items requiring careful alignment and close adjusting. Architects examine this work and some even insist that all screw slots be uniformly set vertically.

Floor-type hinges and door holders fastened to the floor require that recesses be provided or cut into the floor to receive them. These and similar extra labor items are often overlooked by estimators.

Hollow-metal doors require reinforcing at all the hardware locations. This is usually provided for in the door order, but, if not, it must be listed with the hardware intended for such doors.

## Windows, etc.

Double-hung wood windows require sash chains or cords, sash weights and pulleys, or sash balances. These are usually estimated in the carpentry divisions of the estimate rather than under the Hardware heading.

Wood sash are grooved in the mill when they are to receive sash balances, and the balances are usually furnished by the millwork supplier. Sash lifts and sash locks are put under the Hardware heading. The sash balances, stops, and catches are installed after the weather stripping has been completed.

Wood casement windows require hinges and latches and may also have sash adjusters or sash operators. Pairs of casements may require cremone bolts, or one leaf may have top and bottom bolts.

Wood window screens and storm sash require hangers and hooks, and sometimes adjusters are also specified. Shutters and louvers may require hinges, handles, latches, and hooks.

Metal windows generally come fitted complete with hardware, and sometimes with screens.

Transoms which open over doors and windows require hinges or pivots, and catches, and usually also have transom chains and lifts or adjuster rods.

Closets require coat-and-hat hooks and clothes-hanger rods. They may also have shoe racks, hat racks, and other special fittings.

**Cabinets.** Hardware for cabinetwork includes hinges and catches for hinged doors, tracks and pulls for sliding doors, slides and pulls for drawers, and possibly locks for some of the doors and drawers. Fittings for cabinets are often included under the Hardware heading and may consist of shelf supports, shelves, trays, racks, bins, display devices, and many other things. In fine cabinetwork, the hardware may be very expensive and may include concealed hinges and other items requiring more than the usual amount of labor.

CHAPTER

*plumbing*

The term *mechanical trades* is generally applied to the piping, wiring, and machinery lines of work. On regular building work these include plumbing, heating, electrical work, and elevator work. On most jobs these are all separate subcontracts. Some subcontractors, however, handle both plumbing and heating.

The mechanical trades, especially plumbing and heating, are often given out by the owner or the architect separately from the general contract for the construction work. In such cases, the general contractor may theoretically have nothing to do with these lines of work but, because the work is done in the building at the same time that the construction work is being done, conflict of interests may occur. Sometimes the general contractor is given jurisdiction over these separate contracts, even when they are thus separately awarded. Regardless of the specifications covering these lines, the general contractor is governed only by his own contract and the specifications that accompany his contract.

Excavating for the mechanical trades is an item that must always be checked by the general contractor's estimator. Sometimes these trades do all their own excavating and backfilling, but it is more customary for them to do only such excavating, backfilling, and street cutting and patching as is done outside the building.

Trenches for piping and other uses are frequently called for on plans or are specified. It is quite expensive to build these and to provide covers for them and sometimes to provide waterproofing in connection with them.

The hot-water requirements of a building are supplied by a water heater and sometimes involve the use of a hot-water tank. Either or both of these may be specified under Heating or under Plumbing. They may even be specified under both headings. Some of the subcontractors in these lines will omit one or both of these items, thinking that the subcontractors in the other line will include them. Thus, apparently low bids will be received and it is the estimator's duty to straighten out the whole matter and to get sufficient bids in these lines in proper form to enable him to compare prices and arrive at an amount to be entered in his summary.

The plumbing subcontractor takes in the bathtubs, lavatories, and toilets in the bathrooms and toilet rooms, and the sinks and washtubs in the kitchens and cellars. Showers and shower curtains are part of the plumbing work also, and sometimes the glass shower doors are included. Sheet lead, put under the tile floors of standing showers and in special watertight floor construction, is installed by plumbers and listed under Plumbing. Hot-water storage tanks and heaters come under Plumbing, except when they are an integral part of the main heating plant installed by the steamfitters. In any event, all the water supply and water connections are handled by the plumbers. All the water-supply and drainage lines, both inside the building and out to the mains in the street, are in this division. A plumbing contractor is required to have a city license.

Temporary water and toilet facilities for the job may involve the plumbing contractor. If, for example, a large amount of water is required for the construction work, it may prove advantageous to have the plumbers install iron pipes, instead of the workers' depending upon the more customary hose lines. This is especially true if the run to the supply source is long or if other conditions make the use of iron pipe worthwhile. If it is possible to make use of a sewer connection or other sewage-disposal connection, it is worthwhile on a large job to install regular toilets for the workmen, from the beginning of the job. After the toilet roughing is in, the building contractors generally do install such toilets, even having the plumbing contractor furnish second-hand toilets for the purpose. The plumbing contractor is generally called upon to provide a temporary water connection for the job use. The cost of all these items must be considered by the estimator and provided for in the plumbing division or in the general job-expense division of the estimate.

Sewers, catch basins, septic tanks, and manholes, if they are required, are generally specified under Plumbing. The plumbing contractors figuring the job, however, may or may not include all of these. Sometimes they include all but the manholes or all but the manholes and the covers for the catch basins. These are additional items for the general contractor's estimator to check carefully. Several manholes on a job may alone cost $1,000.

Toilet and bathroom accessories, such as holders for soap and paper, mirrors, cabinets, etc., are sometimes specified under Plumbing. Regardless of this, many plumbing contractors choose to omit them, especially the cabinets, and may so state in their bids. Even if they should not state this in their bids, there would be a question as to whether they included these items. This would be especially so if, by chance, their bids read "For all the Plumbing work," instead of "For all the work noted in the Plumbing Section of the specifications."

## Costs

Regular plumbing fixtures cost the general contractor between $110 and $140 each, installed, with the piping and trimmings included. A building with many bathrooms may run between $325 and $400 per bathroom, complete. Small jobs of only a few fixtures cost much more per unit than the same fixtures would cost in a large job. Street sewer and water connections cost several hundred dollars each, with the incidental excavating and replacing of pavements.

Figure 26-1 shows a typical plumbing section for a four-story building. Drawings of this nature are purely diagrammatic and cannot be scaled. This one shows all the plumbing fixtures and the drainage and vent piping.

The following list of definitions will assist the student in becoming familiar with some of the more common plumbing terms and their meanings. They are from the New York City Building Code.

**public sewer.** A sewer constructed or operated by the City.

**private sewer.** A sewer not constructed by the City but complying with the Code of Ordinances.

**sanitary sewer.** A sewer designed or used to carry liquid or water-borne wastes from plumbing fixtures.

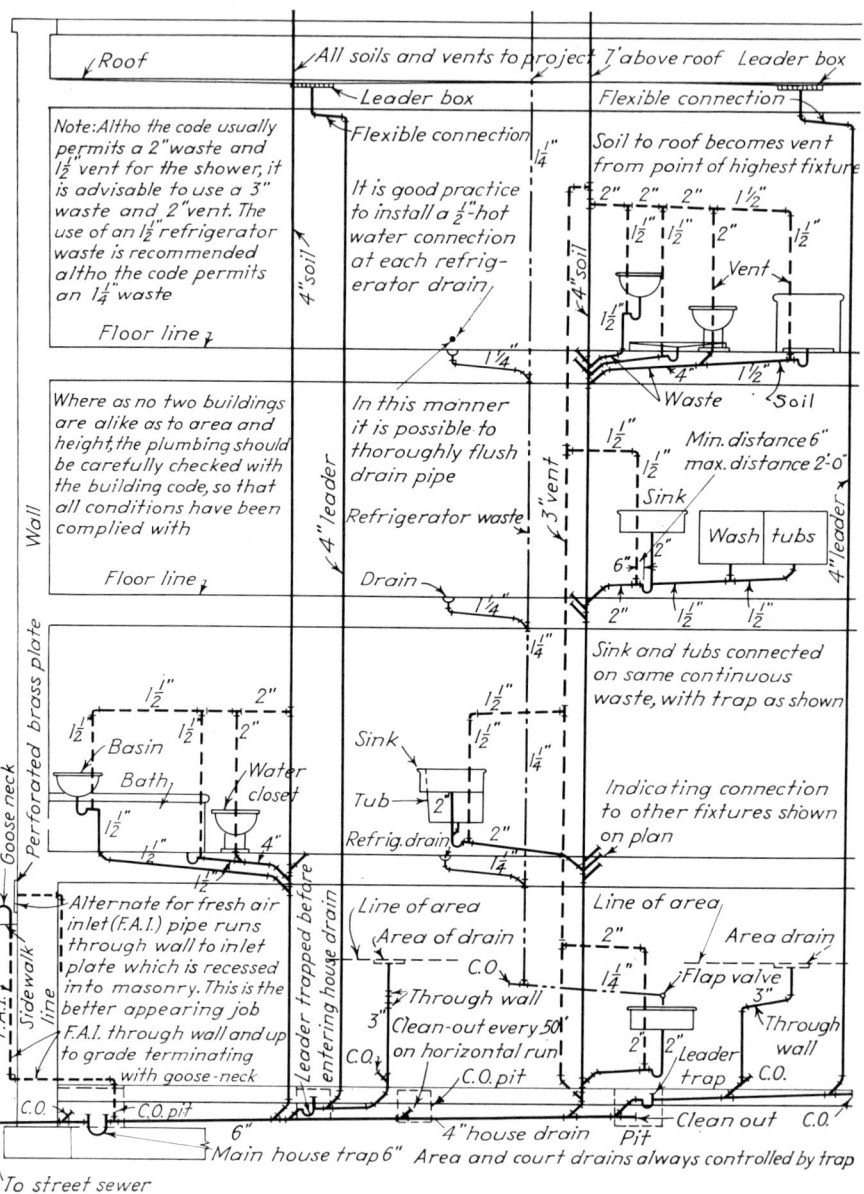

**Fig. 26-1** Plumbing piping.

**storm sewer.** A sewer carrying rain or subsurface water.

**house drain.** That part of the lowest piping of a house drainage system which receives the discharge from soil, waste and other drainage pipes and conveys it by gravity to the house sewer. It ends at the outside of the front wall.

**soil pipe.** Any pipe which conveys to the house drain the discharge of water closets, with or without the discharge from other fixtures.

**waste pipe.** Any pipe which receives the discharge of any fixture except water closets, and conveys it to soil or waste stacks or to the house drain.

**vent pipe.** Any pipe provided to ventilate a house drainage system and to prevent trap siphonage and back pressure.

**stack.** A general term for any vertical line of soil, waste or vent piping.

**leader.** A general term for any vertical line of storm water piping.

**riser.** A water supply pipe which extends vertically to convey water to branches or fixtures.

**branch.** That part of a plumbing system extending from the main to fixtures on not more than two stories.

**trap.** A fitting or device to provide a liquid seal to prevent passage of air while allowing the flow of liquid.

**water service.** That portion of the water pipe which supplies one or more structures and which extends from the public or private main in the street to a main stopcock or valve inside the structure.

**main.** This term, when applied to any system of horizontal, vertical or continuous piping, shall mean that part of such system in which fixtures are connected directly or through branch pipes.

**plumbing fixture.** A receptacle intended to receive and discharge water, liquid, or water-carried waste into a drainage system.

## Requirements

Plumbing systems should be designed so as to guard against fouling and clogging. Pure water, with sufficient volume and pressure, must be provided to enable the fixtures to function satisfactorily under all normal conditions.

Hot-water lines should be provided with safety devices arranged to relieve hazardous pressures and excessive temperatures.

Standpipe and automatic-sprinkler lines should have water in sufficient volume and pressure to enable them to function satisfactorily.

Acids and other damaging wastes should be chemically treated to neutralize their effect, before being discharged into the plumbing system. Where grease in quantities that would produce pipe stoppage is to be handled, an approved device should be provided for intercepting and separating it from the liquid wastes before it is discharged into the plumbing system.

The drainage system should be designed so as to provide adequate circulation of air in the pipes in order that siphonage or pressure will not cause a loss of trap seal. Each vent line should extend to the outer air and be installed so as to minimize the possibility of clogging, frost closure, and the return of foul air to the building.

Adequate clean-outs should be provided and arranged so that the pipes may be readily cleaned.

The following minimum fixtures should be provided:

**Business Buildings.** One toilet for 1 to 15 persons, two for 16 to 35, three for 36 to 55, four for 56 to 80, five for 81 to 110. One-third of the toilets in the men's toilet rooms may be replaced with urinals. One lavatory for 1 to 20 persons is recommended, two for 21 to 40, three for 41 to 60, four for 61 to 80, five for 81 to 100.

**Schools.** One toilet for every 100 boys and one urinal for every 30 boys, one toilet for every 35 girls, and one lavatory for every 50 pupils.

**Hospitals.** One toilet and one lavatory for every 10 patients, and one shower or bathtub for every 20 patients. Fixtures for employees, equal to the numbers for business buildings, should also be provided.

## *House plumbing*

The water-supply main for one-family houses is usually of $\frac{3}{4}''$ or $1''$ galvanized steel or $\frac{3}{4}''$ copper. For business buildings it may be $2''$ or larger. Flush-valve toilets require $1''$ branches. Branches to other ordinary fixtures are generally $\frac{1}{2}''$ pipe or $\frac{3}{8}''$ copper tubing.

Each fixture or group of fixtures is provided with a shutoff valve for repairs or changes.

Cold-water lines are covered with antisweat covering and hot-water lines with heat-insulation covering.

Toilets and toilet tanks and the higher grades of lavatories are made of vitreous china, which is a clay product baked at high temperatures. Other lavatories, and sinks and bathtubs, are usually of enameled iron or pressed steel. Stainless-steel sinks are also available.

The metal trimmings, such as faucets, drains, and shower fixtures, do not come with the sinks, lavatories, and tubs but are supplied and installed by the plumbing subcontractor. Architects frequently specify that all the fixtures shall be of one manufacture and sometimes also specify that the trimmings shall be of the same manufacture.

Sinks are often set into kitchen work counters. In these cases the plumbing subcontractor usually furnishes the sink, but someone else sets it into the counter. The plumbers then connect the sink to the plumbing lines and install the faucets.

When the fixtures are grouped, a saving in material and labor is effected. If the bathroom in a house adjoins the kitchen, for example, or is over the kitchen, one soil-and-vent stack may serve both rooms, whereas two stacks would be required if the rooms were far apart. If there are two other bathrooms, or a bathroom and a toilet room, a saving is similarly made if they adjoin each other. In like manner, a minimum of piping is required if the fixtures are placed along one wall of a room.

**Specifications.** Figure 26-2 is a plumbing section for the house shown in Chap. 6, for which the following specifications apply:

## Plumbing

Provide a complete plumbing system, with all material and workmanship in strict accord with the regulations of all authorities having jurisdiction. The plumbing subcontractor shall obtain all the necessary permits and pay all fees.

All materials shall be new, high-grade and suitable, and the workmanship shall be best union work.

The work under this section includes street sewer and water connections, including excavating and backfilling and street patching, and a complete drainage and water system in the building.

The house drain and soil line shall be XHCI. All other waste lines and the vent lines shall be galvanized steel. All hot- and cold-water lines shall be of copper with brazed joints, copper fittings, and brass valves, and of proper size. Provide copper ¾" water service and water meter. Provide ¾" hose bibs in the garage and outside on the rear wall. Provide water-supply line to the boiler, complete with traps and fittings, and connect the hot-water line to the heater in the boiler. Provide chrome-plated shut-off valves under all fixtures. Provide black iron gas lines from the gas meter to the kitchen range and connect the range.

The kitchen counter, the main bathroom vanity, the dishwashing machine, and the clothes washer will be furnished by others and the plumbers shall connect the fixtures to the plumbing system.

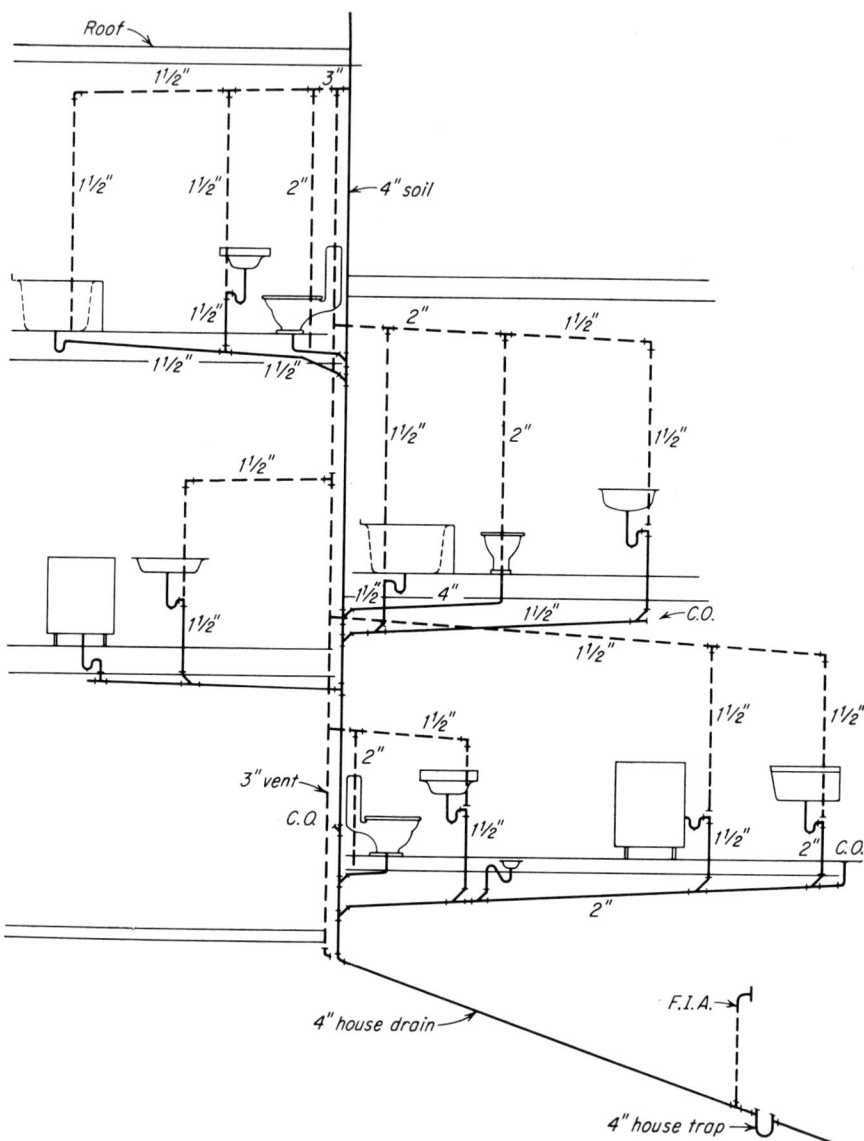

*Fig. 26-2* Plumbing section.

The plumbing subcontractor shall furnish a written warranty to the owner, covering the materials and workmanship for a period of 1 year from final acceptance, and shall furnish the owner with all the customary certificates of approval.

Furnish and install the following American-Standard fixtures, or Crane fixtures of equal quality and design, of colors selected:

Three water closets, "Master one-piece," with Church seats.
One lavatory, "Comrade" 24" × 20", in second-floor bathroom
One lavatory, "Marledge" 20" × 14", in basement
Two bathtubs, "Master Pembroke" 5'6" recess type, with "Waldorf" Chromard fittings and showers
One kitchen sink, 30" × 21", with Chromard fittings, swinging spout, soap dish, and spray
One lavatory, 24" × 20", for main bathroom vanity, with Chromard "Waldorf" trim
One laundry tray, "Lakeview" 24" × 20"
One areaway shower, with Chromard "Waldorf" trim, and Chromard shutoff valves in adjoining toilet room
One Josam floor drain in areaway, with removable top

■ EXERCISES

1. What is a soil line?
2. What is a waste line?
3. What are the so-called mechanical trades in building work?

CHAPTER 27

# heating and air conditioning

The heating boiler, mains, risers, branches, radiators, and return lines, together with the incidental valves and supplies, are in the work of the heating subcontractor. In hot-air heating systems the ducts are also included in this work. Oil burners and fuel-oil tanks are part of the heating work. Heating contractors also install refrigeration systems and special piping and equipment in hospitals, factories, industrial buildings, and processing plants. Some plumbing contractors also do heating work. Air-conditioning and ventilating systems are usually part of the heating work although some subcontractors handle only air conditioning and are not particularly interested in plain heating jobs.

It is necessary for the general contractor's estimator to coordinate the estimates of subcontractors in all lines and to see that everything is properly provided for.

Heating systems are usually tested before they are accepted by the architect. As the testing involves the use of fuel, the estimator must see that the heating contractors figuring the job provide for this, or else he must make an entry here or under General Job Expense to cover the fuel. Likewise, the heating system may be placed in operation before the building is complete, in order to provide temporary heat. If possible, the estimator should have the heating contractor whose bid he plans to use quote also on furnishing the temporary fuel and the radiators and other material that may be required, and on maintaining this temporary heat for the job. If not, he will have to estimate it himself and make an entry here or in the general job-expense division of the estimate, to cover the cost. In some sections, the steamfitters' union requires that a steamfitter's helper be kept on the job while temporary heat is being used.

## Omissions

Boiler pits and foundations and other mason work in connection with heating plants are sometimes specified under Heating. Despite this, the heating contractors very seldom include this work. They may state in their bids that they incude all the heating work, but unless they have stated plainly that they included all the work that is noted in the heating division of the specifications, it would be questionable whether they had these items in their bids.

Ventilating and air-conditioning systems are often included with the heating work. In such cases, the estimator has to check with the sheet-metal estimates and the electrical estimates to see that all the required work in all three lines is included in the bids that he expects to use in his own summary. Much overlapping is possible. Occasionally, it is found that a few of the items involved are included by two different trades, such as fans in roof ventilators and the motors and controls for various fans.

Architects are often vague regarding mechanical work and do not write clear-cut specifications. Sometimes they mention items under two or more headings in the specifications, instead of under only one heading. Work is frequently shown on the plans, or is obviously necessary, but it is not noted at all in the specifications. Plans seldom show the mechanical lines of work completely, and the use of uniform symbols and other indications is not yet well established.

Many new systems of heating, ventilating, and air conditioning are now being introduced, and estimators find that great care is required in analyzing and comparing the bids for quality of work and for possible omissions. Some items that may be omitted by the subcontractors include pipe covering, duct covering, toilet-room vent ducts and registers, roof ventilators, kitchen fans, electric heaters, grilles on fresh-air openings, return-air grilles, meters, oil-tank excavating or enclosures, water heaters, hot-water tanks, water-supply mains, boiler and tank water connections, radiator and convector enclosures, thermostats, controls, electrical and gas connections, machinery foundations, cooling-tower supports, and many others.

These items may or may not be provided for in other divisions of the estimate. Toilet-room and roof vents may be in the roofing division. Water heaters or tanks, or both, may be under Plumbing. Water connections to heating, ventilating, and air-conditioning equipment may be under Plumbing Work. The electrical connections to some or all of the equipment may be under Electrical Work, as may also the thermostats and some of the other control devices. Some of the grilles and enclosures may be under Carpentry or under a metal heading. The machine foundations, supports, and hangers may be in the concrete, masonry, or steel divisions of the estimate.

## Heating systems

The fuel used in most heating systems is either oil or gas. However, the type of fuel used has nothing to do with the method by which heat is brought to rooms.

Oil heating systems have an oil-storage tank which is connected to the boiler or furnace by a small pipeline. Gas heating systems have a direct gas connection. Fuel-oil tanks are placed inside the building, usually in the cellar, or are buried in the ground outside. One or two 275-gal tanks are generally used for indoors. Outside tanks are of 550-gal capacity and sometimes larger. Building codes and Board of Fire Underwriters' requirements must be complied with in regard to tank sizes and locations.

An electrically operated burner is used in connection with either the oil or gas system, to control the boiler or furnace. The burner is activated by a thermostat, which is a type of electric switch controlled by the temperature of the air where the thermostat is located. It turns the heat on or off to suit the desired degree of temperature at which it is set. The dials on the simplest thermostats are set manually. More elaborate thermostats contain a clock which can be set to accommodate different day and night settings. A still more elaborate one is coupled with an outdoor thermostat. The theory of this is that cold, windy weather necessitates a higher indoor temperature to keep the building comfortable. Two-zone and multizone heating employs additional thermostats which activate a pump or blower at the boiler or furnace, sending heat only to the zone desired.

Radiators radiate heat; the heated air rises naturally in a room and, as it cools and drops, recirculates around the radiators and is again heated. Radiators are made of cast iron.

Convectors are like radiators, but they are composed of tubes and fins which have a greater surface than radiators. They heat the air by the force of convection. The cool air enters the lower openings of the convector enclosure and the construction of the enclosure, which forms a flue, increases the speed of the heated air coming from it. Convectors are usually made of copper and this also adds to their efficiency.

Copper tubing is sometimes embedded in floors and convectors are sometimes installed in baseboard enclosures, in place of radiators or regular convectors.

In the *one-pipe system*, steam from the boiler flows through the radiators or convectors consecutively and returns to the boiler in the same pipe, the steam flowing in the upper part of the horizontal pipes and the condensate water in the lower part. The stream of return water is very small, about the size of a lead pencil. Larger than normal radiators are necessary near the end of the line to offset the drop in the temperature of the steam. A gravity one-pipe steam heating system is the least expensive to install, but it is not the most desirable. It is suitable, however, for small buildings.

In the *two-pipe system*, steam or hot water from the boiler flows or is pumped through one pipe to branches leading individually to each radiator or convector, and therefore normal-size radiators or convectors are used throughout. It returns to the boiler through another pipe connected with separate branches leading individually from each radiator or convector.

Other systems of heating are also employed. These include various types of hot-air or air-conditioning systems, vacuum-pumped-steam systems, and special systems of many varieties.

## *Air conditioning*

There are many ways of conditioning air. Some only blow the air around or bring in fresh air and therefore merely employ ventilating devices or systems. Others are for summer use to cool and dehumidify the air. More elaborate systems fully condition the air all year by heating in winter and cooling and dehumidifying in summer.

Ventilating mechanically is accomplished by the use of electric fans or blowers, with or without the use of ducts. The subcontractor doing the heating work in a building may or may not be involved in the ventilating work. In some cases the fans or blowers are estimated under the Electrical Work heading. If there are ducts, these may be provided for under Roofing and Sheet-metal Work.

Small air-conditioning units are for summer cooling only, although the fans in them may be used to circulate air at any time. They are installed in windows or wall openings and are usually of ½-ton or ¾-ton capacity. They take up no floor space and no plumbing connections are necessary, and they may readily be moved to other rooms. This is usually the cheapest way to air-condition a room, but the units give an uneven and drafty cooling effect. The heating contractor might not be involved, as these may be provided for in the electrical work or some other division of the estimate. These so-called window units are easily installed but require proper electric outlets nearby.

Free-standing air-conditioning units are made in capacities of 1 to 5 tons, usually in complete package cabinets. They are easily installed but require water and drainage connections as well as proper electrical connections. Ducts are sometimes used with this type to better serve a small complete building, such as a house, or a large room or several rooms in a larger building.

Central systems are used for large requirements. The conditioned air in them is distributed through ducts to individual rooms. If properly designed and properly constructed, this is an efficient system which provides positive air distribution and fair individual-room control. In the most expensive systems, for very high-grade buildings, the air is distributed at high velocity through pipes. Large ducts and complicated machinery are involved, however, and it is such work that requires very careful study when estimating. Many trades, in addition to those directly connected with the air-conditioning work, are involved. See the section above, headed Omissions, and the section following, headed Building Code.

## Building code

Local building codes, fire insurance underwriters' requirements, and other regulations govern heating and air-conditioning work. The following general notes from the New York City Building Code are of present interest:

It shall be unlawful to make any contact between steam or hot water pipes and any woodwork or other combustible material.

Steam or hot water pipes shall have a minimum clearance from any combustible material of one-half inch.

Where steam or hot water pipes are located within one inch of any combustible material, such material shall be protected by a metal casing or lining and where such pipes pass through stock shelving, they shall be covered with at least one-half inch of insulating material.

Concealed hot water piping may be installed in an outer wall in any structure only when amply protected against freezing.

Heating pipes shall be so installed as to provide safely for all expansion and contraction.

Distributing pipes connected to warm air furnaces shall be kept at least one inch away from any woodwork, and if less than two inches away, the woodwork shall be protected by sheet metal covering or other incombustible material.

A clear working space of at least 18" on the sides and 24" on the top shall be provided around all furnaces and boilers. Such separation shall be maintained with respect to walls as well as pumps and other apparatus used in connection with the heating plant.

Every boiler generating steam shall be equipped with a safety valve adjusted to open under a lesser pressure than the maximum working pressure for which the boiler was designed.

Every closed hot water heating system shall be equipped with an approved pressure relief valve adjusted to open at a pressure slightly higher than the normal operating pressure of the hot water heating system.

Rooms in which boilers or furnaces are located shall have adequate ventilation to supply fresh air for proper combustion. It shall be unlawful to make any direct connection of air inlets to the ash pits or combustion chambers of boilers or furnaces, except where forced draft is employed.

## ■ EXERCISES

1. Name 10 items that may be specified by an architect under Heating Work and yet may be omitted by subcontractors bidding on this work.
2. What trades, other than those directly connected with air-conditioning work, may be required for the installation of a complete air-conditioning system in a large building?
3. What are the customary sizes of fuel-oil tanks?

CHAPTER 28

# *electrical work*

All electric wiring done at the job for lighting, power-supply, and signal purposes comes under the heading of Electrical Work. Special machinery, such as elevator machines, pumps, fans, etc., are installed under other headings, but the electrical connections are made by electricians and the cost for these connections is included under Electrical Work. Plain and special lighting outlets and reflectors, also, are always put under this heading, but the lighting fixtures are sometimes given a special heading. Electrical contractors are licensed by the city, and their work also comes under the inspection of the Board of Fire Underwriters organized by the fire insurance companies.

Electrical symbols are shown in Fig. 28-1 as they appear on plans.

The electrical work in a building gives trouble to the general contractor's estimator mainly in regard to the machinery, fans, and other such equipment. Concerning the main wiring and the lighting system it is usually quite well established as to who furnishes what. Motors attached to fans and unit heaters, elevators, oil burners, and other motor-driven equipment are the things that must be watched. Unless these are very clearly specified, it is likely that neither the subcontractor furnishing the motor nor the electrical contractor will have included the control switches and connections at the motors. In preparing the estimate, it is necessary to see that one or the other does include the cost of this work. Heavy controls are expensive. Sometimes even the motors are left out, no one having included them. The sheet-metal contractor, for example, may take in a lot of exhaust fans and omit the motors for them. The electrical contractors may assume that these will come with the fans and so merely include wiring to the motors, even omitting the switches or other forms of controls.

Another item of electrical work that gives trouble is the temporary lighting or power that the job may require. Several of the subcontractors, requiring electric power to operate their equipment on the job, may specify in their bids that they assume the general contractor will furnish them with suitable current connections at proper locations in the building and at their shanties. The general contractor also may require power for equipment and nearly always has to provide temporary lighting in and about the job. Owing partly to regulations of the electrical unions (something of a "racket"), this temporary electrical service has become an expensive item, sometimes to the extent that a job must employ one or more electricians on full time and with considerable overtime, merely to turn switches off and on morning and evening. This matter must be investigated for every job and properly provided for in the estimate. The electrical contractors fight shy of it, and usually the estimator has to use his own judgment in estimating the cost of installation, maintenance, overtime, and current consumption for the entire job requirements.

If the electrical contractor whose bid is to be used states that he includes only the work inside the building, or that extending to a point a few feet outside the building, the requirements of the job must be checked thoroughly, as there may be much outside work, such as poles, transformers, outside lighting, signal work, etc. Even if the electrical utility company is to do some of the outside work— running the service to the building, furnishing transformers, etc.—it may be left to the general contractor to pay for this service.

Electrical work sometimes involves the construction of a transformer vault, machine foundations, supports for motors and fixtures, etc. Items like these require that the estimator look into the question as to who furnishes them, since they are not electrical work and, although they are specified in the electrical portion of the specifications, the subcontractor who states in his bid that he includes "all the electrical work" may take the stand that because these are not electrical work he does not include them, even though they may have been so designated in the specifications.

Lighting fixtures and other items of electrical equpiment made in nonunion shops will almost surely cause trouble with the electricians working on the job. If items of this nature are to be supplied by the owner, it is the general contractor's place to notify him, through a statement in the bid for the job, that it is assumed any items to be thus furnished will be satisfactory in every way to the union requirements on the job. Some electrical unions insist upon removing all the wiring from lighting fixtures at the job because they were wired by nonunion shopmen or by members of another union, and then rewiring them at the job, all at the expense of somebody, who may be the general contractor.

Some types of lighting fixtures require considerable carpentry work to be done at the job to accommodate them—boxes for recessed lights, etc. This work is usually done by the carpenters on the job, and, unless it is actually included in the electrical bids, it should be provided for under the heading Carpentry.

If a sum of money is specified for the purchase of lighting fixtures or any other items on the job, it is the task of the estimator to ascertain just how the money is intended to apply, whether as the list price for such items or as the actual price to be paid out by the subcontractors. Otherwise, there may be a dispute when the items are being purchased.

## Costs

Wiring and outlets cost between $6 and $9 per outlet for lighting work, if flexible cable is used, and between $9 and $14 per outlet if rigid pipe conduit is used. Ordinary lighting fixtures cost between $1.50 and $8.00 each, delivered, and about $2 to $4 each for installing. These are the prices that the general contractor is charged by the electrical subcontractor. Switches count as outlets. Letter boxes also come under the heading of Electrical Work, as they are commonly installed in connection with the bell system.

## Regulations

Electrical work is generally done in accordance with the requirements of the National Board of Fire Underwriters in addition to the regulations of the local authorities having jurisdiction.

# ELECTRICAL WORK

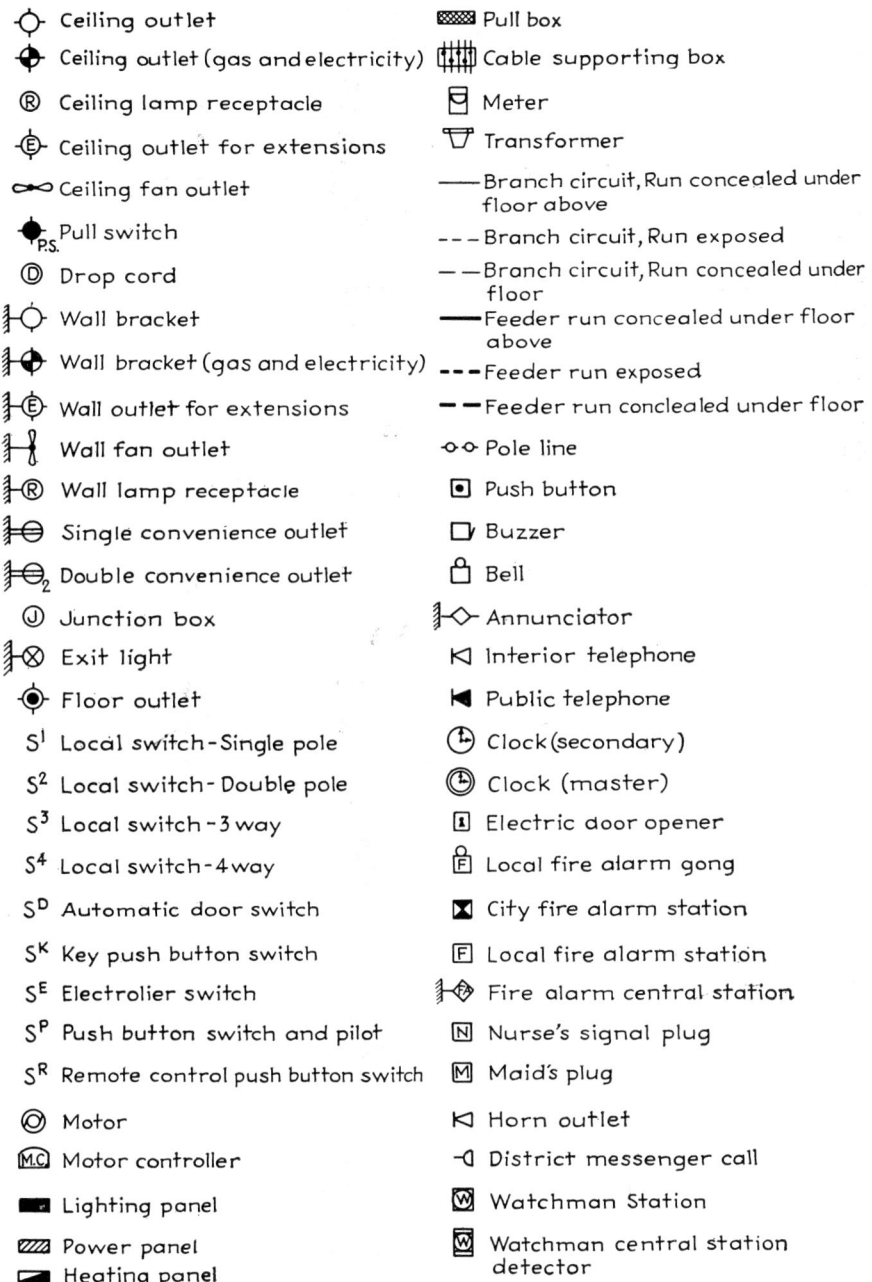

Fig. 28-1  *Electrical symbols.*

The local electrical utility company also has regulations governing electrical work. The following, from one company, will introduce the student to such requirements:

**Service.** The company has adopted as its standard the three phase, four wire, alternating current system of distribution, at approximately 60 cycles with 120 and 208 volts, in the interest of a standardized, unified and economical system.

The standard service comprises: three phase, four wire, 120/208 volt service; or single phase, two wire, 120 volt service; or three wire, 120/208 volt service, with two conductors and a neutral of the three phase, four wire system.

Three phase, four wire, 265/460 volt service will be designated by the company, subject to the customer's concurrence, for supply to buildings when warranted by the magnitude or location of the load or other physical conditions.

**Connections.** Electric service will be supplied to each building or premises through a single service lateral, except where, for reasons of company economy, conditions on the company's distribution system, improvements of service conditions, or magnitude of the customer's load, the company elects to install more than one service lateral. The company reserves the right to determine the location of any service lateral.

If the company designates overhead service, the company will install its service conductors from its street system to the first point of attachment on or near the front face of the building or to the first intermediate supporting structure on the customer's property which, in such case, shall be the point of service termination.

If the company designates underground service, the company will install its service conduit and cables from its street system to the property line or suitable sub-sidewalk space, which, in such case, shall be the point of service termination.

**Transformers.** Where the company considers transformers and associated equipment reasonably necessary for the adequate supply of service to a customer or a customer's premises, the customer shall provide suitable space and reasonable access thereto. To facilitate access and ventilation, such space shall, wherever practicable, be adjacent to the property line and should be outside the building and immediately below the street grade.

At the request of the customer, the company's transformers and associated equipment may be installed by the customer at one or more points in his building or premises on the same or different levels, provided that the entire service installation within his premises, including the installation of, or connections to the company's transformers and associated equipment, or replacements thereof, is made at the customer's expense in accordance with the company's specifications.

**Construction.** Where service is requested for construction purposes and where the facilities installed therefor will not be used for permanent supply, the customer will be required to pay in advance to the company a sum of money, as determined by the company and endorsed upon the agreement for service, which shall be the estimated non-recoverable cost of furnishing and installing all necessary additional facilities of the company and the removal thereof.

**Meters.** The company will install, upon the request of the customer, as many meters as he shall desire, provided the circuit or circuits connected to each meter are kept separate from all other circuits.

The customer shall furnish, install and maintain all wiring and equipment, including standpipes, conduits, fittings, wires, cables, fuses, end boxes, service switch, meter equipment and meter wiring, beginning with the point of service termination. The customer shall install and connect metering transformers on initial installation and upon subsequent alteration to the main cable or bus circuit.

The company will not supply service until the customer's installation shall have fulfilled the company's requirements and shall have been approved by the authorities having jurisdiction over the same. The final connection for making the service alive shall be made only by the company.

## *House wiring*

The electrical service may be either a simple two-wire service or a three-wire service. The three-wire service is preferred as it provides the regular voltage for lighting and also double voltage for power requirements.

In the three-wire service three wires enter the building. One is a neutral wire which is connected to one leg of each of the building circuits. The remaining legs are connected to the other two service wires, one half of the legs being connected to each, thus providing for 110- to 120-volt circuits. Where a 230-volt circuit is required, the neutral wire is connected to one leg and the two live wires are connected to the other leg. The 230-volt wiring is used in circuits to motors and heaters that operate at twice the lamp voltage. Most small appliances, however, operate at 110 to 120 volts.

Metallic tubing, or "thin-wall" conduit, is commonly used for good work. It is like rigid pipe conduit but is lightweight and easy to bend and cut. The ends are not threaded but are fastened with clamp-type couplings.

Rigid conduit, or steel "pipe conduit," comes in 10-ft lengths with a coupling screwed on one end. It is cut with a hack saw, and the cut ends are then threaded to fit the couplings. Both the thin-wall and pipe conduits are empty and the wires are pulled through them. Pipe conduit is used on very high-grade jobs and in certain cases where the applying regulations require its use.

Flexible spiral armored cable, or BX wiring, comes in coils with the wires already contained in it. This is used in ordinary work, and for unimportant branch wiring in connection with jobs having pipe conduit or thin-wall conduit.

TABLE 28-1

| | | | |
|---|---|---|---|
| Oil burner | 255 | Built-in oven | 4,000 |
| Furnace fan | 229 | Dishwasher | 1,165 |
| Room air conditioner | 940 | Food waste disposer | 345 |
| Attic fan | 370 | Refrigerator | 205 |
| Dehumidifier | 210 | Home freezer | 270 |
| Bathroom heater | 1,140 | Kitchen fan | 75 |
| Water heater | 2,430 | Clothes washer | 1,200 |
| Range | 10,910 | Clothes dryer | 4,455 |
| Cooking top | 6,700 | Ironer | 1,495 |

Outlets are controlled from two locations, when so desired, by the use of two 3-way switches. For a greater number of locations, a 4-way switch is used at each extra location.

The lighting circuits are employed for lighting and for clocks, radios, small fans, portable lamps, etc.

Appliance circuits are used for hand irons, mixers, refrigerators, and small appliances.

Individual circuits, of double voltage, are used for ranges, water heaters, clothes washers, and other equipment requiring heavy-capacity wiring. Several such circuits may be required, especially if electric heating is to be provided or if a workshop with electric machinery is involved.

Table 28-1 gives the average wattage required for household equipment and will serve to point out the need for ample wiring.

Push buttons are installed at front and back doors to ring bells, buzzers, or chimes located in the hall or kitchen. An electric door opener, operated by a push button on the second floor, is sometimes used.

Conduits and outlet boxes for telephone wires, to one or more locations, are often specified.

Television connections, consisting of nonmetallic outlet boxes located in the attic and in several of the rooms and connected with a suitable transmission line, are provided in modern houses.

### ■ EXERCISES

1. Draw 10 electrical symbols, and name them.
2. What are pipe conduits and thin-wall conduits?
3. Prepare an estimate of the electrical work for the house shown in Chap. 6. Figure 28-2 is an electrical plan of the first floor. The specifications are as follows:

   ELECTRICAL WORK

   Provide a complete electrical sytsem, with all materials and workmanship in strict accord with the recommendations of the National Board of Fire Underwriters and the regulations of all authorities having jurisdiction.

   All materials shall be new, high-grade, and suitable, and all work shall be union work.

   The work under this section includes the necessary service equipment, feeders, panels, power circuits, lighting circuits, outlets, switches, porcelain pull-chain receptacles, television connection outlets, bell and buzzer system, and the installation of the owner's fixtures and equipment requiring electrical connections.

   The service equipment shall include a 120-amp, three-wire, safety-type service switch connected with three No. 2 service wires.

   Install a 4-circuit, three-wire panel box in the garage. Run the power wiring, three-wire, from this panel, in steel pipe conduit.

   Install a 10-circuit, three-wire panel box in the garage. Run the lighting wiring and the wiring to the minor appliances from this panel, in BX conduit, properly distributed, and providing for 100-watt capacity in each outlet throughout the building.

   All wiring, except in the basement and attic, shall be concealed. All exterior wiring and outlets shall be waterproof.

   All switches shall be the silent, mercury type. All base and wall receptacles shall be duplex and the face plates on these and on the switches shall be ivory-finished.

   Provide four separate power circuits and connections to serve the clothes washer and dryer in the laundry and the dishwashing machine and toaster outlet in the kitchen.

   Provide proper electrical connection to the oil burner and install a thermostat for this in the living room and a remote-control switch at the top of the basement stairs.

Install a kitchen 10" exhaust fan, of good make, with aluminum shutters outside, a sheet-metal duct, and a chrome-plated grille on the kitchen side.

Install nonmetallic television outlet boxes in the entrance foyer and in the basement playroom. Run a suitable transmission line from these to the attic, with a slack length of 20 ft in the attic.

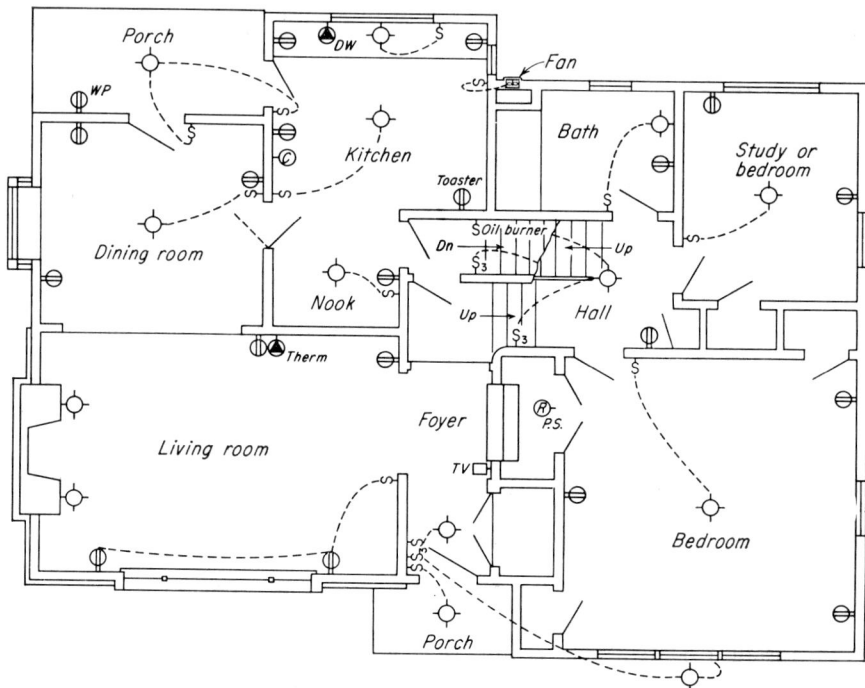

*Fig. 28-2  First-floor electrical plan.*

Install the electric outlets and wiring shown on the first-floor electrical plan.

Install in the basement: 8 ceiling-light outlets, 2 wall-light outlets, 5 wall switches, 6 wall convenience outlets, 1 pull-chain outlet with porcelain lamp receptacle, a flexible-cable convenience outlet in the garage, and also 3-way switches for the garage and stairway lights.

Install in the second floor and attic: 5 ceiling-light outlets, 1 wall-light outlet, 6 wall switches, 8 base outlets, 1 wall convenience outlet, 1 pull-chain outlet with porcelain lamp receptacle, and a 3-way switch for the lower hall light.

Install the laundry and kitchen equipment and the lighting fixtures. These will be furnished by the owner.

Install a complete signaling system, including transformers, push buttons at the front and kitchen doors, and a combination bell and buzzer in the kitchen and in the second-floor hall.

The electrical subcontractor shall furnish a written warranty to the owner covering the materials and workmanship for a period of 1 year from final acceptance, and shall furnish the owner with all the customary certificates of approval.

CHAPTER

# *are you an estimator?*

This book was planned as a complete course in the work of the building contractor's estimator. If you have conscientiously studied everything in it, you are at least well on the way toward becoming an estimator. Practical experience is needed, however, just as in any other line of work. When a young man completes a course in chemistry or in bookkeeping, he may call himself a chemist or a bookkeeper, but he is hardly qualified to take charge of a chemical laboratory or a bookkeeping department. In the same way, if you have completed the course outlined in this book, you may call yourself an estimator, but you should not expect to have full charge of preparing complete estimates until you have had considerable practical experience in building work.

This chapter contains a comprehensive examination, which will test your ability in estimating. If you have done the work called for in the book and receive a mark of 70 per cent or better in this examination, you may feel sure of embarking on actual estimating work. If your mark is between 80 and 90 per cent, you are better than the ordinary assistant in estimating. If you have really studied and worked satisfactorily during the whole course and earn between 90 and 100 per cent in the examination, especially if you have reached a mark in this range in each of the two sections of the examination, you need have no hesitancy in calling yourself an estimator. Put it this way: You may classify yourself as a junior assistant in estimating if your final grade is between 70 and 80 per cent, as a junior estimator if it is between 80 and 90 per cent, and as an estimator if it is between 90 and 100 per cent.

Before taking the second part of the examination, get as much practice as possible by borrowing several sets of plans and specifications from architects and builders and going through the entire procedure of making complete and accurate detailed estimates for all of them.

If your mark is over 90 per cent in each of the two sections of the examination, take this book with you when you are seeking a position, and tell your prospective employer that you know everything in it. Carry along all your work sheets and your notebook for the entire course. If your interviewer is at all fair, he will have to admit that you have a good knowledge of the subject.

Before taking the examination, review all the chapters from the beginning. Chapter 1 calls your attention to the importance of the work of the estimator. It is work that requires patience and a methodical way of doing things. It is not an undertaking for a careless person to attempt. Never vary your way of working in an attempt to please somebody who is looking for speed; you will not be thanked for it if there is a mistake in your work. If it is not prepared with your usual methodical care, there will be mistakes—always remember that. The estimator is usually the center of activity in a well-organized office, and this means that his work is the important work of the office, the work that means the difference between a profit or a loss on the job! Such work is not always properly valued by employers or by other members of a contracting organization; but the estimator finally learns that his reputation as a careful man is one of his main assets.

Chapters 2 to 4 deal with the people you will meet in building work. They are all interesting, and you will enjoy working with them. One striking characteristic of your work is the constant change that takes place, both in the kind and location of jobs and in the steady flow of different people—architects, engineers, inspectors, superintendents, draftsmen, foremen, subcontractors, material men, and workmen. Here is as great a variety as you could wish for; surely your work will never become monotonous. Learn your work well and really enjoy the company of other trained men, who, in turn, will enjoy working with you. Learn also to judge men, for you will find that not all are as sincere as you are. Keep your own character and reputation clean throughout all the trials and tribulations you encounter, however, for you will find truth in the statement made in Chap. 4—that character and reputation of the right sort contribute to your happiness.

Chapers 5 to 7 discuss plans and specifications and the legal side of building work. Plans and specifications have an important legal standing themselves, as they form the portion of the contract that explains just what is to be done. Everything shown on the plans or specified in the specifications is to be furnished or done by the contractor—the general conditions or the contract itself usually states the matter that way. Advise against signing any contract that you feel is not entirely clear and fair. Make your own subcontracts and purchase orders fair, and thus uphold your reputation. Bear in mind that the recommendations given in Chap. 8 for a sort of estimating code are not actually in vogue. Perhaps the time will come when the building industry will develop and use certain standards of bidding that will eliminate some of the abuses of today. In the meantime, we shall have to make the best of the existing methods. Abide by the law in all things; remember that building codes are only schedules of minimum requirements and try to do better than is called for in these minimum requirements.

Chapters 8 to 10 give a broad view of the preparation of estimates and of the general expense involved in the carrying on of building operations. Some contractors make up estimates hastily, sometimes without the aid of an estimator trained as you have been. Such estimates can only be termed approximations, and, as is stated in Chap. 8, approximate estimates have a bad habit of not being correct and of bringing disappointment to those who use them. There is nothing more satisfying to a regular estimator than to see his carefully prepared estimate being used all through the life of a job and serving all purposes faithfully. Remember to picture in your mind, as you estimate, the actual working conditions, especially when you are gauging the unit prices to be applied. Remember to check all the large items and the unit prices you apply to them; they are the ones that will have the greatest effect on the total cost of the job. Do not let others influence your judgment too much. Look into every item thoroughly, yourself, so that you can be confident of your own judgment. Keep your estimates and quantity sheets in good order at all times; arrange them so that you or anybody else will be able to find any item quickly and to understand just what every item is for. Remember that work that is out of the ordinary in character is always more expensive to build than that with which every workman is familiar. Remember that work to be perfomed at a great distance from the center of business areas involves particular attention and care in arranging for workmen, deliveries of material, etc., and may require expensive transportation of both men and materials. Winter work usually means a whole list of items of extra expense.

Chapters 11 to 18 give detailed instructions for estimating the work that is usually done by the general contractor's own men. Only regular work has been considered, and you are therefore warned to watch out for unusual work or conditions and to study and analyze them to the best of your ability if you are to estimate on them. Excavating that you do not feel entirely confident in handling should be figured for you by subcontractors. Experienced estimators never hesitate to say that an excavating problem is too much for their firm to handle—if they really think such is the case. Building contractors should not be expected to be experts in difficult excavation or foundation work. Concrete work, likewise, may stump you, especially very complicated form-work. If you attempt to handle any complicated work in any line, measure and list the items with extra care and do not hesitate to put a much higher unit price on such work than you ordinarily would for the same work when it is not complicated. Your high price may prove to be low enough when the job is done, and you will not get any thanks for anything else.

Chapters 19 to 28 treat of the work usually done by subcontractors. Unless your firm has had experience in actually doing a number of jobs in these lines, it is always better to rely on dependable subcontractors than to attempt to do or even to estimate this work. The mechanical trades, especially, are tricky, and estimators who have thrown in prices on them have often been sorry for having done so. Learn the ways of subcontractors. Learn to judge them. Beware of very young estimators who are sometimes used by subcontractors; they may not have had the training that you have and may lead both you and themselves into trouble. Do not measure plans or give figures for any subcontractors; let them get and be responsible for their own figures. Even if they forgive you for unwittingly misleading them, they will always feel that they have a moral hold on you and expect you somehow to fix things so they can regain what they lost or what they say that they lost. Be businesslike and perfectly honest with the subcontractors, but do not allow them to develop a personal friendship and then take advantage of such friendship. Be as impersonal as possible, and keep your reputation clean. Be loyal to the firm you work for. Work only for good people.

When you take this examination, do so with the knowledge that you will fool nobody more than yourself if you get a high mark without deserving it, by not making it a test that really proves to you yourself that you are an estimator. Fail in the test willingly, rather than start your career with a stain on your reputation!

The examination is divided into two parts, the first covering Chaps. 1 to 10 and the second, Chaps. 11 to 28. If it is desired, Part I may be given immediately after the classwork on Chaps. 1 to 10 has been completed. This is recommended especially when these chapters form the work of the first year in a two-year course.

The students are to do one of the exercises from each of the groups given in the chapters. These are to be selected by the instructor or the school administration and announced or issued when the examination begins.

In addition to the exercises taken from the chapter groups, the following questions and problems are also to be included. To conserve time, note that in all problems involving the measuring of plans, only the names of the items and the dimensions of them are required, no time being taken for making any extensions of figures. The numbers noted in parentheses after some of the questions indicate the number of words used by the average student properly to answer the questions in the time allowed.

## *Examination*

PART I

*20 questions and problems at 5% each*

**1–9.** These are exercises taken from the lists at the end of Chaps. 1 to 10, and will be announced by the instructor.

**10.** Building estimating requires a working knowledge of all phases of building work. Why? (50 to 75 words.)

**11.** The estimator, in a well-organized office, is usually the center of activity in the office. Why? (50 to 75 words.)

**12.** Before the work at the job can begin, the estimator makes up the various schedules, etc. What are these, and how are they used? (30 to 50 words.)

**13.** When the job gets under way, the estimator gradually relinquishes hold. How should he turn the work over to the superintendent? (40 to 60 words.)

**14.** The construction of a building involves many kinds of administrative and technical skills. Name 10 positions of administrative and technical skill that are found in the actual construction of a building.

**15.** How does a contractor go about getting plans for new jobs that are to be estimated? (40 to 60 words.)

**16.** What information should a good estimate contain? (50 to 75 words.)

**17.** On a regular building job costing about $200,000 there are usually many items of general job expense. Make a list of these for such a job and show after each the approximate cost that might be involved.

**18.** Imagine that the job in Prob. 17 is actually under way. Make a daily report such as the job superintendent would send to his home office for a day when the job is well organized and about half completed. Include remarks regarding a dispute between the superintendent and the architect's inspector about the quality of some material.

**19.** Make a diagram and describe the relationship between the men usually concerned with building work, similar to the treatment of this topic in Chap. 2, Construction Relations. (60 to 80 words and diagram.)

**20.** Write a descriptive composition of 100 to 150 words on Plan Reading, using the plans in Figs. 29-1 to 29-13 to illustrate the points brought out in your composition.

PART II

**1–15.** These are exercises taken from Chaps. 11 to 28, and will be announced by the instructor. (15 at 3% each.)

**16.** List every tenth term from the Index, to a total of 20 terms. Every student is to start with a different term. One student starts with the first term in the Index, the next student with the second term, etc., and then each tenth term thereafter. State the meaning and use of each term clearly in 20 to 30 words. Sketches will be accepted in lieu of 10 to 20 of the words. (20 terms at 1% each.)

**17.** Make a complete detailed estimate for the plans given in Figs. 29-1 to 29-13, omitting the extensions of quantities, to save time, but including roughly approximate totals for the purpose of pricing. Extend and total all money amounts. Use assumed subcontractors' figures for the plumbing, heating, and elevator work only. (35%.)

## Bank and office building outline specifications

**General.** The intention and meaning of these specifications and the accompanying drawings is to provide and secure a new complete building 30'0" × 50'0" containing cellar, three floors, and mezzanine. The work includes the structure and the plumbing, heating, and elevator work. The electric work, signals, and electric fixtures are not included in this contract and will be done by others under a separate contract.

The standard form of the "General Conditions of the Contract," copyrighted 1937, as established by the American Institute of Architects, is hereby included in and made a part of these specifications, whether or not attached hereto. A copy of these "General Conditions" is on file at the office of the architect where same may be inspected by contractors figuring on the work.

The reference to "Supervision" in the Standard Form shall mean that the Contractor shall provide a superintendent having at least twenty years' experience on building construction (at least ten years as Superintendent), a job engineer having at least ten years' experience on building construction work (at least five years as Engineer), a job clerk having at least eight years' experience on building construction work (at least five years as Chief Job Clerk) and as many assistants as required to properly carry on and expedite the work. The three "key" men shall be on duty at the job at least forty hours per week from the start to the final completion and acceptance and shall not be replaced without the written consent of the Owner and the Architect.

The Contractor shall provide a job office, not less than 10 × 16 ft, with not less than two rooms and fitted with electric lights, telephone, desks, chairs, plan tables, racks, files, typewriter, engineers' instruments, etc., and shall maintain same in good order solely for office purposes.

**Excavating.** Do all excavating of every description and whatever substance encountered. Should the excavation through accident or otherwise be taken out below the levels shown on the drawings the contractor shall fill in the resulting excess excavation with 1:2:4 stone or gravel concrete at no extra cost.

**Concrete Work.** All materials shall be new, clean and of substantial quality and quantity. Cement shall be portland conforming to the latest standard specifications of the American Society for Testing Materials. Sand shall be coarse and washed. The coarse aggregate for all concrete below the finish first floor level shall be washed gravel or crushed stone uniformly graded from ¼" to 1". The coarse aggregate above the first floor shall be clean steam cinders.

Provide substantial forms, oiled for all exposed concrete surfaces, for all the concrete work including the footings.

All concrete below the finish first floor level shall be 1:2:4 stone or gravel concrete. Provide 4" × 4" × No. 6 wire mesh reinforcing in the first floor slab. Set and grout all anchor bolts and steel bearing plates. Grout to be 1:3.

All concrete above the first floor shall be 1:2:5 cinder concrete. Provide 4" × 4" × No. 8 wire mesh reinforcing in all the second floor and hall and toilet floor slabs. Provide standard steel soffit clips on all steel beams. Slabs are 4" thick.

Fill the iron stair treads with 1:3 mortar, troweled smooth, with light wire mesh embedded in it.

Run a 6" coved cement base throughout the second floor and in all the halls, toilets and slop sink closets, except in the cellar.

Lay a 5" slab throughout the cellar and in the elevator pit and apply a 1" finish of 1:3 on it and troweled to a smooth and perfectly level finish.

Patch the sidewalk neatly, adjoining the building, with concrete and finish and mark off to match the present work.

Lay 1:3 cement finish, troweled smooth and perfectly level, on the slabs throughout the second floor and in all the toilets and the slop sink closets, and in all the halls of the second and third floors and roof bulkhead.

**Masonwork.** All materials shall be new, clean, and of substantial quality and quantity. Cement shall be portland conforming to the latest specifications of the Society for Testing Materials. Lime shall be an approved hydrated lime. Sand shall be fine sand suitable for brick masonry. Brick shall be hard burned red common brick of uniform size and with true shape. The two walls returning from the fronts shall be faced for a distance back of 10 ft with smooth face brick matching in color the limestone. Terra cotta blocks shall be Natco or equal. All mortar shall be 1:1:6 and all joints shall be full shoved joints not more than ½" thick and shall be uniform and straight.

All piers shall be well bonded within themselves and to the adjoining walls.

All cellar partitions shall be of smooth finish terra cotta blocks. All other terra cotta blocks shall be scored to receive plaster, those at the shafts being smooth on the shaft side.

Clean and point all exposed interior and exterior brick and terra cotta with muriatic acid and water.

Install terra cotta copings on the two rear walls, except over the face brick. Install 12" × 16" terra cotta flue lining.

Set, bed, and grout all bearing plates, window frames, beam anchors, brackets, bolts, flashings, etc.

Provide cast iron cleanout door and frame and terra cotta smoke pipe thimble in the boiler room.

Install approved reinforced gypsum structural plank flooring on the mezzanine floor and on the mezzanine stair landing. This shall be $2\frac{1}{2}''$ thick and provided with rabbeted edges to fit between the structural steel angles.

**Stonework.** All granite shall be Deer Isle granite. This consists of the base course, on the two street fronts, extending 2" below the sidewalk grade. It shall have "eight-point" finish. The pier bases shall be in one piece 6" thick. Also included are the two entrance steps which shall be in one piece each, 8" thick, 14" wide, with wash on top.

All limestone shall be standard buff genuine Indiana limestone. This consists of the wall facings as shown and also the copings and return copings, the window sills, and the chimney cap. All exposed surfaces shall be smooth planed. The name: FIRST NATIONAL BANK, shall be neatly and accurately carved in the stone over the bank entrance as shown. The general ashlar shall be installed in proper bonding with the brick backing and to comply with all local building ordinances. All bond stones shall be 4" thicker than the normal thickness of the facing. Greater dimensions shall be provided where required for jambs, reveals, lintels, architraves and sills. No stone shall have a bed less than 4".

All exterior marble shall be White Alabama marble $1\frac{1}{4}''$ thick with polished finish.

Submit shop drawings showing the bedding, jointing, and anchoring of all stonework. The entire two street fronts and the exposed portions of all the openings in them shall be covered with stone. Provide openings for the cellar vents and for the plumbing fresh air inlet. Provide all anchors required. Set all stone and parge back of all stone with nonstaining cement mortar. Rake out all joints $\frac{3}{4}''$ deep and fill with pointing mortar colored to match the stone. Clean down all stonework upon completion of the job. Protect all stone with wood covering.

**Steel and Iron.** All steelwork shall conform to the latest edition of the standard specifications for structural steel for buildings as adopted by the American Institute of Steel Construction. All steel shall have a shop coat of red lead and linseed oil paint before it is delivered to the job. All shopwork shall be riveted and all field connections bolted with the threads upset.

All ironwork shall be substantial, neat and perfect for the purposes intended. All joints shall be smooth and close. All material shall have a shop coat of red lead and linseed oil paint before being delivered to the job.

The ironwork includes standard stairs and railings of office building type. The treads are to be pan type for cement fill and the flight from the main hall is to have a decorative cast-iron newel about $4'' \times 4''$ with a bronze cap and decorative iron balusters, and this flight is to be set to receive $1\frac{1}{8}''$ marble treads placed on top of the cement fill. Also included are the standard cast-iron fluted saddles for the elevator shaft door openings and bronze fluted saddle for the door between the entrance hall and the bank. Provide also a simple railing with moulded cap along the edge of the mezzanine, and a pair of flush checkered steel trap doors and frame in the sidewalk.

Submit complete shop and setting drawings for all the steel and ironwork.

Include two $12'' \times 31$ lb sheave beams in the elevator machine room. Include substantial framing for support of elevator bulkhead, and iron stair to machine room. Include standard bearing plates and anchors for all steel beams and columns.

**Carpentry.** All materials shall be new, clean, and of substantial quality and quantity. All floor beams shall be Dense Select Structural Douglas Fir or Dense Structural Southern Pine, officially grade-marked. Roof beams shall be No. 1 Common Douglas Fir. Finish flooring for third-floor office space shall be No. 2 plain red oak, $1\frac{3}{16}'' \times 2\frac{1}{4}''$ face, D&M and end matched, underlaid with asphalt-saturated felt. Subflooring and roof sheathing shall be No. 2 common pine or fir, $\frac{7}{8}'' \times 6''$ T&G. Cross-bridging shall be $1'' \times 3''$. Fur all exterior walls with $1'' \times 2''$ pine furring strips $16''$ on centers, except in cellar and shafts.

Provide anchors for every fourth beam end. Provide roof blocking and cant strips as required. Provide curbs for the skylights and vent. Install $2'' \times 6''$ roof beams over the stair bulkhead, resting on a $2'' \times 4''$ plate anchored to the block walls. Provide $1\frac{1}{2}''$ round hardwood handrails on the iron stair railings. Provide $1'' \times 6''$ white pine or poplar baseboard and two mouldings, in the workroom and mezzanine portions of the bank and in the third floor office space.

All the doors indicated to be of wood shall be of No. 1 Door Stock White Pine or Ponderosa Pine, $1\frac{3}{4}''$ thick. Provide cylinder locks, ball-bearing butts, liquid door closers, and stops with hooks for all office doors, with six keys for each door and two master keys. Provide bit key locks and plain butts for all the other wood doors, with six keys for each door and four master keys.

Provide double-hung sash and frames for all windows on the second and third floors, and stools, aprons and casings. The sash shall be $1\frac{3}{4}''$ thick white pine. Stools shall be $1\frac{1}{8}''$ thick, aprons $\frac{3}{4}'' \times 4\frac{1}{4}''$ housed into stools, and casings $\frac{3}{4}'' \times 4\frac{1}{4}''$ moulded and provided with backbands.

Provide a standard steel sash, stationary type, in the stairway bulkhead, anchored to the blocks. Caulk all windows.

Provide a paper holder, a coat hook, a glass lavatory shelf, and a chrome-framed $18'' \times 20''$ mirror in each toilet room.

**Hollow Metal and Kalamein.** Provide pressed steel frames for all wood and metal doors, except the elevator doors, which are specified under "Elevator Work."

Provide standard hollow metal bronze doors for the two openings leading from the halls to the bank spaces. These shall be one-panel doors and shall be provided with cylinder locks having solid bronze knobs and escutcheons. They shall also have solid bronze ball-bearing butts and burglarproof dead locks with solid bronze escutcheons, and heavy liquid door closers.

Provide standard kalamein doors where indicated, one-panel type, fitted with butts and bit-key locks. Provide burglarproof dead locks also on the two doors to storage spaces A and B and also to the storage room C. The two roof doors shall be flush construction on the outside. Provide six keys for each door, and three master keys for the bit-key locks only.

**Architectural Metal.** This division of the work shall be done by a subcontractor approved by the Architect and the Owner. To meet approval, the subcontractor must have a record of many satisfactory jobs of high grade and must have a well-equipped shop.

The work includes all the door and window frames on the two street fronts of the first story and mezzanine and the two sets of entrance doors, together with all the fittings and hardware for these openings. It also includes the grilles on these openings and the frames, sash, and grilles for the cellar vent openings, and the perforated cover on the plumbing fresh-air inlet.

All the construction of this work shall be of heavy type and of the highest grade of workmanship. All exposed parts shall be polished stainless steel. Complete shop drawings shall be submitted to the Architect and shall be revised until they meet his entire approval.

The door saddles may be either aluminum or stainless steel abrasive saddles, eight inches wide, and provided with Rixson door hinges. Heavy cylinder locks with stainless steel escutcheons and handles shall be provided, and the bank entrance shall also have a burglarproof mortise dead bolt. The lock bolts shall have stainless steel ends.

Complete inside trim for all the openings is to be included also, as no other finish will be provided except the marble stools in the window openings.

**Cabinetwork.** This division of the work shall be done by a subcontractor approved by the Architect and the Owner. To meet approval, the subcontractor must have a record of many satisfactory jobs of high grade and must have a well-equipped shop.

The work includes the partition along the edge of the mezzanine floor, the partition and door below the mezzanine, the complete bank screen with counter, drawers, and door, the railing at the manager's office, the bank entrance vestibule and vestibule doors, and the hall entrance vestibule and vestibule doors, together with all fittings and hardware.

All this work must be of the highest type "Cabinetwork." Complete shop drawings shall be submitted to the Architect and shall be revised until they meet his entire approval.

The partitions, bank screen, railing, and the roof of the bank vestibule shall be formed of $2'' \times 3''$ framing and covered on both sides with solid stiles and rails and veneered paneling. The doors and the drawers shall be of veneered wood. All the other wood shall be solid. All exposed wood throughout shall be selected and matched American Walnut and shall be given a high-grade "furniture" finish.

The bank screen shall be 5'0" high. The customers' shelf shall be $1\frac{1}{8}''$ thick and extend through to the back of the screen. The portion above the customers' shelf shall consist of five posts supporting four pieces of $\frac{1}{2}''$ thick plate glass. The top edges of the glass shall be polished and each piece shall have a pass opening formed in it. The return portion of the bank screen shall be similar but with ground plate glass.

The counter top shall be $1\frac{1}{8}''$ thick. A continuous foot rest shall be provided under the counter, and eight drawers shall be installed to occupy the entire length immediately below the counter top. The counter top shall be level and 36" above the floor. The customers' counter shall be 39" above the floor. The railing shall be 39" high with a flat top 5" wide. The drawers shall have roller or suspension guides.

Provide complete hardware for all the doors, with aluminum or chrome-plated brass exposed parts. Include Rixson floor hinges for the active doors. Include cylinder locks with electric openers, and $\frac{3}{16}''$ thick ground plate glass, for the two doors in the bank screen and partition. Include push plates, pulls, and $\frac{3}{16}''$ clear plate glass for the four vestibule doors. Include $\frac{3}{16}''$ ground plate glass for the partition below the mezzanine and $\frac{3}{16}''$ clear plate glass for the partition on the mezzanine and for the transom sash above the hall vestibule doors.

**Plastering.** There will be no lathing or plastering required in the cellar, nor in the shafts, nor on the ceiling under the mezzanine. All iron stair soffits will be exposed except the one in the entrance hall. This one will be lathed and plastered.

Apply 3# metal lath on all the wood furring strips on the exterior walls throughout. Apply a coat of cement mortar on this lath for the marble wainscoting in the entrance hall and in the bank public space and manager's office. Apply three coats of lime plaster on all other portions of this metal lath. Finish the walls in the toilet rooms with Keene's cement plaster and finish the walls elsewhere with a hard white plaster of Paris finish.

Apply two coats of plaster, finished hard white, on all the terra cotta partitions and in the stairway bulkhead, and also on the ceilings in the toilets and in the halls of the mezzanine and second floors, and on the mezzanine ceiling.

Apply 3# metal lath and three coats of plaster, finished hard white, on the ceiling areas in the balance of the second and third floors and in the stairway bulkhead.

Construct hung ceilings over the high portion of the bank space and also in the entrance hall and the entrance hall vestibule. This construction shall meet all the requirements of the local building ordinances. The hangers and runners shall be of substantial iron members well anchored to the concrete construction. Lath shall be 3# metal lath. Plaster shall be three coats with hard white finish. The ceiling in the bank shall be run level directly under the floor beams of the second floor. The ceiling in the entrance hall shall be 12'0" above the floor. The ceiling in the vestibule shall be as high as possible. Close in the vertical spaces above the hall and vestibule ceilings with similar construction.

Run a 12" girth moulded plaster cornice around all four sides of the bank space, and close in the vertical space back of the cornice along the mezzanine side. Run a 4" girth moulded plaster cornice in the entrance hall, with a double face portion across the stair soffit and around the stair well to form a ceiling panel.

**Interior Marble.** Install Gray Tennessee marble wainscoting on the two street walls in the public space and manager's office in the bank, and also on the exterior wall at the back of the manager's office. This shall extend up to the window sill level and shall consist of a base $1\frac{1}{4}$" thick and 9" high, matched flush panels $\frac{7}{8}$" thick and a moulded cap $2'' \times 2''$. Install $1\frac{1}{8}$" thick stools in the window openings, extending to $\frac{1}{2}$" beyond the cap and provided with moulded exposed edges. Install a similar stool in the window below the mezzanine, with an apron $\frac{3}{4}'' \times 4''$ under it.

Install Gray Tennesse marble on the walls of the entrance hall and vestibule, full height, and extending up the stairway to the level of the mezzanine floor, and extending on the wall under the stairway down to the level of the first floor. This shall consist of a base 9" high and $1\frac{1}{4}$" thick, matched flush panels $\frac{7}{8}$" thick and 24" high, a band $1\frac{1}{8}'' \times 3''$, and matched panels above $\frac{7}{8}$" thick.

Install Gray Tennessee marble treads $1\frac{1}{8}$" thick on the stair from the entrance hall to the mezzanine.

Submit complete drawings to the Architect for approval before starting any work.

Clean and polish all marble upon completion of job.

**Terrazzo.** Lay a 1" mortar underbed and a $\frac{1}{2}$" thick terrazzo floor in the entrance hall and vestibule. Provide white brass strips with $\frac{1}{4}$" wide tops to form a modernistic design as will be detailed. The terrazzo will be formed with marble and granite chips of three colors as will be selected by the Architect. The terrazzo shall be machine rubbed and grouted, the grouting coat removed by machine, and the floor then fine-stoned and washed.

**Glass.** Glaze the exterior windows of the first and mezzanine floors with $\frac{1}{4}''$ thick polished plate glass. Glaze the windows of the second and third floors with double-thick A-quality sheet glass. Glaze the steel sash in the stairway bulkhead with $\frac{1}{4}''$ rough wire glass. Glaze the six office doors with $\frac{3}{16}''$ thick Syenite glass. The balance of the glass is specified under other headings.

All glass shall be bedded in putty and be back-puttied. Face puttying shall be run smooth and true. Putty shall be of an approved brand and special putty shall be used in all the metal windows.

**Roofing and Sheet Metal.** Lay composition roofing on the main roof, composed of one layer of roofing paper nailed to the roof sheathing, four layers of roofing felt each embedded in a full coat of asphalt roofing cement, and a surfacing of 400 lb of washed gravel or 300 lb of crushed slag per 100 sq ft.

Install copper flashings and counterflashings at all vertical surfaces, extending up at least 12″ and built into brick joints at the top $1\frac{1}{2}''$.

Cover the stairway and elevator bulkhead roofs with building paper and four layers of roofing felt each embedded in a full coat of asphalt roofing cement, and a surfacing of Ruberoid roofing lapped 17″ and turned down over the edges of the roofs. Cover the outside of the terra cotta block walls of the bulkhead with a coat of creosote oil and a coat of fibrous asphalt mastic troweled smooth.

Dampproof the outside of the elevator pit walls and the outside of the cellar walls up to the grade level by applying a priming coat of creosote oil and a mop coat of hot coal-tar pitch.

Install Barrett-Holt Type 1-LG leader connections.

Furnish a ten-year written guarantee, signed by the roofing contractor and the general contractor, covering all roofing and flashings.

Provide a copper, wind-driven vent, with 24″ throat, of a good standard manufacture and approved by the Architect, over the toilet vent shaft.

Provide copper skylights, with ridge ventilators, over the top floor toilet rooms, glazed with rough wire glass.

Provide Fire Underwriters' type louvered vents and frames for the other toilet rooms, 24″ wide × 24″ high.

Flash the wall above the skylights and vent with copper flashing and counterflashing extending up to and under the coping.

**Painting.** Calcimine all plaster ceilings and cornices throughout.

Apply three coats of linseed oil paint on the ceiling under the mezzanine.

Apply four coats of linseed oil paint on all plastered wall surfaces on the first and mezzanine floors.

Touch up all bare spots of exposed steel and iron with red lead and apply two coats of linseed oil paint to all the exposed steel and iron.

Apply two coats of linseed oil paint on all plaster wall surfaces on the second and third floors and stair bulkhead, and to the wood baseboard.

Apply one coat of cement-lime paint on all exterior wall surfaces in the cellar and the entire height of the elevator and vent shafts, extending this 12″ on to the adjoining surfaces also.

Apply one coat of cold-water paint on all *interior* wall and partition surfaces in the cellar and on all wall and partition surfaces in the elevator and vent shafts, in addition to the painting already specified.

Apply one coat of gloss enamel on all wall surfaces in the toilets (except in cellar), slop sink closets, and for a height of 5 ft in the halls and stairways of the second and third floors, in addition to the painting already specified.

Apply three coats of linseed oil paint to all the pressed-steel door frames, all the kalamein doors, all the metal louvers, and all the wood doors and windows except those covered under "Cabinetwork."

Apply one coat of combination filler and finish to the wood flooring on the third floor.

All surfaces shall be perfectly dry and ready for painting before any painting is started. The painting subcontractor will be held responsible for inspecting and testing the surfaces and producing a guaranteed job.

All materials shall be fresh and of approved makes. Each paragraph above calling for paint shall provide for a different color, at the option of the Architect. Each coat of paint shall be of a slightly different tone so as to provide for easy inspection as to the surfaces being fully covered by each coat. Each coat shall be approved by the Architect before the next coat is started. The painting contractor shall cover the floors throughout the building with building paper and when the work is completed and ready for final inspection, he shall remove this paper and all surplus materials and tools, etc.

**Plumbing.** Install a complete system of plumbing, including water and sewer connections under the street, and all in full accordance with the local building regulations and ordinances. This includes brass hot and cold water piping, 3″ cast-iron leaders with screw joints, and fixtures of Standard Sanitary Mfg. Co. make with chrome-plated brass trimmings and exposed branches. Run water lines to the heating boiler, which will be provided also for water heating. Provide a forty-gallon tank for hot water.

Cover all hot and cold water lines and the tank with approved covering. Provide a water meter and water connection of ample size. Provide a control valve on each water line in the cellar and a control valve under each fixture.

**Heating.** Install a complete two-pipe gravity oil-fired steam heating system guaranteed to heat the entire building to 70° when the outside temperature is 10° below zero, and also to supply hot water during winter and summer.

All materials shall be new, of substantial quality, and of approved standard makes. The boiler shall be designed for a safe steam working pressure of 15 lb gauge and a hydrostatic test pressure of 60 lb per sq in. and shall be complete with all fittings, trimmings, and tools, and connected to the flue with an iron smoke pipe.

Steam and return piping shall be standard-weight wrought steel. Fittings shall be standard-weight cast-iron screwed steam pattern fittings. Valves shall be single wedge standard-weight 125# W.S.P. screwed pattern, brass-bodied, brass-mounted, bronze stem.

Provide ceiling-type radiators in the cellar. Provide a copper convector with finished enclosure under each window throughout the building and one in each vestibule. The sizes of all radiators, convectors, and piping shall be such as will produce an even heating distribution.

The oil burner shall be a Petro or General Electric outfit complete in every respect, with a 450-gal tank.

Provide a Minneapolis-Honeywell thermostat control system with the thermostat located in the bank manager's office.

Cover all steam and return piping with approved covering. Install floor flanges around piping running through floors. Provide heating risers in the toilets, with covering and canvas outer covering on the lower four feet in each toilet.

Provide chrome-plated radiator inlet valves. Provide a chrome-plated brass-body thermostatic radiator trap of the multiple diaphragm type at the return end of each radiator. Provide all other required or recommended valves, traps, strainers, specialties, fittings, supplies, etc., to produce a noiseless and thoroughly efficient installation throughout.

Furnish a written guarantee, signed by the heating contractor and the general contractor, covering all of this work for a period of one year.

**Elevator.** Install a complete passenger-elevator outfit and shaft doors and frames. This work shall be done by Otis, Westinghouse, or an approved equal manufacturer. The elevator shall be as large as the shaft dimensions permit and all the work shall meet the requirements of the local building ordinances.

The car shall serve all five landings and have a speed of approximately

75 fpm. It shall be of the automatic push-button type with all safety devices and floor leveling automatic mechanism and door interlocks. All the materials shall be new and of the latest types. The work includes a car with rubber-tile flooring and decorated baked-enamel steel enclosure, overhead machine, cables, guides, controls, counterweights, limit switches, motor, light, buffers, lubricators, safety and governor, wiring in hatchway, switches, and all other items necessary for proper and complete installation. Also included are hollow steel shaft doors with baked enamel finish and pressed-steel frames, car gate, door closers and interlocks, push-button controls, and car-in-use indicators.

The electrical contractor will furnish an outlet near the mid-point of the shaft for lighting and a power feeder with a knife switch in the machine room. The elevator contractor shall do all other wiring and electrical work.

Provide a written guarantee covering the entire installation for one year and providing for free maintenance and repairs and parts during that period. If any repair or maintenance work or service is found necessary it shall be done promptly and at any time of day or night to suit the Owner's convenience.

**Electric.** The electrical work, including bank and building signals and lighting fixtures, will be done by others under a separate contract. The general contractor and his subcontractors shall cooperate with the electrical contractor in every reasonable way.

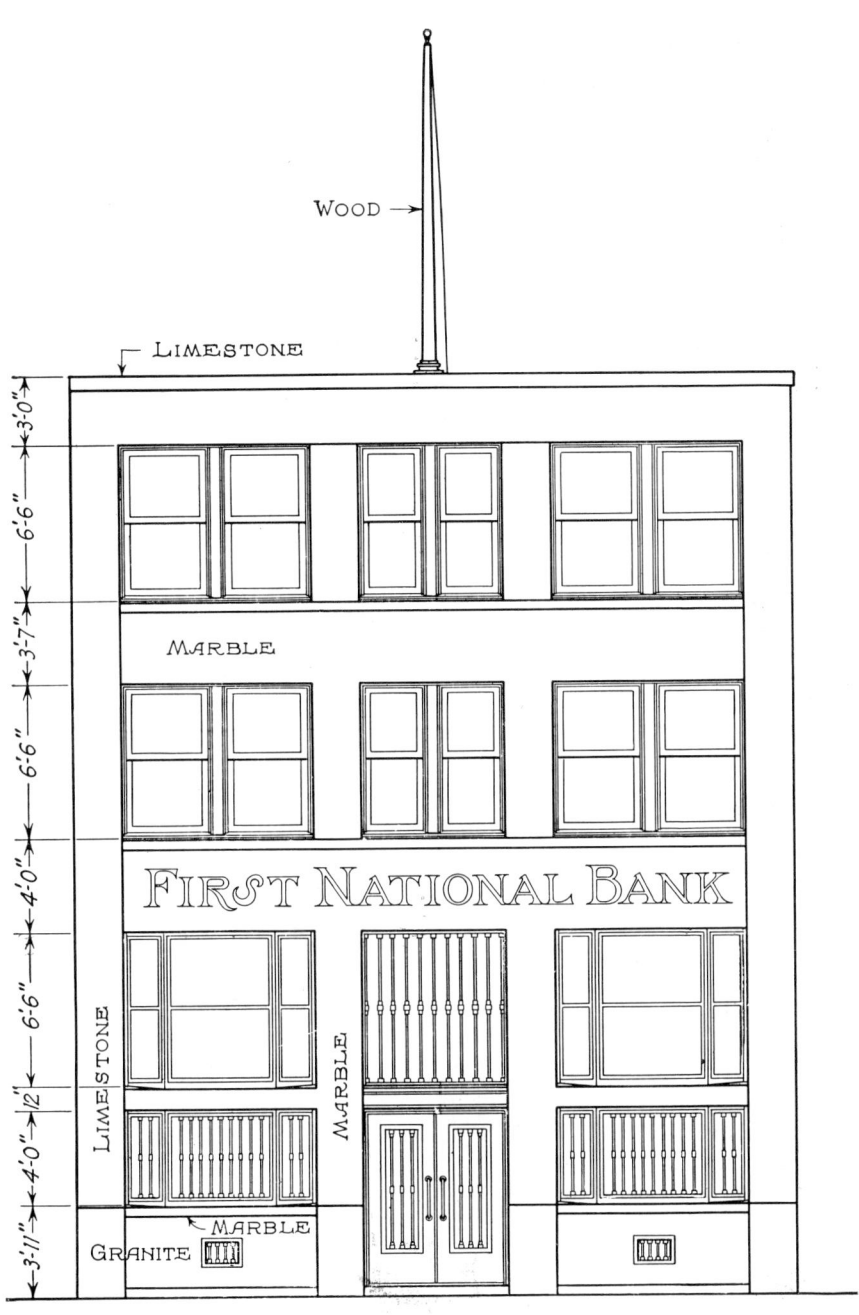

Fig. 29-1.

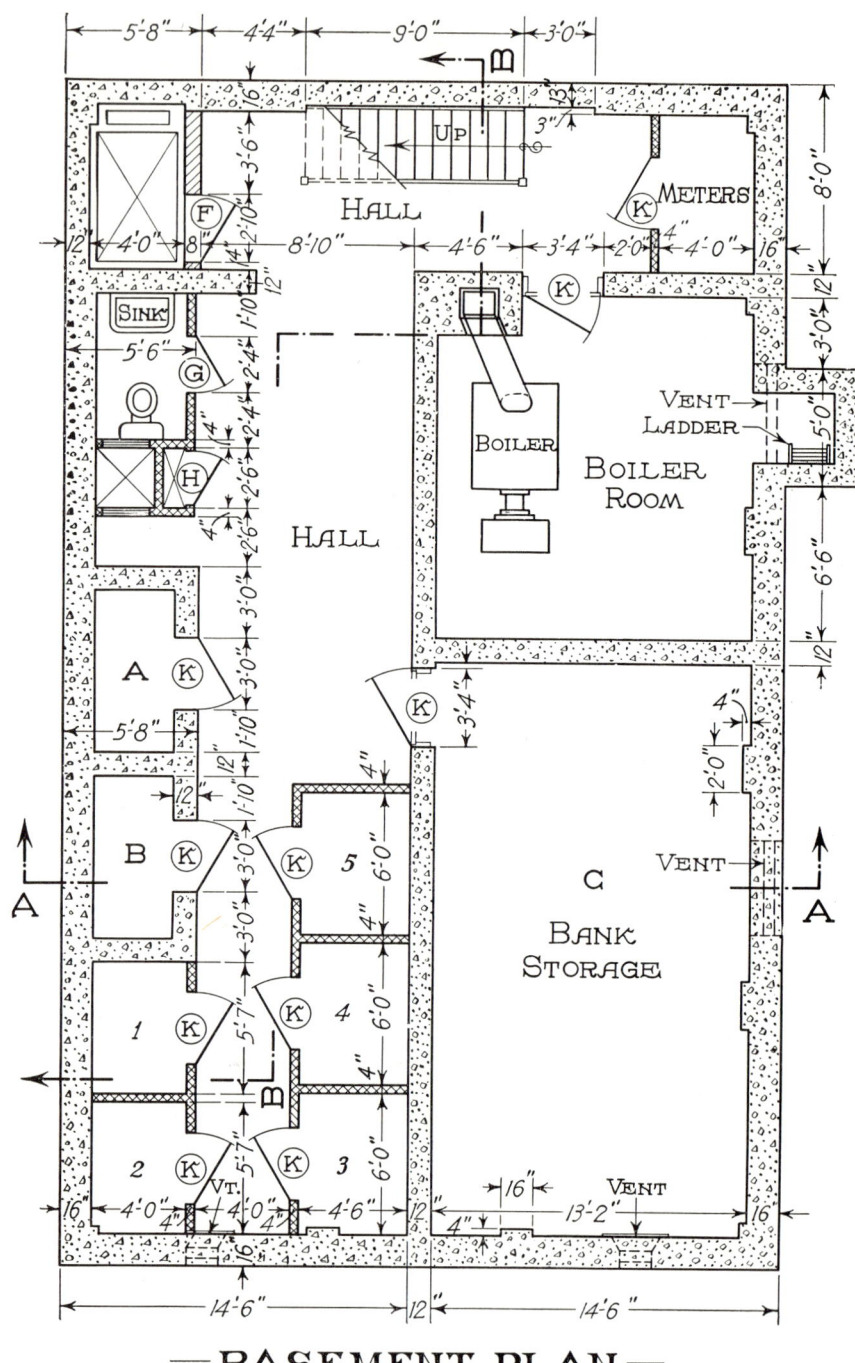

—BASEMENT PLAN—
—SCALE: 1/8"= 1'-0"—

Fig. 29-2.

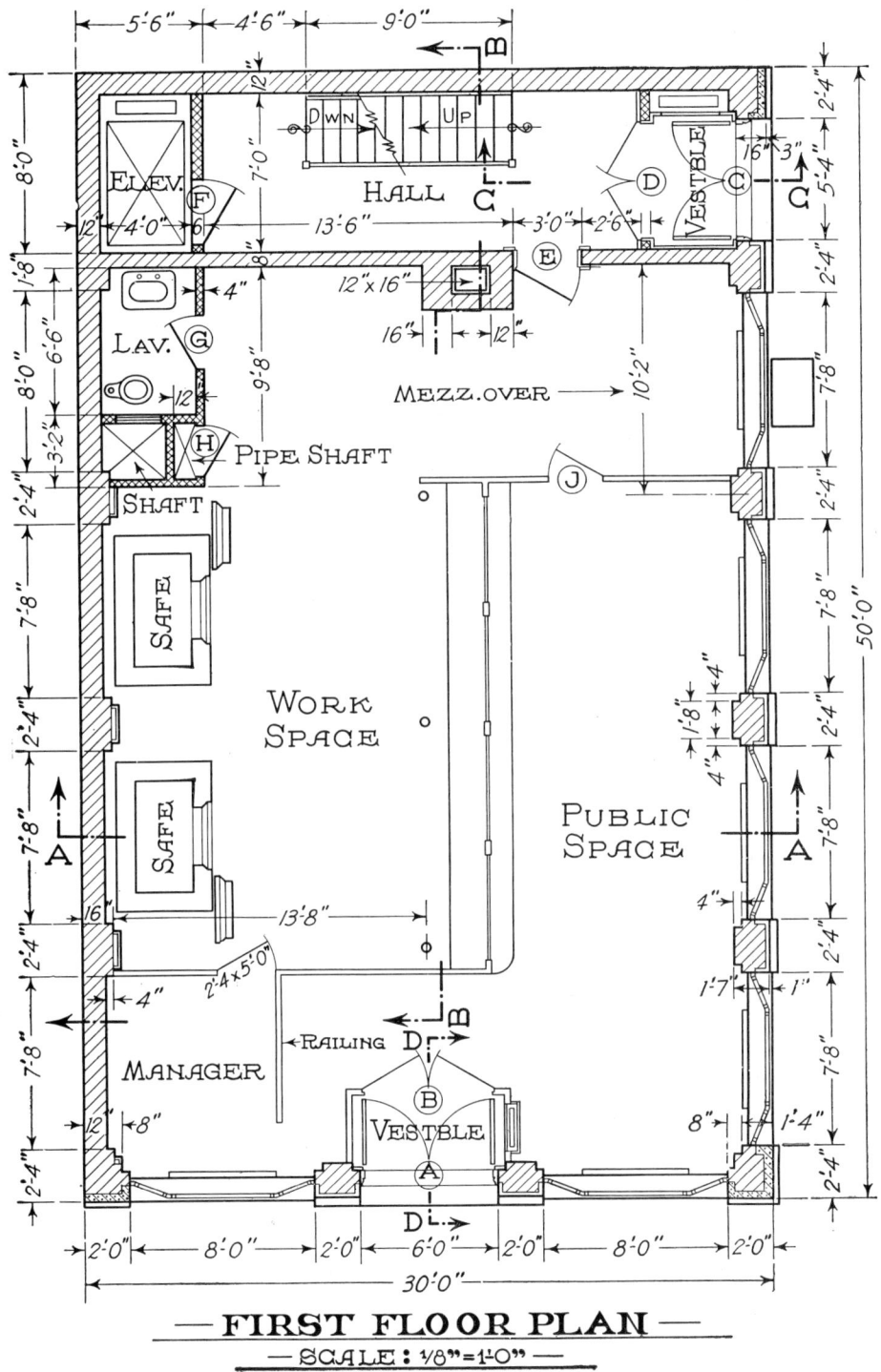

Fig. 29-3.

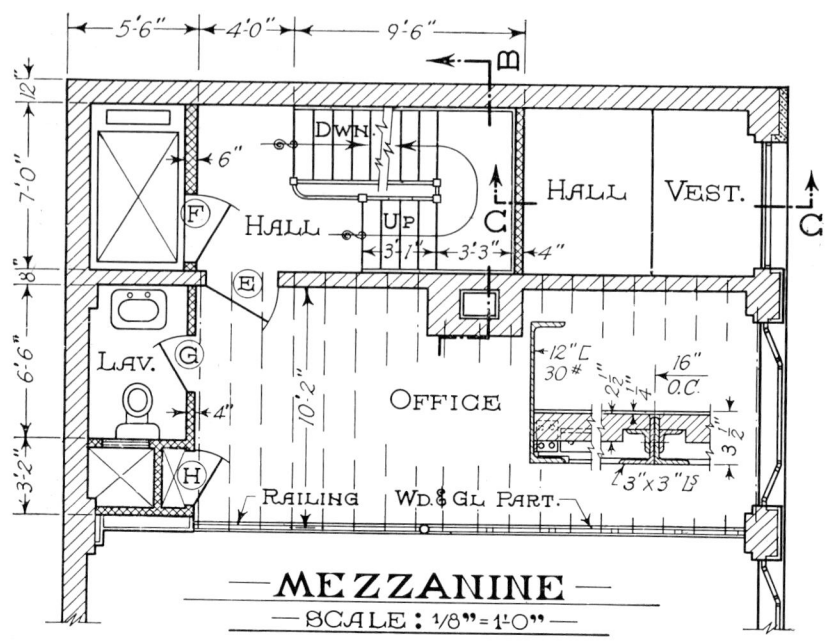

SCALE, ALL PLANS: 1/8" = 1 FOOT

KEY: ▓▓▓▓ = CONC.  ▨▨▨▨ = BRICK  ▩▩▩▩ = T.C.

DOOR SCHEDULE:

| TYPE | SIZE | MATERIAL | REMARKS |
|---|---|---|---|
| A | 6-0 x 15-0 | BRONZE & GLASS | SEE DETAIL |
| B | 5-8 x 7-6 | WOOD & GLASS | WOOD FRAME |
| C | 5-4 x 15-0 | BRONZE & GLASS | SEE DETAIL |
| D | 5-4 x 7-6 | WOOD & GLASS | WOOD FRAME |
| E | 3-0 x 6-8 | HOLLOW METAL | STEEL FRAME |
| F | 2-10 x 6-10 | HOLLOW METAL | STEEL FRAME |
| G | 2-4 x 6-8 | WOOD | LOUVRES BOTT. PAN. |
| H | 2-4 x 6-8 | KALAMEIN | STEEL FRAME |
| J | 2-6 x 7-0 | WOOD & GLASS | WOOD FRAME |
| K | 3-0 x 7-0 | KALAMEIN | STEEL FRAME |
| L | 2-8 x 6-8 | WOOD | STEEL FRAME |
| M | 3-0 x 7-0 | WOOD & GLASS | STEEL FRAME |
| N | 3-0 x 6-6 | FLUSH KAL. | STEEL FRAME |
| P | 3-0 x 5-0 | FLUSH KAL. | STEEL FRAME |

Fig. 29-4.

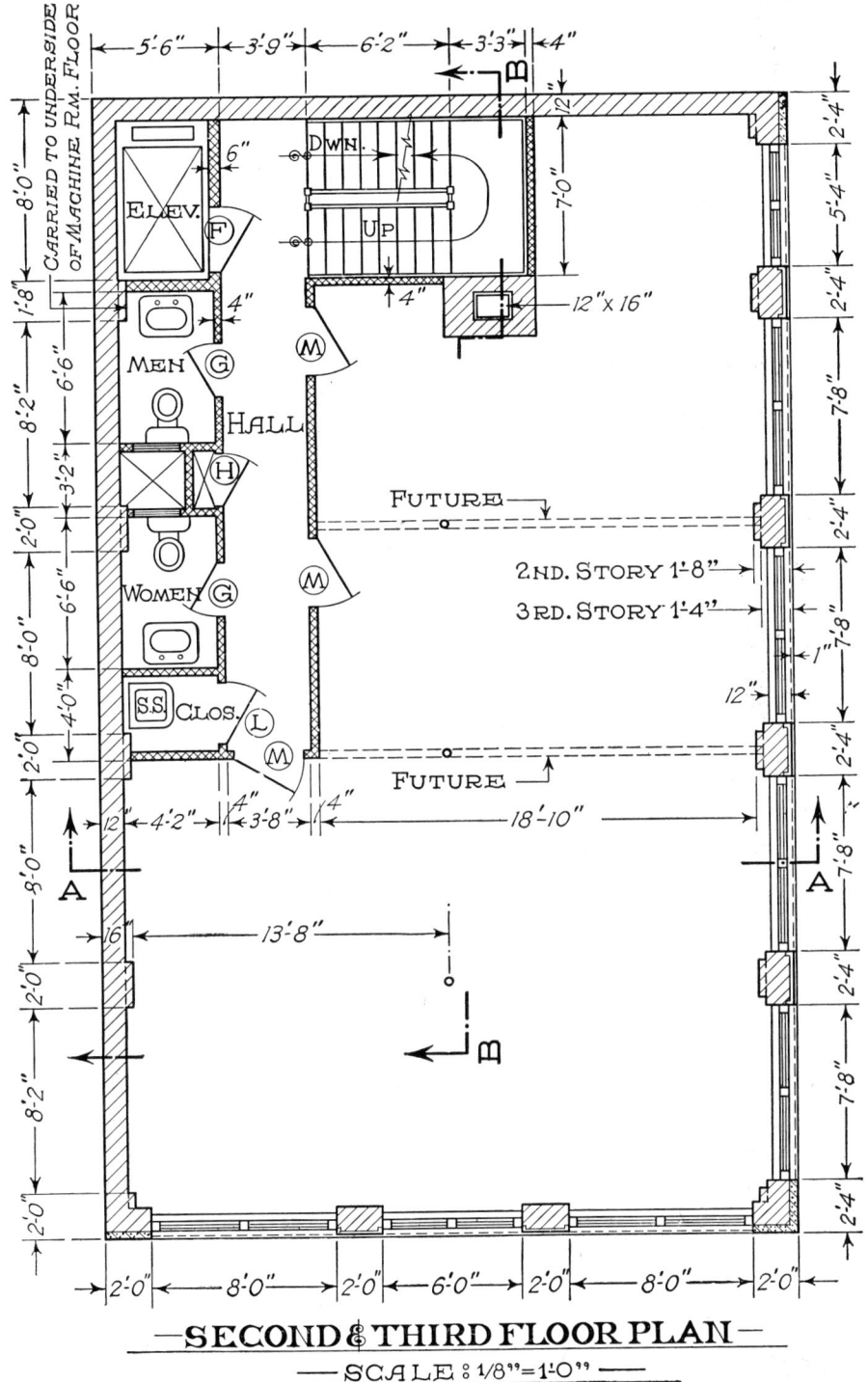

Fig. 29-5.

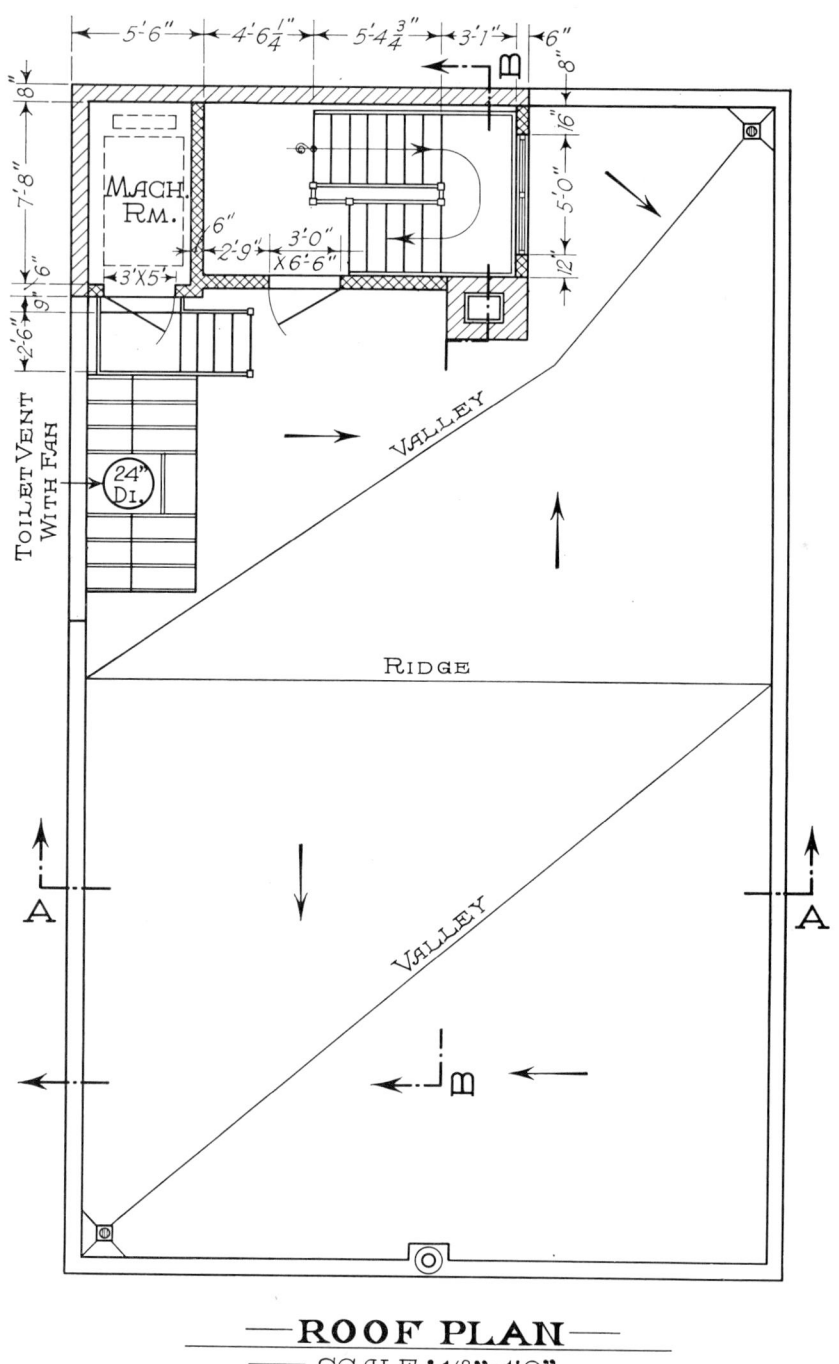

Fig. 29-6.

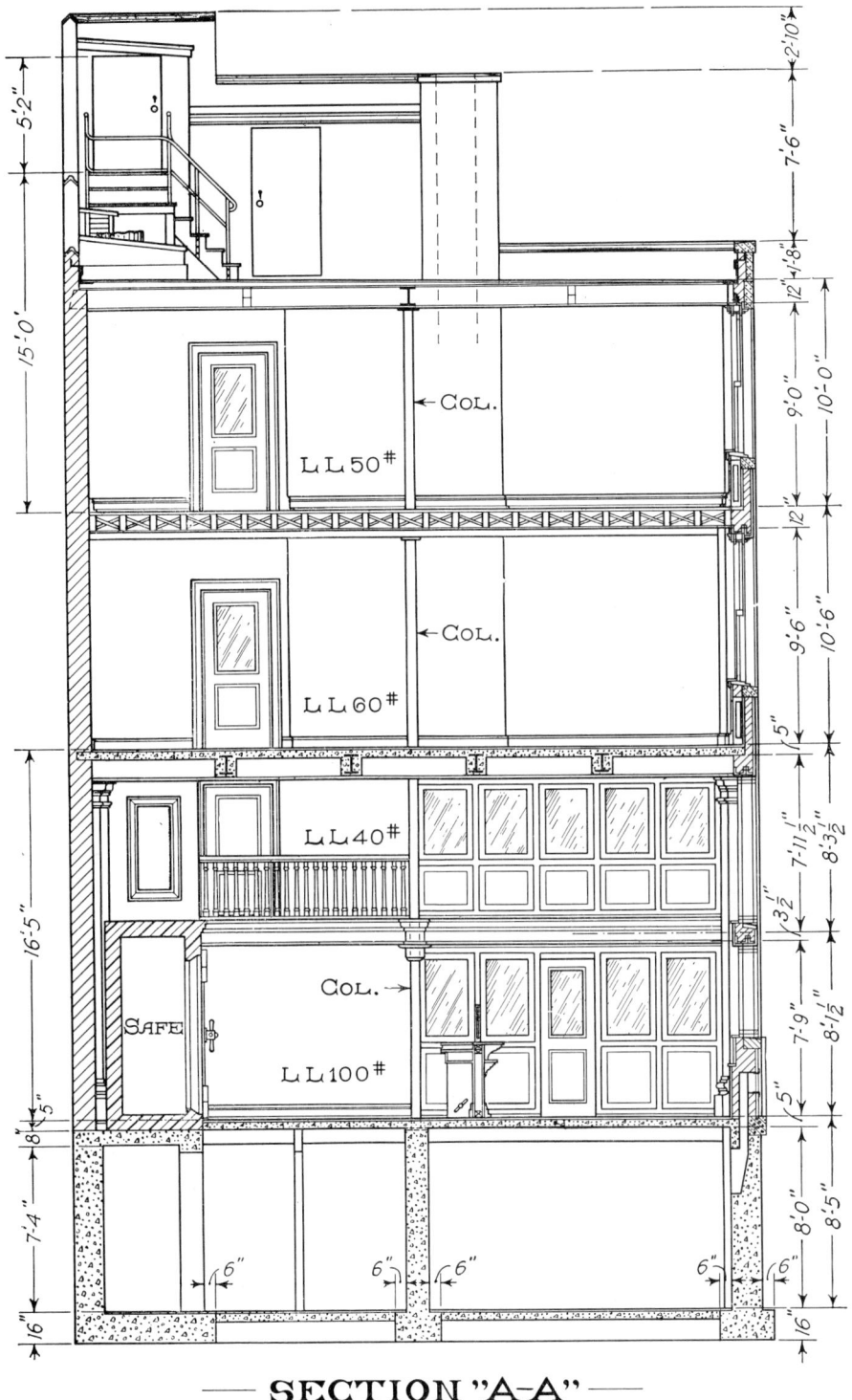

Fig. 29-7.

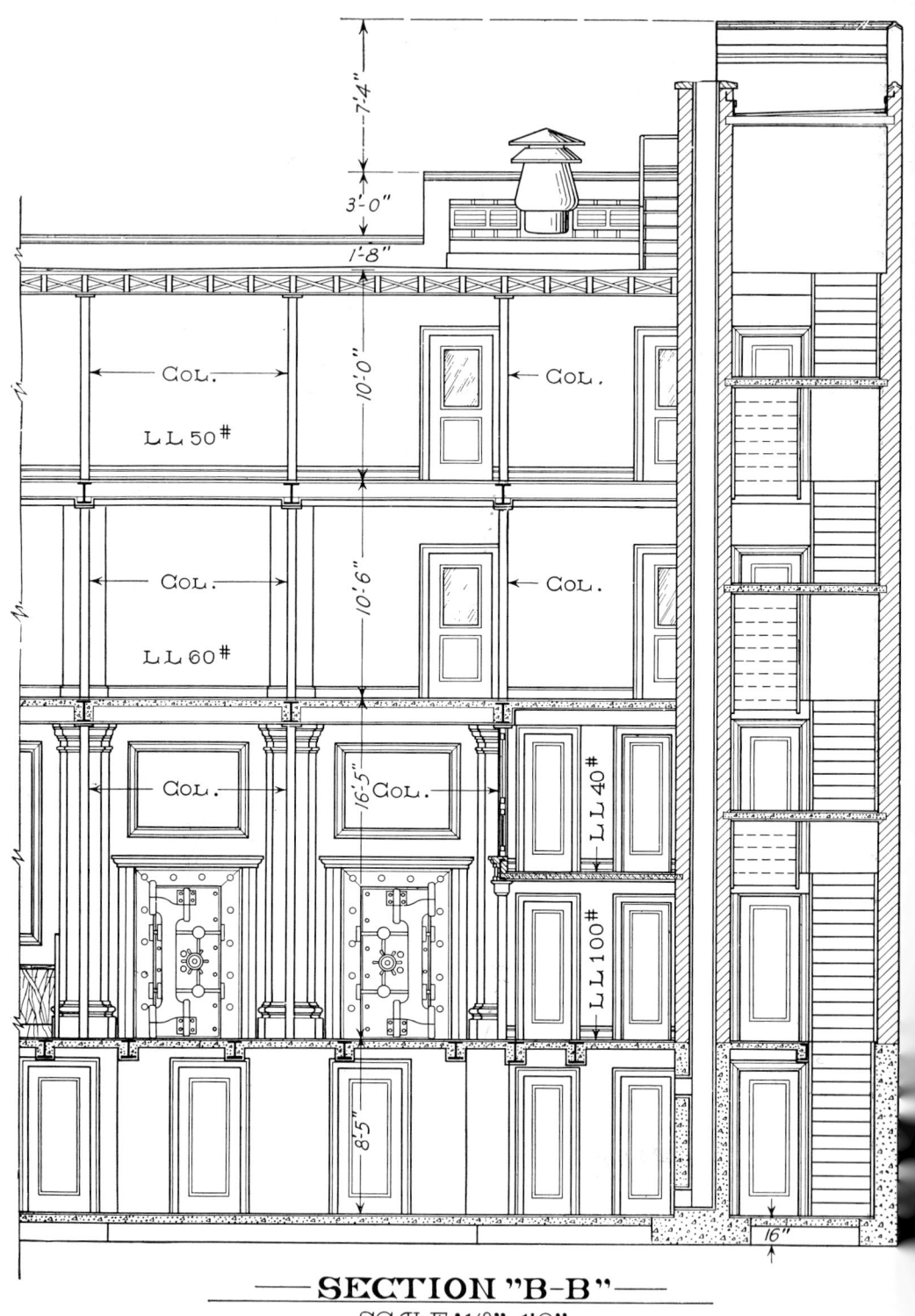

Fig. 29-8.

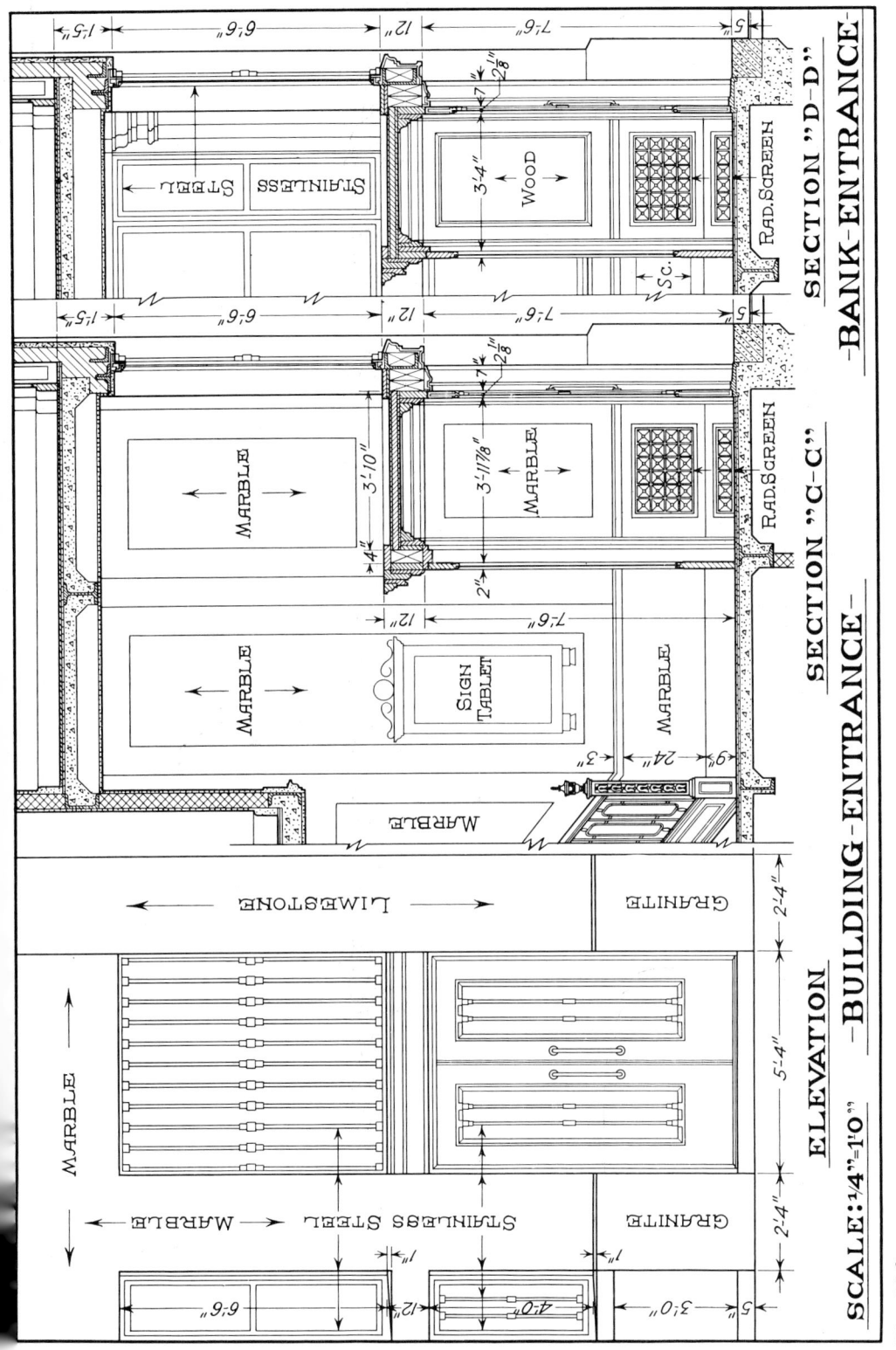

Fig. 29-9.

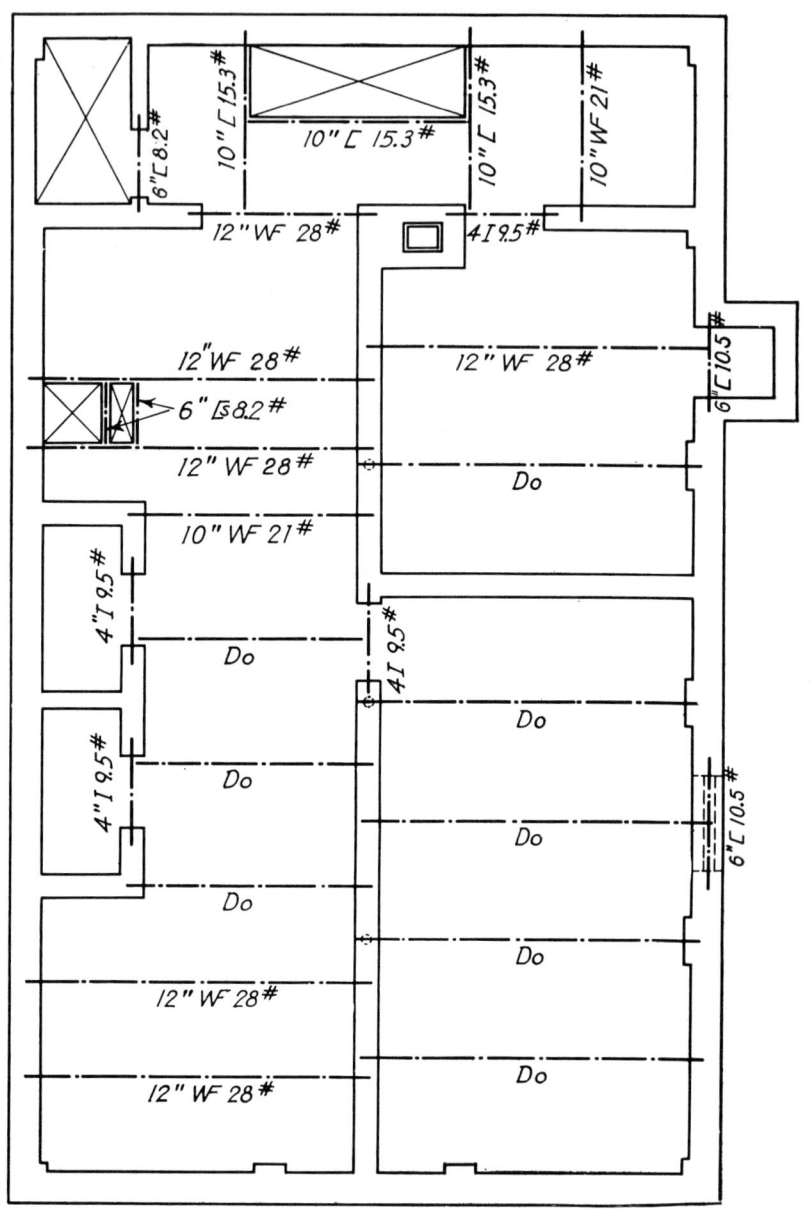

Fig. 29-10.

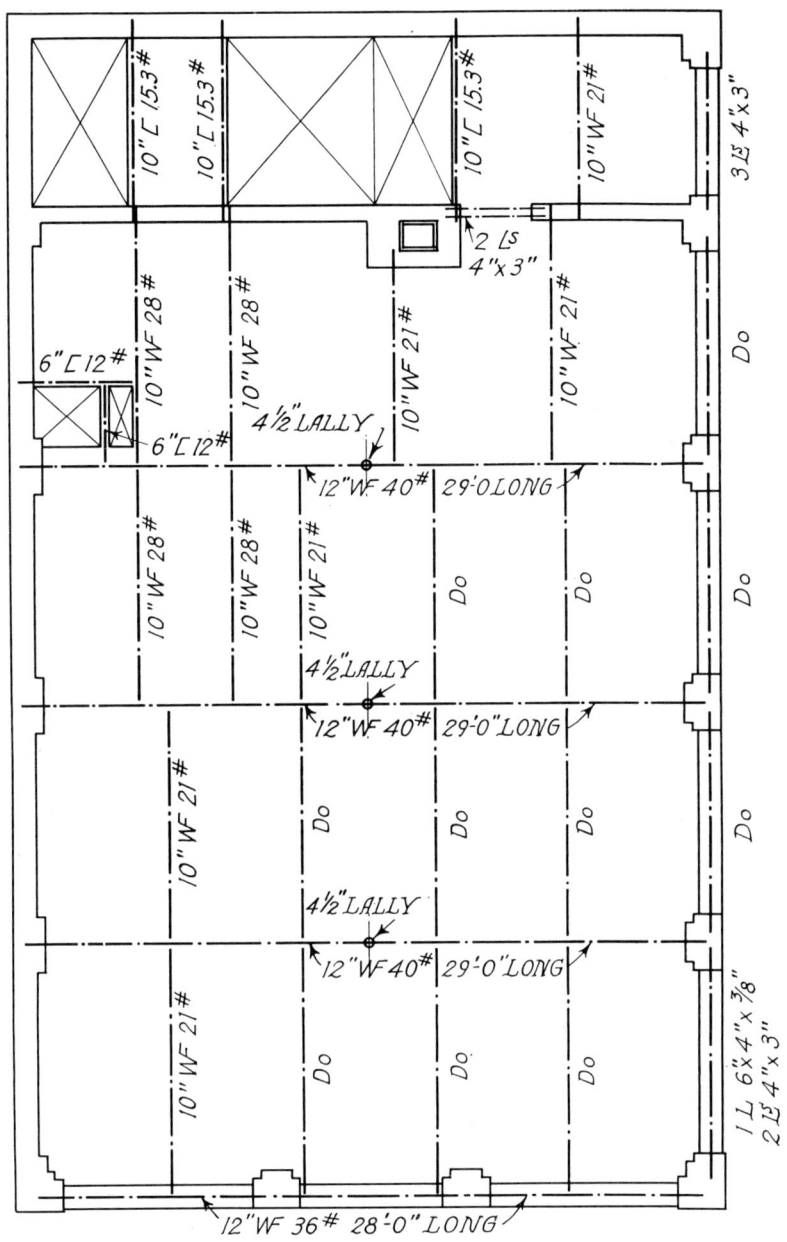

Fig. 29-11.

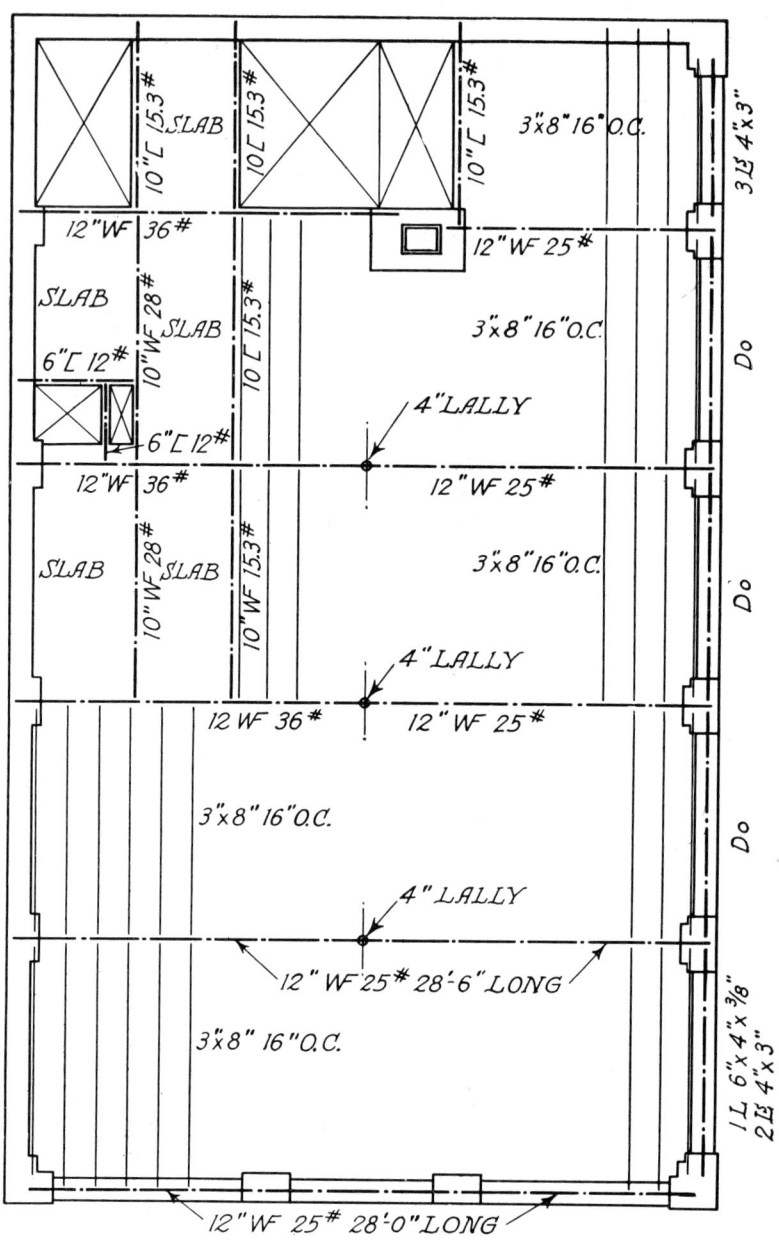

Fig. 29-12.

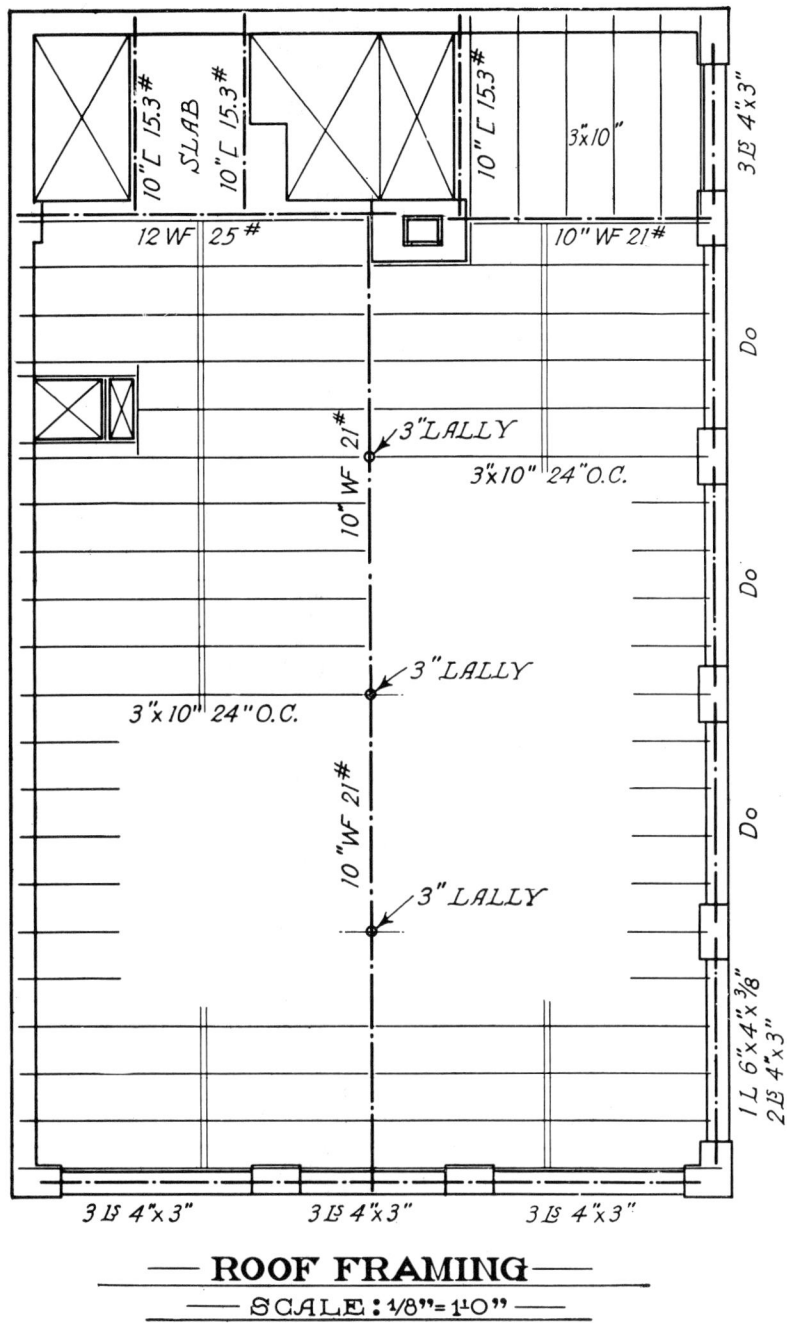

Fig. 29-13.

# *index*

Accessories, bathroom, 318, 339
Acoustic plaster, 257
Aggregates, 150, 154
Agitator, concrete, 150
Agreements, 80
Air conditioning, 349
American bond, 187
Anchors, carpentry, 213, 227, 331
   concrete, 153, 171
   masonry, 199
Angle of repose, 135
Angles, steel, 287
Apron, window, 268, 271
Apron wall, 38
Arbitration, 84
Arches, brick, 187
   concrete, 158
Architect, functions, 11
   license, 14
   office, 12
   peculiarities, 18, 103
   superintendent, 5, 17
Architectural plans, 30, 34
Architectural terra cotta, 194
Areaway, 57, 61
Ashlar, 305
Astragal, 265, 270

Backband, 271

Backfill, 138
Back-up blocks, 189
Balloon framing, 57, 209
Bar spacers and bolsters, 171
Bars, reinforcing, 144, 148
Baseboard, 267, 274
Basket-weave brickwork, 187
Bathroom accessories, 318, 339
Batter boards, 124
Beams, ceiling, 211
   steel, 32, 285, 388
   wood, 205, 215, 227
Bearing partition, 223
Bid summary, 130, 132
Blocking, wood, 274
Blocks, concrete, 176, 192
   gypsum, 176
   terra-cotta, 175, 189
Blueprints, 30
Board measure, 99, 112, 208
Bolsters, bar, 171
Bolts, anchor, 216, 226
Bond, brick, 181, 186
Bonds, surety, 124, 126
Bowstring truss, 228
Box gutter, 300
Brace, corner, 211, 221
Braced frame, 57, 209, 211
Branch, plumbing, 341

Brick, bond, 181, 186
　common, 176
　enameled, 197–199
　face, 175, 181
　joints, 182
　veneer, 45, 189, 202
Bridging, 215, 217, 231
Budget estimates, 90
Building code, 38, 169, 198, 231
Building department, 6, 37
Built-up roofing, 296
Bullnose brick, 199

Carpentry specifications, 231, 279, 371
Casement window, 53, 270, 313
Cast stone, 305
Caulking, 186
Ceiling, hung, 250, 251, 256
Ceiling beam, 211
Cement, 187
Ceramic tile, 318
Chair rail, 274
Channels, steel, 286
Cinder concrete arches, 158
Cinder fill, 159, 165, 172
City departments, 6, 37
Claw-plate connectors, 222, 230
Clipped bond, 198
Closures, brick, 180
Coarse aggregate, 150, 154
Collar beam, 53
Column forms, 163
Columns, steel, 284
Comb-grain flooring, 275
Combination walls, 189
Common bond, 187
Common brick, 176
Composition roofing, 296
Concrete, controlled, 39, 169
　plain, 150
　reinforced, 150, 169
　reinforcing, 144
　specifications for, 368
Concrete arches, 158
Concrete blocks, 176, 192
Concrete forms, 144
Concrete foundations, 143
Concrete inserts, 153, 171
Concrete joints, 152
Concrete mix, 150, 237

Concrete planks, 166, 168
Concrete stairs and steps, 173, 242
Connectors, timber, 222, 230
Construction joints, 152
Contingencies, 100
Contract, arbitration, 84
　definition, 80
　financing, 82
　lump-sum type, 81, 84
　management type, 81, 87
　subcontract, 101
　unfair dealings, 103
Contractor, duties, 4, 20
　office, 21, 26
　personnel, 21, 28, 116
Controlled concrete, 39, 169
Convectors, 349
Conventional indications, 33
Copper flashing, 296
Corner beads, 249
Corner braces, 211, 221
Cornices, stone, 307
　wood, 272
Cost-plus contracts, 81
Cost records, 24
Counterflashing, 299
Cow stall, 242
Cricket, roof, 299
Cross bridging, 215, 217, 231
Cross section, 73, 395
Curbs, concrete, 241
Curtain wall, 38

Dampproofing, 184
Datum, 41
Detail drawings, 36, 77, 387
Diamond mesh lath, 256, 258
Doors, designation of, 267, 334
　fire, 312
　fireproof, 310
　kalamein, 310
　rolling steel, 315
　tinclad, 312
　wood, 263
Dormer framing, 225
Double-hung window, 45, 268
Douglas fir, 206
Dovetail, 274
Driveways, concrete, 241
Drop girt, 211

## INDEX 395

Ducts, 303, 350
Dutch bond, 187

Edge-grain flooring, 275
Electrical plan, 360
Electrical service, 356
   temporary, 122, 354
Electrical symbols, 355
Electrical work, 352
   regulations, 354
   specifications, 359, 377
   wattage table, 358
   wiring, 357
Elevations, 33, 53, 379
Elevator specifications, 377
Employee classifications, 8, 116
Enameled brick, 197–199
Enclosure wall, 38
Engineers, designing, 14, 34
English bond, 187
Estimate, arrangement, 90
   bidding, 90
   divisions, 92
   preliminary, 90
   summary, 130
   working, 91
Ethics of contracting, 22, 84, 102, 284
Excavating, 134
   mechanical trades, 337
   specifications, 142
Expansion joint, 152

Fabric dampproofing, 185
Face brick, 175, 181
Federal laws, 37
Financing contracts, 82
Fine aggregate, 150, 154
Finish carpentry, 262
Finish grade, 41
Finish hardware, 332
Fink truss, 228, 288
Fir, 206
Fire doors, 312
Fire partitions, 38
Fire Underwriters, 313, 354
Fire walls, 38
Fire windows, 38
Flashing, chimney, 299
   roof, 296
   wall, 298

Flashing, window, 269
Flat arch, brick, 188
Flat rib lath, 258
Flat slab construction, 162
Flemish bond, 187
Float finish, cement, 236
Floor arches, 158
Floor bridging, 215, 217
Flooring, cement, 235
   tile and terrazzo, 318
   wood, 66, 112, 275
Flue lining, 184
Formwork, 144
Four-way system, 170
Framing, types, 57, 209
Frieze, 300
Furring, wood, 270, 272

Garage plan, 113
Girt, 211
Glazed terra cotta, 194
Glazing specifications, 331, 375
Grade marking, 207
Granite, 305
Gravel roofing, 296
Grillage beams, 301
Grounds, 245, 274
Guarantees, 124
Gutters, 300
Gypsum blocks, 176
Gypsum board, 279
Gypsum plaster, 251

Hardeners, floor, 236
Hardware, 332
Header beam, 215
Headers, brick, 180
Heating controls, 348
Heating specifications, 377
Heating systems, 348
   temporary, 122
Herringbone brickwork, 188
Hip rafter, 210, 224
Hollow block floors, 165, 172
Hollow metal, 310
House drain, 341, 344
Howe truss, 288
Hung ceiling, 250, 251, 256

I-beams, 285
Inserts, 154

Inspector, architect's, 5, 16
  municipal department, 6
Insulation, 223
Insurance, 125
Integral cement finish, 236

Job expense, 115, 133
Job organization, 22, 28, 120
Joints, brick, 182
  construction, 152
  expansion, 152
  wood, 274
Joist, wood, 209

Kalamein, 310
Keene's cement, 256
King post truss, 288

Labor laws, 37, 129
Lally column, 61
Laminated floor, 212
Lathing, 248
Laws, Federal, 37
  municipal, 37
  state, 37
Licensing, architects, 14
  electrical contractors, 352
Lighting fixtures, 354
Limestone, 305
Lines and batters, 124
Live loads, 73
Long-span roof slab, 167
Lot line, 42
Lumber, grading, 207
  sizes, 205
Lump-sum contracts, 81, 84

Maple flooring, 275, 278
Marble, 318
Marble specifications, 374
Mason materials, 175
Masonry specifications, 198, 203, 369
Material men, 8
Mechanic, 8
Mechanical plan, 35
Mechanical trades, 3, 37
Meeting rail, 268
Membrane waterproofing, 184, 301
Mesh, reinforcing, 161
Metal lathing, 248

Metal pan system, 164
Mill construction, 57, 209
Millwork, 262
Monolithic finish, 236
Mortar, 187
Mouldings, wood, 274
Mullion, 57, 268
Municipal departments, 6
Muntin, 57, 268

Nail sizes, 220
N.C. pine, 206

Oak flooring, 66, 112, 275
Original grade, 41
Ornamental iron, 289
Outlookers, 272
Overhead expense, 127

Painting specifications, 375
Pan system, 164
Panel wall, 38
Parapet wall, 38
Pavements, concrete, 241
Pine, 206
Piping, 340, 343
Plancia, 300
Planks, concrete, 166, 168
Plans, 30, 32
  architectural, 30, 41, 379
  bank, 379
  garage, 113
  house, 41
  mechanical, 35
  office building, 379
  structural, 34
Plaster boards, 245
Plastering specifications, 260, 373
Plate, roof, 210, 218
Plate anchor, 272
Platform framing, 57
Plot plan, 41
Plumbing, 337
  requirements, 341
  specifications, 343, 376
Plumbing section, 340, 344
Plywood, 279
Poplar, 208
Portland cement, 187
Pratt truss, 228, 288

Precast planks, 166, 168
Precast slabs, 166
Preliminary estimate, 90
Present grade, 41
Progress schedule, 22
Project manager, 28, 116
Protection, temporary, 119
Protective assembly, 38

Radiators, 349
Rafter, 210, 225
Raised girt, 211
Raked joint, 182
Rectangular mesh, 161
Reinforced concrete, 150, 169
Reinforcing, concrete, 148
Reinforcing mesh, 161
Reinforcing rods, 149
Relieving arch, 188
Ribbon, 210, 218
Ridge, 224
Rolling steel doors, 315
Roof framing, 224
Roof trusses, 228, 229, 288
Roofing specifications, 303, 375
Rowlock arch, 188
Rowlock brick, 187
Rubble stone, 305
Running bond, 181

Sand, 150, 154
Scaffolds, 182
Scale rule, 108
Section, 33, 73, 385
Sheathing, 219, 222, 226
Sheet metal, 296
Sheet piling, 135
Shiplap, 266
Shop drawing, 11
Shortleaf pine, 206
Show windows, 315
Sidewalk shed, 119, 122
Sidewalks, 240
Siding, 266
Site examination, 96
Skeleton sheeting, 135
Skewback, 188
Skylights, 302
Slab, concrete, 162
    flat, 162, 170

Slab, forms, 162
    inserts, 171
    long-span, 167
    reinforcing, 161
Slag roofing, 302
Slate work, 318
Slot anchors, 153
Soffit clips, 160
Soil pipe, 341
Soldier brick, 186
Spandrel, 38
Split rings, 222, 230
Spruce, 206
Stack, plumbing, 341
State laws, 37, 129
State licenses, 14
Steel doors, 315
Steel joists, 291
Steel sash, 313
Steps, concrete, 173, 242
Stone specifications, 309, 370
Stool, window, 268
Store fronts, 315
Stretchers, brick, 186
Structural plans, 32, 388
Stucco, 245
Studs, metal, 253
    wood, 209, 213, 223
Subflooring, 223
Subcontractor, approved, 103
    definition, 7
    estimates, 101
    responsibility, 365
    substitutions, 102
Summary sheet, 130, 132
Sun room, 273
Surety bonds, 124, 126
Surveys, 42, 123
Suspended ceiling, 250, 251, 256
Symbols, plan, 36
    electrical, 355

Tail beam, 215, 231
Temporary construction, 98, 119, 155
    enclosures, 119
    heating, 100, 122
    job office, 119
    light and power, 12, 353
    sidewalk shed, 119, 122
    toilets, 119, 338

Temporary construction, water lines, 122, 338
Terra cotta, 175, 189, 194
Terrazzo, 318
Tile specifications, 321
Timber connectors, 22, 230
Tinclad doors, 312
Topsoil, 41
Triangle mesh, 161
Trimmer beam, 215
Trusses, 228, 229, 288
Two-way floors, 169

Underflooring, 223
Unfair dealings, 103, 130
Union regulations, 120, 353
Union wage rates, 8, 98

Veneer, brick, 45, 180, 202
Vent pipe, 340
Ventilating, 349

Wage rates, 8, 98
Walks, cement, 240
Warren truss, 228, 288

Waste, carpentry, 99
 concrete and masonry, 98
 element of cost, 98
 wood flooring, 100
Waste pipe, 341
Water tables, 26, 306
Waterproofing, 184, 301
Wide flange beams, 285
Windows, casement, 53, 270, 313
 fire, 38
 fireproof, 310
 steel, 313
 wood, 268
Wire gauges, 291
Wire work, 292
Wiring, electrical, 357
Wood, cornices, 272
 doors, 263
 flooring, 112, 276
 framing, 207
 grading, 208
 trusses, 228, 229
 windows, 263
Working estimates, 91